I0043098

The Practical Guide to
STATISTICS
Basic Concepts, Methods and Meaning

The Practical Guide to STATISTICS

Basic Concepts, Methods and Meaning

*Applications with R, Excel,
and OpenOffice Calc*

Kilem Li Gwet, Ph.D.

Advanced Analytics, LLC
P.O. Box 2696
Gaithersburg, MD 20886-2696
USA

Copyright © 2010 by Kilem Li Gwet, Ph.D. All rights reserved.

Published by Advanced Analytics, LLC .

No part of this document or the related files may be reproduced or transmitted in any form, by any means (electronic, photocopying, recording, or otherwise) without the prior written permission of the publisher.

Advanced Analytics, LLC
PO BOX 2696,
Gaithersburg, MD 20886-2696
e-mail : info@advancedanalyticsllc.com

This publication is designed to provide accurate and authoritative information in regard of the subject matter covered. However, it is sold with the understanding that the publisher assumes no responsibility for errors, inaccuracies or omissions. The publisher is not engaged in rendering any professional services. A competent professional person should be sought for expert assistance.

Publisher's Cataloguing in Publication Data :

Gwet, Kilem Li
The Practical Guide to STATISTICS
  Basic Concepts, Methods and Meaning/ By Kilem Li Gwet
    p. cm.
    Includes bibliographical references and index.
  1. Statistics
  2. Statistical Methods
  3. Research Methods - Study - Learning. I. Title.
  ISBN 978-0-9708062-9-1

Preface

This book presents basic statistical concepts, and methods with a particular emphasis on their meaning, and practicality in real life. Many introductory statistics texts do not provide the practical motivations behind the techniques, nor the reasons why the techniques were formulated the way they are. The focus in these texts appears to be on describing various statistical methods in a very simple language (sometimes in plain English) to make them accessible to the general public. The problem with this approach is that implementing a statistical technique even properly does not tell much about the value of the statistical solution.

Many students in my statistics classes experience difficulties for various reasons. Some of them came to my class with a rather limited experience with statistical concepts in their prior education. Others had their algebra too rusty to be of any help, or could not use the latest technology such as Microsoft Excel to implement the methods. However difficulties that stem from the reasons just mentioned are not the most serious since they are easily identifiable. In my opinion, the most fundamental issue struggling students had in common was a lack of general statistical literacy to make any sense of what I was talking about. What I mean by "lack of general statistical literacy" is the inability to see why a statistical solution is even called a solution in the first place. In this context, for a Statistics professor to be telling students to ask questions when they do not understand does not really help. It does not help because students will not want to ask their professor why they are being taught statistics, which is often the question some have in mind.

Professionals and students tend to deal with their frustrations with statistics by rushing to a local bookstore or to Amazon.com in order to purchase the first "Statistics for Dummies" book they can find. I can see two problems with this reaction. The first problem is that treating oneself as dummy is not known to be a good path to empowerment. Secondly if you are a dummy (I mean a real one) then Statistics might not be for you. Statistics requires high-level thinking, and a simple cooking recipe may help you get a decent grade from a statistics professor who may not care that much about your statistical literacy. The more effective approach is to first realize that when you do not understand something, you must focus on articulating good fundamental questions about what you are learning, rather than seeking small (and sometimes misleading) answers from books written for dummies or for the "utterly confused." I do believe that any person with some college-level education should be able to acquire general statistical literacy from a text offering a clear exposition of the fundamental ideas.

Throughout this book, I will constantly attempt to motivate the statistical concepts and methods being described. The main purpose of the methods will be kept before the reader, and used as often as needed to justify each step taken. A particular effort is made towards reconciling statistical logic and common sense, with the objective of gaining insight into the real value of the statistical solution. Statistics is not made more comprehensible by trading its native language for an easier one. Instead, statistics becomes more accessible by getting down to the subtle job of teasing apart abstract and concrete facts, theory and practice, absolute truth and pragmatism. The oversimplification of concepts in many introductory "for-dummy" statistics texts does not just make statistics simpler. It makes it simpler and useless.

It may kill our pain and frustrations for a short-lived satisfaction. I would rather acquire a sharp tool to crack the nut. This book aims to help researchers and students see what the most commonly-used statistical procedures are made up of.

Kilem Li Gwet, Ph.D.

Contents

Contents

CHAPTER $\boxed{1}$

Introduction

OBJECTIVE

The purpose of this introductory chapter is to cover basic notions that I thought you will need in subsequent chapters, including a broad and limited overview of the field of statistics. I review some software products that I recommend for processing your data, and end the chapter with short survey of the concept of probability

CONTENTS

1.1. Statistics and Abstraction

Students in my statistics classes often ask me the question "Why do have to learn this?" or "What is this going to be used for?" or "What are the concrete applications of this technique?" Rather than asking questions, other students will make bold statements such as "We need concrete solutions, and not statistical theory," or that "I will probably never use this stuff in practice." In my opinion all these questions and statements actually conceal a more fundamental issue, which is resistance to abstraction. Many of my students refuse to study statistics by developing basic technical skills first before applying them to solve real-world problems. They seem to prefer statistics to emerge in the middle of a discussion about a specific concrete matter. This sounds to me as if instead of teaching a child to count from 1 to 10, you decide to teach that child how to count cars, then dogs, then trees, and so on. That child will certainly learn to count all those things, but without establishing any relationship among them, without capturing the essential notion of number, abstracted from any concrete identifiable object.

I do believe that a minimum level of abstraction is necessary if you want to learn one technique, and use it to solve many different practical problems. The technique I have used for dealing with resistance to abstraction, and which I apply in this book is to use motivational concrete framework. I always start by describing a practical business or social problem to motivate the statistical investigation. My next step is to describe what the statistical solution looks like, and justify why it can be seen as solution, and finally start by developing the statistical apparatus needed to go from the problem to the solution. I must admit that, it could be quite difficult to remain motivated in studying a statistical technique over a certain period of time without knowing where it is going to lead us.

A discussion over the nature of a statistical solution to a specific problem, can indeed be quite passionate if well conducted. And such a discussion can be done before the technical concepts are introduced. Its purpose is not to replace the statistics course, and it cannot. Instead, its purpose is to have a conversation about our destination before embarking on the journey. You may expect as much as I do, that this conversation will make the frustrations and all sort of inconveniences associated with the journey more acceptable.

Practitioners and students tend to insist on the concreteness of what they learn, not simply as a rejection of abstraction, but also because they want to be busy producing numbers. But insisting too much of the production of numbers will probably overemphasize the importance of these numbers why neglecting the more fundamental question of their intrinsic value in the context of research. Practitioners do not only need to know how to do a test of hypothesis, or how to construct a confidence interval. Having a good grasp of the very nature of the progress the statistical solution will help us achieve, makes more sense to me. You need to know whether the statistical solution is even worth the effort, and what importance should you put on it in your investigative effort.

1.2. The Field of Statistics

The field of statistics has become so vast that there is no one single point of entry that can lead to all of its branches. This situation has made the study of statistics confusing. Consequently, before deciding what statistics class to take or what statistics book to buy or what statistical consultant to hire, a broad and high-level overview of the field of statistics is necessary.

The basic statistical activity consists of organizing and summarizing a data series. Sales data in a department store could be presented in the form of monthly, quarterly, or annual total sales. The statistical measure in this context will be total or the sum. The same data is often displayed graphically with pie and bar charts. Some creative approaches for displaying statistical data are powerful exploration techniques, which have allowed analysts and statisticians to extract useful information from databases. This activity that consists of organizing and summarizing data is referred to as *Exploratory Statistics*[1] . Its role is limited to providing practitioners with metrics and graphical methods to show progress, to evaluate risk, to compare magnitudes. However, descriptive statistics does not provide the tools to answer important and broad research questions, when the researcher's interest goes beyond what observed data can tell. For example "How is sales revenue affected by the level of advertising expenses in general ?" or "How likely a male aged 15 to 25 is to be a smoker ?" These are two general research questions that do not refer to a specific data set. Therefore the data set can only be used as a stepping stone for exploring a whole universe of possibilities, or for inferring from observed data. Here you are entering the domain of *"Inferential Statistics."*

Inferential Statistics makes data speak through various modeling techniques. These techniques are not always very precise, but remain useful as the only means for studying hidden relationships and invisible parameters. But inferential statistics itself has many sub-branches with different aims. I will confined myself to mentioning only some of these sub-branches that ordinary scientists are likely to encounter in their professional life. The 2 branches of inferential statistics I like to mention are *Classical (or Mainstream) Inferential Statistics* and *Finite Population Sampling and Inference.* You will later see that each of these branches of infe-

---

[1]Some texts use the term *Descriptive Statistics*. However, some of the techniques used go beyond a mere description of data, and perform further exploration to extract information

rential statistics is further subdivided into the parametric[2] and nonparametric[3] branches with different aims. But for now, I like to show the difference between *Classical (or Mainstream) Inferential Statistics* and *Finite Population Sampling and Inference*.

Finite Population Sampling and Inference

An automobile insurance company may want to know about the likelihood for a driver to have one accident or more in a given year. If the interest is limited to a specific group of drivers who can be located (e.g. all drivers who live in the state of Maryland in the US), then the likelihood for a driver to have an accident represents the actual *proportion* (or relative number) of Maryland drivers who got into a car accident in any given reference year. This proportion is a concrete measure with a well-understood operational definition[4]. The number of Maryland drivers is known, and all of them constitute a finite population. Estimating quantities in this context appeals to the techniques of Finite Population Sampling and Inference. The quality of our estimations will also be evaluated with respect to the specific concrete population that is being targeted in our study.

Classical Inferential Statistics

If you do not want to confine yourself to a particular geographic location, you could look at the driver as an abstract person who could live anywhere (an abstract person does not have a real residence after all). In this context, the likelihood for a driver to have an accident is not associated with any well-defined operational

---

[2]Parametric methods are based on hypothesized models

[3]Nonparametric methods are based on the processes that generated the data being analyzed

[4]The operational definition associated with a concept is seen here as the step-by-step procedure for quantifying it

definition. It does not represent a concrete measure. Instead, it is a theoretical construct that is often referred to as the *Probability* that a driver will have an accident. Calculating this probability appeals to the techniques of Classical Inferential Statistics.

Of these two branches of inferential statistics, classical inferential statistics is the oldest one. Its development started several centuries ago to support scientific research. Finite population sampling and inference was invented primarily to meet governments' needs. Since classical inferential statistics does not refer to any specific population of interest, it is with no surprise that government officials have shown little interest in it. Given the limited resources they have at their disposal and the fact that there is generally a specific group of people they have to look out for, such a broad framework is inappropriate. Until today, Finite population sampling and inference has been used primarily in government-sponsored survey projects, and the importance of government databases to researchers is the primary reason for including this inferential framework in this book.

1.3. Statistics and the Notion of Variable

Statistics in general can be seen as the study and management of variability. Houses and cars for example have different prices. Even the price associated with a particular car model may change monthly, or quarterly depending on various factors such as government tax regulations or gas price. This variability in car or house prices explains the need to compile statistics to see what happened. Without variability there is no need for statistics. *Exploratory Statistics* is more concerned with the description of that variability, while *Inferential Statistics* will be concerned with both the description and the management of variability.

The field of exploratory statistics can be divided into the following three components :

▶ The descriptive data analysis

▶ The summary measures of quantitative variables,

▶ The exploratory data analysis

In statistics, characteristics of interest such as *age, educational attainment, height, gender* are generally coded or recoded with numeric values. The recoded characteristics are called *variables*. Let me consider educational attainment for example. This characteristic may be defined as {Some High School with no degree, High school graduate, Some college with no degree, College degree or more}. It could be recoded as {1, 2, 3, 4}, where 1 represents the lowest level (high school with no degree), and 4 the highest (college degree or more). This characteristic, which is recoded with numeric values with only an indirect connection to education, has now become an abstract variable that I could refer to as X (capital X). Such abstraction makes it easier to identify an appropriate existing statistical technique that could be applied to the study of educational attainment. Without such abstraction, it would be necessary to develop a technique for educational attainment, and another technique for a similar characteristic such as high school grade levels (1=freshman, 2=sophomore, 3=junior, 4=senior or more). Moreover, one may not even realize that both techniques are identical, a knowledge that eliminates the need to continuously reinvent the wheel. Throughout this book, most techniques will be described in terms of abstract variables such as X, Y, or Z. When applying them to solve a problem, it will be up to the researcher to recode the raw characteristic so as to match the variable so defined to a specific technique.

The use of variables in statistics is sometimes confusing to some students. It should not be. In fact using abstract variables instead of the more concrete characteristics is an exercise that we all do on a daily basis without even realizing it. When you purchase two items from a groceries store at the cost of $5.00, and \$6.00, you know that the total to pay is \$11.00. You implicitly took the abstract numbers 5 and 6 (with no dollar sign) and remembered learning in elementary school how to add them to obtain 11, before putting back the dollar sign. This way you do not need to learn how to add dollars, pens, houses or people independently. You first learned to add abstract numbers before applying these skills in a concrete situation. But skills are developed more effectively in an abstract context. Statistical methods will often be described in terms of abstract variables taking numeric values. It will be up to the analyst to consider the 4 levels of the educational attainment characteristic independently of the concept of education to create an abstract variable X (actually how the 4 levels are defined has no statistical value). The variable X will be processed with the statistical technique, and the result will then be associated with the concept of educational attainment to formulate the research finding. You will later see in this book that the transition from the concrete to the abstract to back to the concrete is not always as smooth as we may want it to be.

Table 1.1 shows a small dataset of 9 professionals along with their gender and educational attainment, both expressed as characteristics then as variables. The 2 variables X, and Y contain all the information needed for analysis, while only the characteristics contain all the information needed to interpret the results. In addition to facilitating the development of statistical techniques, the variables present an auxiliary benefit. The numeric values they take make it easier to refer to specific groups of subjects. In the group of professionals of Table 1.1 for example, rather than saying "all professionals with a high school degree at the minimum," we could say $(X \geq 2)$ *(read "X greater than or equal 2")*. In subsequent chapters, you will see that such groups are subject to numerous manipulations, which makes frequent references to long sentences particularly unwieldy.

Table 1.1:

Educational Attainment and Gender of 9 Professionals

Individual	*Characteristics*		*Variables*	
	Education	Gender	X	Y
1	Some High School	Male	1	1
2	Some College	Female	3	0
3	Some College	Female	3	0
4	College Degree or More	Male	4	1
5	Some High School	Female	1	0
6	College Degree or More	Male	4	1
7	High School Graduate	Female	2	0
8	High School Graduate	Male	2	1
9	College Degree or More	Male	4	1

The coding of characteristics should be done so as to maximize the usefulness of the created variables. For example, educational attainment, which is an *Ordinal* characteristic[5] must be coded in such a way that its lowest level (i.e. some high school) receives the smallest of the 4 numbers, while its highest level (i.e. college degree or more) receives the highest number. Although we have the luxury of assigning any number to any level, coding "College degree or more" as 1 and "Some high school" as 4 clearly contradicts our intuition, while leading to a statistical analysis that will be harder to read. Likewise, gender can be coded as 3 for male and 4 for female or vice versa. In this case, which gender type receives the smaller number is irrelevant because Gender is a Nominal[6] characteristic. However, a more effective coding scheme for dichotomous (or binary) characteristics such as gender, is to assign the number 0 to one of the gender

---

[5]An *ordinal* characteristic is one whose levels can be ranked for low to high

[6]A *nominal* characteristic is one whose values are simply labels, identifiers, or attributes to identify subjects

type and 1 to the other. Its main advantage is that all the codes will sum to the count of individuals with the gender type that receives the code of 1.

1.4. Measurement Types

Throughout this book, I will describe various statistical techniques. Each of these procedures will only be valid for variables of a certain type. Therefore, knowing the variable type will be essential for identifying the proper procedure to use. The two main variable types I am concerned about are the *Categorical Variables* and the *Measurement Variables*.

1.4.1 Categorical Variables

A variable is called categorical when it takes a limited number of values. The simplest of all categorical variables are dichotomous variables (also called binary variables) such as gender that may be arbitrarily coded as 0 for male and 1 for female. If you conduct a telephone survey of 100 individuals, some of them will agree to participate while others will decline. You may defined a categorical variable X where $X_i = 1$ if individual i is a respondent, and $X_i = 0$ if individual i is a nonrespondent.

Dichotomous variables are commonly coded using 0 and 1. Although any pair of numbers would be suitable for coding the two groups of a dichotomous variable, the 0-1 coding scheme has the major advantage mentioned in section 1.3 that you cannot overlook. That is, all coded values sum to the number of cases coded as 1. Moreover, the arithmetic mean of the coded values equals the proportion of cases coded as one, which by itself is a quantity of interest. Because some statistical procedures are complex, any simplification in the coding scheme or in the notations will pay off.

Other examples of categorical variables include motorcycle manufacturers (i.e. $X = $ *Motorcycle Manufacturer*) that can be coded as 1 for Honda, 2 for Yamaha, 3 for Kawasaki, 4 for Suzuki, 5 for Hartley-Davidson, and 5 for other. That is $X = 1, 2, 3, 4, 5$, and these codes represent labels or identifiers, and cannot be part of an arithmetic calculation in any meaningful way. You may distribute the number of motorcycles sold in a given year across manufacturers to determine market share. Note that no ranking of motorcycle manufacturers is possible based on the codes assigned to them, nor is any ranking possible among the values (0 and 1) of the gender variable. Random variables not offering any ranking possibility form a special class of categorical variables called *nominal scale* variables. These variables are used is some inferential procedures involving proportions such as the chi-square test to be discussed in subsequent chapters.

Other categorical variables allow for ranking and are called *ordinal scale* variables. This class of variables includes for example *Educational Attainment* defined in Table 1.1 and taking the values 1,2,3, and 4. There is a natural order in these numbers because "high school graduate" is normally superior to "some high school." Does it really matter whether you code this variable as 1,2,3, and 4 as oppose to 4, 7, 23,and 31 ? The answer is no, it does not matter, because ordinal scale variables are typically ranked first and the obtained ranks are further processed with a special statistical technique (e.g. non-parametric tests of hypothesis to be discussed later in the book). Consequently, the only thing that matters when coding educational attainment for example is to code "high school graduate" with a value higher than that used to code "Some high school" in order to preserve the natural order of things. Otherwise, the analysis will be impossible to interpret.

1.4.2   Measurement Variables

Length, Width, or *Height* are examples of *Measurement variables.* A measurement variable can be defined as a variable that takes values produced by a measuring instrument. The length takes values produced by a yardstick, while the weight takes values produced by a weighting scale. Note that a measuring instrument is not necessarily a hard physical equipment such as the weighting scale, it could also be one of these survey instruments used in psychological assessment, or a scoring model such as those used by credit card companies.

If the values that the measurement variable takes have numerical increments that correspond to equal differences in the physical entity over the entire range of measurement, then this variable is said to be an *Interval scale* variable. Measurements aimed at quantifying opinions would generally not produce interval scale variables. For example a variable measuring the level of satisfaction as "very dissatisfied" (coded as 1), "dissatisfied" (coded as 2), "neutral" (coded as 3), "satisfied" (coded as 4), and "very satisfied" (coded as 5) cannot be of interval type, since a numerical increment of 1 from very dissatisfied to dissatisfied will certainly not translate into the same chance in satisfaction level as the an increment of 1 from satisfied to very satisfied. Examples of interval data include, temperature (in Celsius or Fahrenheit), year (e.g. 1980, 1981, 1982, etc...), and the different psychological test or credit scores. Note that the interpretation of a credit score for example may be debatable, but it is measured in such a way that a numerical increment will generally correspond to equal differences in abilities to reimburse debts. Several statistical procedures discussed in chapters 8 and 10 will be valid for interval scale data, but not for categorical data.

Interval scale variables such as the temperature or the credit score have a fundamental difference with other measurement variables such as the height or the weight, which is the location and the meaning of the numerical origin. The height and the weight have both a numerical origin of 0, which corresponds to the total absence of any physical matter. For height and weight, 0 is a natural origin, which provides a clear-cut point where the measurement of the physical magnitudes begins. A 0 temperature on the other hand does not represent a total absence of temperature (very few people will dare wear t-shirts in a zero-degree temperature, especially if expressed in the Fahrenheit scale). Where does the measurement of temperature begin? Variables which take values with a natural origin are called *Ratio scale* variables.

The existence of a natural origin for ratio scale measurements has an important implication in practice, which is the possibility of making ratio comparisons. For example weighting 200 pounds means that you are weighting twice heavier than a 100-pound person. Such a ratio comparison would be impossible with interval scale variables. When the temperature is 40^0C, you will certainly not feel twice warmer than when the temperature is 20^0C. You should probably not put more emphasis on the difference between interval and ratio scale variables more than it is necessary. Most statistical techniques that are valid for one of these 2 types will also be valid for the other.

In many statistics textbooks you will often see the terms *discrete variables* or *continuous variables*. These 2 terms belong primarily to the language of mathematics. They represent notions that are relevant for a rigorous formulation of statistical theories based on the language of mathematics. The continuous variable is the mathematical idealization of a measurement variable, and is assumed to take all possible values in a continuum of possible values. The discrete variable is any variable that is not continuous. The set of possible values of a discrete variable if often assumed to be countable. Practitioners should talk about measurement and categorical variables, while mathematical statisticians can talk about continuous and discrete variables.

A measurement variable such as height as used in practice is not really continuous, since it will generally be rounded either to the nearest integer or to a single digit after the decimal point. Still the rounded height is seen as a rough approximation of a hypothetical and more exact value that belongs to a continuum. Statistical results that are valid for continuous variables should nevertheless be applied to these pseudo-continuous numbers. There is nothing wrong with that, and it is unnecessary to waste time wondering whether your variable is continuous enough. The ride from the drawing board of theory to the messy world of hard can be bumpy. People with good judgement under those circumstances will have an edge.

1.5. The Software

If you are a practitioner, the study of statistical methods and techniques for you, is a first step towards achieving other scientific goals. Because soon or later you will need to take a dataset and actually implement those techniques you learned. This will be achieved only if you have a software product that you master reasonably well, and which you can use to process your data before interpreting the output. Unless you are dealing with a very small dataset, and perhaps wanting to implement a basic technique for producing simple descriptive statistics such as averages, you are generally not going to succeed with a manual manipulation of your data. The era of statistical analysis by hand and handheld electronic calculators is over, and has in fact been over for decades. A computer, and a software product (ideally simple to learn) are mandatory if you are going to be serious about statistics.

What software? and for what purpose? It is difficult to recommend a particular software that is suitable to all needs. Such a software does not exist. I personally use multiple software packages for my consulting projects depending on the nature of the task. Among others, I used the R package, Excel, SAS, OpenOffice Calc, and only occasionally Stata, and SPSS. All examples in this book are given in the MS Excel for Windows (2007 & 2010), R, or OpenOffice Calc. My choice of these 3 products is justified by the fact that they are either free or widely available for personable use. R and OpenOffice Calc are completely free, while Microsoft Excel, although not free is widely available and often come loaded in most new computers running the Windows operating system. I will now briefly discuss the merits and limitations of each of these 3 products.

1.5.1   MS Excel (Shipped with the MS Office Suite)

Some statisticians have advised against the use of Excel for statistical analysis. I beg to disagree on this, and strongly advise using Excel when appropriate. There are indeed some odd datasets full of unusual numbers that Excel may not be able to process adequately. But it is highly unlikely that you will ever have to deal with such datasets in business or social research. However, if you are conducting high-level scientific research that requires robust algorithms to adequately handle unusual data, then Excel is certainly not for you. Excel in my opinion, is an excellent compromise between statistical capability and ease of use. I myself use Excel each time I need to perform a quick analysis on a small to medium size dataset.

Excel may not be a very efficient production tool for practitioners wanting to perform a large number of analyzes of the same type using perhaps many different datasets. If you are able to develop Excel macros then you can automate several tasks and transform Excel into a good production tool. But investing in learning macro development is probably not justified unless you are already a heavy Excel user.

Which Version of Excel?

If you want to use Excel for statistical analysis, and your version is older than Excel 2010, I would advise that you upgrade to upgrade to Excel 2010, which is the latest version at the time of the writing of this book. In Excel 2010, the implementation of statistical functions has improved dramatically from what it was in the earlier version 2007. Excel developers have done a incredibly good job making the statistical functions work the way most practicing statisticians expect them to work. This is particularly true for the different functions used to evaluate the probability distribution functions discussed in chapter 2. Section E.8 of Appendix E describes the statistical functions used in this book.

Having said that, Excel 2007 still implements reasonably well most of the statistical techniques you may care about. If you already own Excel 2007 you may continue using it until you see the need to upgrade. You may well not need. However, a version older than 2007 may simply not be particularly useful.

Excel Add-Ins

An Excel Add-In is an external file that Excel can load when it starts up. The file contains a program that adds additional functionality to Excel, usually in the form of new functions or modules. Excel is shipped with a variety of Add-Ins ready for you to load and start using, and many third-party Add-Ins are available. The two Excel Add-Ins that I use in this book are the "Analysis ToolPak" and "Solver". Both come with MS Excel, but must be activated before they can be used for the first time. Appendix E provides all the instructions for setting up the Analysis

ToolPak. These instructions are to be used for setting up Solver as well (you would select Solver Add-In instead of Analysis ToolPak).

The Analysis ToolPak contains a long list of statistical modules each of which implements a specific statistical procedure. Solver on the other hand, is a power tool for solving optimization problems. The reason I introduced solver in this book, is to allow you to use many important statistical techniques that were previously accessible only to individuals with a good grasp of Calculus. With Solver, you no longer need Calculus to understand and implement advanced statistical techniques such as nonlinear regression, maximum likelihood estimation and more.

This book does not teach you how to use Excel. However, I will show you what can be done with Excel, with the Analysis ToolPak, and Excel Solver. But feel free to further explore these tools if you are going to use them to perform your own analysis.

1.5.2   OpenOffice Calc (it's free)

OpenOffice is a free fully-fledged Office suite that comprises among other products a text processor called Writer, and a spreadsheet called Calc that I will be concerned about. In this book, I use OpenOffice 3.2 that can be downloaded at,

`http://www.openoffice.org`

OpenOffice Calc 3.2 mimics the 2007 version of Excel very closely at least as far as statistical functions are concerned. The statistical functions in both Calc 3.2 and Excel 2007 generally have the same names. The only exception is that the function arguments in Excel are separated with commas, while they are separated with semi colons (;) in Calc.

Note that with OpenOffice Calc, you will not have access to powerful add-ins such as the Analysis ToolPak and Solver of Excel. Again, you may want to proceed with the reading of the book until you see what you need and what you do not need. I have used Calc on many occasions, and I was satisfied with what I was able to do with it.

1.5.3   The R Package (it's free)

The R package has become an immensely popular statistical package across the world. If you are going to do statistical analysis on a regular basis for many years, and you do not know which statistical software to learn, this is the one to get. No doubt. You will enjoy the support of an extended online support group where you will be able to ask questions. Moreover, the product is entirely free, and numerous books have been published to help practitioners and scientists learn this product.

The R package can be downloaded at,

`http://www.r-project.com`

Furthermore, the PDF file "Using R for Introductory Statistics" by John Verzani, which provides a short and friendly introduction to the R package, and a good overview of its capabilities can be downloaded at,

`http://cran.r-project.org/doc/contrib/Verzani-SimpleR.pdf`

R is an interactive computing environment that makes a large collection of statistical functions available to you. Using R is about finding the right function and learning how to use it. I provide several examples in this book that uses R. If you have never previously used R, you may want to proceed with the reading of this book to see what one can do with this package before deciding whether you want to use it or not. R gives you the opportunity to develop your own functions for performing routine tasks as well as develop completely new packages for advanced users.

Appendix F describes some of the R functions most frequently used in this book. In order to avoid filling the body of the book with R material that non R users are not interested in, I will regularly refer to Appendix F where interested readers could find more details on the use of R.

1.6. Basic Exploratory Statistics

In the beginning of this chapter, I indicated that exploratory statistics was concerned about the description of the variability associated with random variables. Although variability at the population level will often be your primary interest, the sample is what you will often get. Therefore the description of variability will be done based on sample data. I now like to formally introduce the important notions of population and sample.

Populations and Samples

Statistics is often made difficult either because the study population has been ill-defined or because defining it has been downright omitted. Occasionally, the study population will be well defined then ignored afterwards, when in reality it should be driving the formulation and the implementation of the statistical procedure. The way you define your population must reflect your interests at the time of the investigation. This shows how critical it is for you to take the time to clearly envision what you want to focus on. This phase of the investigation is not statistical. It is about setting specific goals.

If you want to study income for example, you will have two

types of populations you can define. The first population will be the specific group of individuals whose income levels are of interest to you. If you investigate income as it relates to residents of the city of Los Angeles, it will be wise to consider defining your population as all residents of Los Angeles who are included in the labor force, and to see income as the characteristic of interest. This population is a *Population of Units*. If your goal is to study income as it is affected by educational attainment with no reference to a specific group of individuals, then you will want to define a *Numerical population*, where one member is an income figure such as $46,900.00. Your numerical population will then be a large collection possible numbers.

Populations of units are finite by nature. That is they represent a finite group of units that are the focus of the investigation. A correct and rigorous investigation of such populations may require the use of survey sampling methods that are discussed in chapter 12. Numerical populations on the other hand, have an unspecified number of numerical values that you will probably not be interested in, and which can be arbitrarily large. The mathematical idealization of these numerical populations consists of saying that these are *infinite populations*, which can take any values in a continuum. This idealization facilitates the formulation of statistical theories that only mathematical statisticians should care about, not practitioners.

To conduct your investigation you will generally focus on a small portion of your population called the **Sample**. Because the numbers in the numerical population are not tied to specific physical units, these populations are abstract in nature. The selection of samples from these populations is generally simple, since you do not have to worry about reflecting a particular structure that may be inherent to them. The only requirement being to select a random sample in order to remove any possible selection bias from the process. Most techniques presented in standard statistics textbooks assume that you are dealing with numerical populations. If you are dealing with a concrete finite population of units, then sampling should be carried out carefully.

The remaining portion of this chapter is devoted to the study of basic exploratory techniques that you would apply to the sample data and initiate your statistical analysis. These techniques revolve around a graphical description of sample frequency distributions as a rough approximation of the underlying probability distributions, and the use of sample summary statistics as numerical approximations of the probability distribution parameters.

1.6.1 Frequency Distributions

Studying the frequency distribution of categorical and measurement variables requires different approaches. I will start with categorical variables and will discuss measurement variables afterwards.

Frequency Distribution of Categorical Variables

For categorical variables, the description of variability is essentially done with the *Frequency table* and the *Bar graph*. Consider for example Table 1.2 that shows the distribution of 500 subscribers of a newspaper by the type of community where the subscriber resides. The variable of interest is $X = $ *Type of community*, which takes 3 possible values, {city, suburb, rural} conveniently coded as 1, 2, and 3 respectively. Table 1.2 shows the raw count of subscribers (also called frequency)as well as the relative frequency (expressed in percentages) representing the ratio of the frequency to the total number of subscribers.

The column of relative frequencies is the most important since it describes the distribution of subscribers without being much affected by the number of subscribers. Relative frequencies are likely to remain stable in other studies based on a different number of respondents. This is stable information you can rely upon during a decision-making process.

Table 1.2 : Distribution of subscribers by community type

Type of Community	Coded Value	Number of Subscribers	Relative Frequency (%)
City	1	190	38
Suburb	2	170	34
Rural	3	140	28
Total		500	100

Figure 1.1 shows a bar graph that depicts the distribution of subscribers by type of the community of residence. Its main advantage is its ability to provide a quick and visual comparison of the different levels of the categorical variable X.

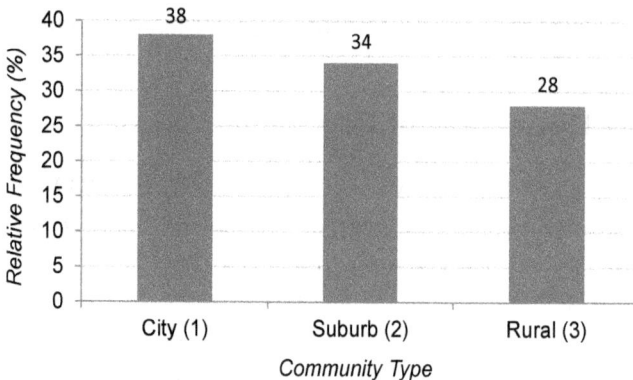

Figure 1.1. Bar graph for categorical variable "Type of Community"

The ordering of categories in the graph is arbitrary as are the codes 1,2, and 3 assigned to them. However, ordinal categorical variables will generally suggest a more natural ordering.

I created Figure 1.1 using Excel, although the same bar graph may be produced with R by typing the appropriate commands in the R console as shown in Figure 1.2. The formatting of this type of charts is generally more convenient with Excel. However, if you are already an R user, you may continue using it for the purpose of creating charts as well.

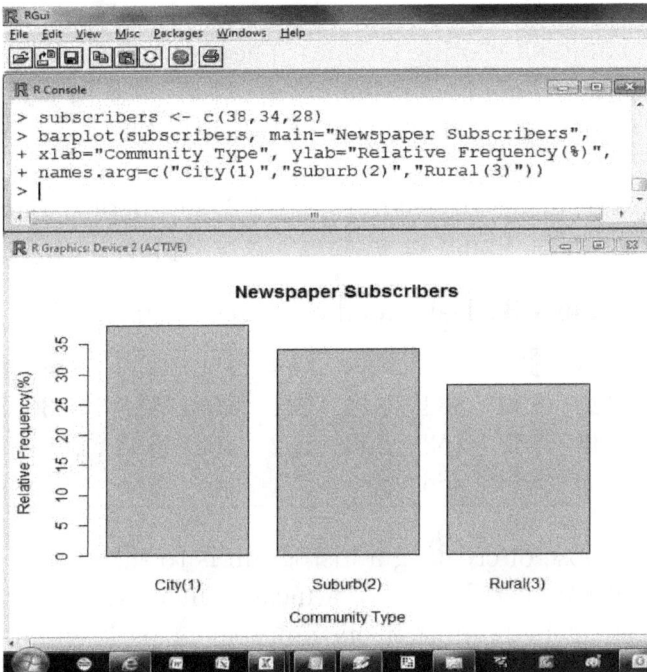

Figure 1.2. Bar graph for categorical variable "Type of Community" Using R

Frequency Distribution of Measurement Variables

Analyzing the frequency distribution of a measurement variable requires that you first categorize it before creating a frequency table and a *histogram*. The frequency table shows counts of observations within each of the categories called *bins*. The histogram is a graphical representation of the distribution of measurement data. It is similar to the bar graph, with the exception that the bars are adjacent and joined at the class boundaries to reflect the continuous nature of measurement data.

Table 1.3 shows the dollar amount that 40 families spent on food on a given day in an amusement park. Such a flat list of numbers does not allow us to tell any useful story regarding the spending habits of visiting families. The measurement variable of interest is $X = Daily\ dollar\ amount\ spent\ on\ food\ by\ a\ family$. In order to explore the distribution of this variable, I used the **His-togram** tool of Excel's Analysis ToolPak to create the frequency table 1.4 and the histogram in Figure 1.3. The details for creating Figure 1.3 can be found in section E.8 of Appendix E.

Table 1.3 : Daily food expenses of 40 families

$77	$18	$63	$84	$38	$54	$50	$59	$54	$56
$36	$26	$50	$34	$44	$41	$58	$58	$53	$51
$62	$43	$52	$53	$63	$62	$62	$65	$61	$52
$60	$60	$45	$66	$83	$71	$63	$58	$61	$71

The purpose of creating a histogram is to take a first look at the form of the distribution of a measurement variable. Are the amounts spent by families in the amusement park symmetrically distributed? Are some of these amounts unduly high? Unduly low? Is the distribution skewed to right? Skewed to left? Is there a particular value around which most amounts are concentrated?

There is no unique number of classes (or bins) that will allow you to best explore all aspects of the frequency distribution. To determine the number of bins, most statistics book recommend the use of the so-called "2 to the k rule" that recommends to use as number of bins, the smallest integer value k for which 2^{k-1} exceeds the number of observations. This rule is also known as the Sturges' rule because it was proposed by Sturges (1926).

Excel appears to use the simpler rule that recommends the number of bins to equal the square root of the number of observations, rounded up to the nearest integer. It does not really matter much which rule you use, you may still need to modify the bins to obtain a better look at the distribution. For example, if your data series contains outliers (these are values that are located far away from the bulk of the data), you will need to have additional bins (some of which will have 0 observation) in order to have a more detailed look at the middle of the distribution. This is due to the fact that the bin width is calculated as (*Max Value* − *Min Value*)$/(k-1)$ where k is the number of bins. Therefore, a large extreme value will increase the bin width dramatically creating a high concentration of observations in the middle of the distribution, which will conceal the form of the distribution at that location.

Table 1.4: Frequency Table of daily food expense data

Bin	Bin Code[a]	Frequency
$0 to $18	18	1
$18 to $29	29	1
$29 to $40	40	3
$40 to $51	51	7
$51 to $62	62	18
$62 to $73	73	7
$73 to $84	More	3
Total		40

[a]The bin code is all Excel produces. The associated bin definition as interval is in the first column.

If the construction of histograms is an issue that inspires you, you may want to look at other rules that have been suggested in the literature such as those of Freedman and Diaconis (1981), and Scott (1979) .

Histogram

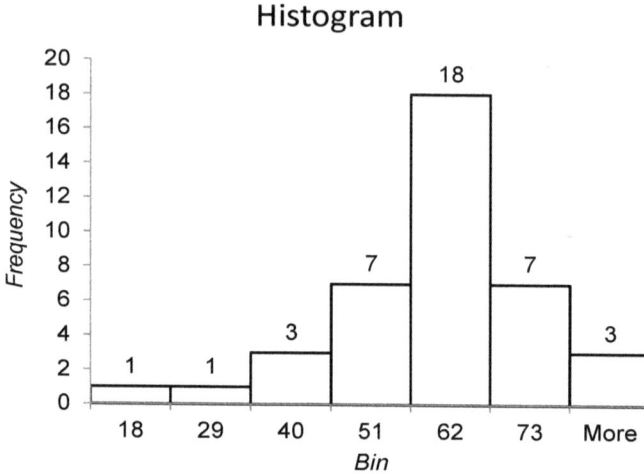

Figure 1.3. Histogram of Table 1.3 data

Figure 1.3 shows a rather regular smoothed distribution, which does not appear to have outliers. There is probably no need to to have both bins 18 and 29. Both may well be collapsed into a single bin 29 to obtain a smoother histogram. If a family had spent about $6 or $150 for example these would have been outliers resulting in a distortion of the histogram.

The R package too, offers an easy way to plot histograms with the function `hist()`. In its simplest form, this function will take a single argument representing the vector containing the data series. Numerous other parameters can be specified to have other features added to the graph (see the R documentation for more details).

1.6.2 Summary Statistics

In the last section, I discussed two of the most commonly-used graphical methods for exploring the probability distribution of a random variable X, which are the bar graphs for categorical variables, and the histograms for measurement variables. There are other graphical techniques such as the boxplot, the pie chart, or the frequency polygon that I did not discuss due to their limited implications in statistical inference. In this section, I want to add to these visualization techniques, some numerical summary measures that will describe specific aspects of the variable distribution in a more precise way.

Two types of summary measures will be described in this section. These are the measures of location that will inform you about the location of the bulk of the data, and the measures of dispersion intended to inform you about the spread of the possible values of the characteristic of interest. A too wide range of values is an indication that any given value would be expected to be situated far away form the middle of the distribution. This situation will require a massive collection of data to obtain reliable information about the population being investigated.

Measures of Location

The measures of location that I like to mention are the mean, and the median. Each of these quantities has a population version and a sample version.

The Mean

When your population of interest is a finite and concrete population of N identifiable units, then the population mean of a variable X will be the average of all values $X_1, X_2, \cdots, X_N$ taken by the units. This population mean is denoted by $\overline{X}$ (read capital X bar), and has the following algebraic expression:

$$\overline{X} = \frac{1}{N} \sum_{i=1}^{N} X_i. \qquad (1.1)$$

If $X =$ *annual income* and your target population is made up of the 300,000 residents of a given city, then $\overline{X}$ as defined by equation 1.1 will be the population mean income, and will be seen as the finite population parameter

If the population of interest is an abstract numerical population as is often the case in most textbooks on classical statistics, the mean is actually the expected value of characteristic X and will be denoted by μ (read mu) or $E(X)$. That is $\mu = E(X)$. This expected value can be computed only if the theoretical probability distribution of X is specified. Several theoretical probability distributions will be discussed in chapter 2 that could be used for that purpose.

Whether the population parameter is a concrete (although unknown) quantity such as $\overline{X}$ or a theoretical construct such as μ, you will be able to approximate it numerically by selecting a sample of size n. The unknown population mean will then be approximated by the arithmetic mean (also called the sample mean) of the n sample values $x_1, x_2, \cdots, x_n$, which will be denoted by $\overline{x}$ (read small x bar). The equation of the sample mean is given by:

$$\overline{x} = \frac{1}{n} \sum_{i=1}^{n} x_i. \qquad (1.2)$$

The Median

Along with the mean, the median is another popular measure of central tendency that represents the middle of the distribution of a characteristic X. For a finite population of size N, the population median of X is the numeric value $\widetilde{X}$ that splits the N population values $X_1, X_2, \cdots, X_N$ in half. For a numerical population (viewed mathematically as *infinite*), the median denoted by $\widetilde{\mu}$ is a theoretical construct, which the random variable X will exceed with a probability of 0.5. The population median ($\widetilde{\mu}$ or $\widetilde{X}$) will generally be approximated in practice by the sample median $\widetilde{x}$, a quantity that splits the n sample observations $x_1, x_2, \cdots, x_n$ in half.

If your sample contains 5 values $\{2,5,4,3,1\}$ for example, its median is obtained by first sorting it as $\{1,2,3,4,5\}$, and by taking it middle value, which in this case is 3. That is, $\widetilde{x} = 3$. When the sample size is an odd number (e.g. 5), the median will always be one of the numbers in the sample. The situation is different when the sample size is an even number. If $\{2,5,3,1\}$ represents your sample, then its sorted version will be $\{1,2,3,5\}$, which has 2 middle points 2 and 3. The median in this case will be the average of the 2 middle points. That is, $\widetilde{x} = (2+3)/2 = 2.5$.

If the mean is already available as a measure of central tendency, why would you need the median as an alternative measure? The reason is that the mean is highly sensitive to the presence of outliers in the sample, making the sample median the preferred choice when the distribution of data is skewed. The mean income calculated from a sample $\{\$55,000, \$50,000, \$75,000, \$63,000, \$67,000\}$ is $\overline{x} = (\$55,000+\$50,000+\$75,000+\$63,000+\$67,000)/5 = \$62,000$. However, replacing $\$67,000$ with much larger number such as $\$280,000$ will yield a dramatically high mean income of $\$104,600$, which does not provide a good representation of the bulk of your data. By the way, the median in both samples remains at the same level of $\widetilde{x} = \$63,000$.

Proportions and Probabilities

The mean and the median are typically calculated for measurement variables. With categorical variables, you generally want to calculate proportions or probabilities. The *proportion* of students who receive an A grade in the last statistics represents the relative number of students to have received that grade, and will be denoted by p.

Suppose that you want to evaluate the chance of a new graduate to get a job within 3 months after graduation. In this context, you have not specified any particular population of students in any school or any city, nor have you specified any particular timeframe for your inquiry. You are then dealing with an abstract population to which you cannot associate a concrete measure such as a proportion. Therefore, you will not talk about the proportion of new graduates to get a job within 3 months after graduation. Instead, you will talk about the *probability* for a new graduate to get a job. The probability represents a population parameter that you will denote by π (read pi).

Measures of Dispersion

The measures of dispersion that I like to present are the variance, the standard deviation, and the percentiles. If you have n sample data points $\{x_1, x_2, \cdots, x_n\}$, the *sample variance* denoted by s^2 represents the mean squared difference from the sample observations to the overall mean. This measure summarizes the spread of the sample observations around the sample mean, and is mathematically formulated as,

$$s^2 = \frac{1}{n-1} \sum_{i=1}^{n} (x_i - \overline{x})^2. \tag{1.3}$$

If you are seeing the above expression for the first time, you may be intrigued by its denominator, which is $n-1$ as opposed to the n often used in the calculation of averages. In practice using n or $n-1$ in the denominator will not affect the numerical values in a noticeable way. However, using $n-1$ gives the sample variance a mathematical property called unbiasedness, which mathematical statisticians seem to care about very much.

Example 1.1 _____

If your sample contains the following numbers {$55,000, $50,000, $75,000, $63,000, $67,000} then the sample mean will be $\bar{x} =$ $62,000$, and the sample variance calculated as follows:

$$s^2 = \left[(55000 - 62000)^2 + (50000 - 62000)^2 + (75000 - 62000)^2 \right.$$
$$\left. + (63000 - 62000)^2 + (67000 - 62000)^2\right]/(5 - 1),$$
$$= 97,000,000.$$

The difficulty with the use of sample variance stems from the fact that it is not expressed in the same original units as your sample data. In the above example 97,000,000 does not represents a dollar amount because the original values were squared to obtain the sample variance. To correct this problem, it is recommended to use the *sample standard deviation*, which represents the square root of the sample variance. The standard deviation associated with the sample variance of example 1.1 is $s = \sqrt{97,000,000} = \$9,848.86$, which compares very well with the initial sample values.

At the population level, the *Population Variance* measures how large you expect the squared difference of any given observation to its expected value to be. It is a population parameter that is denoted by S^2 when the population is finite and concrete, and is denoted by σ^2 (read sigma square) for abstract numerical populations. The population variance is generally unknown and should be estimated from the sample. The population standard deviation is the square root of the population variance and is denoted by S or σ for finite and abstract populations respectively.

Percentiles

Just as the median divides the sample data into two equal parts, the *Quartiles* divide the sample data into 4 equal parts, the *Deciles* into 10 parts, and the *Percentiles* into 100 equal parts. There are 3 quartiles denoted by Q_1, Q_2, and Q_3, which respectively represent the 25*th*, 50*th*,and the 75*th* percentiles. Likewise, the 2*nd* decile is actually the 20*th* percentile. Consequently, knowing how to calculate percentiles is sufficient for calculating the quartiles and the deciles.

The general approach for calculating the pth percentile is to first sort your n sample data points $x_1, x_2, \cdots, x_n$ in ascending order. Let L_p be the location (or the rank) of the pth percentile on the sorted list. Therefore,

$$L_p = Integer\ part\ of\ \left((n+1)\frac{p}{100}\right). \qquad (1.4)$$

Example 1.2 _____

Consider the following bi-weekly salaries of 15 entry-level accountants: $2,038; $1,758; $1,721; $1,637; $2,097; $2,047; $2,205; $1,787; $2,287; $1,940; $2,311; $2,054; $2,406; $1,471; $1,460. To calculate the third quartile Q_3, I need to first sort the series in ascending order as follows: $1,460; $1,471; $1,637; $1,721; $1,758; $1,787; $1,940; $2,038; $2,047; $2,054; $2,097; $2,205; $2,287; $2,311; $2,406. Therefore, the location of the third quartile (i.e. 75*th* percentile is,

$$L_{75} = Integer\ part\ of\ \left((15+1) \times \frac{75}{100}\right),$$

$$= 12.$$

Consequently the 12*th* number on the sorted list, which $2,205 is the third quartile ($Q_3 = \$2,205$).

_____

Percentiles are useful statistics, since they allow you to identify the top 10% units in a given group for example. Note that p^{th} percentile of an abstract numerical population for which only the probability distribution might be available, will generally be defined as the number x_p that will be exceeded with a probability of $1 - p$. I will further discuss these notions in subsequent chapters.

MS Excel can be used for producing basic descriptive statistics. One option would be to use the "Descriptive Statistics" module of the Analysis ToolPak. You will want to ensure that the "Summary Statistics" check box of the dialog form is checked. I have not discussed some of the summary statistics that Excel produces. One of them is the *Standard error*. It represents the standard deviation of the sample mean[7], and is generally obtained as the ratio of the simple standard deviation to the square root of the sample size. You will also want to know that Excel has functions for computing the percentiles, such as PERCENTILE.EXC, PERCENTILE.INC, PERCENTRANK.EXC, and PERCENTRANK.INC for Excel 2010, and PERCENTILE, and PERCENTRANK for Excel 2007. However, these functions do not compute the percentiles according to the algorithm presented earlier. I still advise to use them unless you are conducting a high-level scientific research where any tiny loss of precision is a big loss.

The R package too offers the summary() function that generates some limited basic summary statistics. Many R packages have been developed, and can be installed to produce a wider range of summary statistics. See the R documentation for more details.

---

[7]Note that as an average of n random variables, the sample mean itself is a random variable, which has a mean and a variance of its own.

1.7. The Calculus of Probability

I previously indicated that statistics in general can be seen as the study and management of variability. One notion that happened to play a pivotal role in the study of variability is the law of probability. It is the quantitative representation of variability, and the foundation of statistical inference.

Suppose you conduct telephone interviews for a survey research firm at the cost of $50 per completed interview, and the response rate is known to be 70%. If potential respondents are selected randomly, then the dollar amount spent per interview is a two-value random variable that I will call X and that takes the 2 values $5 (if respondent declines the interview) and $50 otherwise. The law of probability associated with X is defined as follows:

$$P(X = \$50) = 0.70 \; and \; P(X = \$5) = 0.30, \qquad (1.5)$$

where $P(X = \$50)$ and $P(X = \$5)$ represent the probability for the interview cost to be $50 and $5 respectively, and 0.30 is obtained as $1 - 0.70$. Once you know the law of probability of the variable of interest, you can use it to derive various statistical measures such as the mean, the median, the variance, and even conduct tests of hypotheses and more. Therefore, it should not come as a surprise if several chapters are often devoted to the notion of probability in many statistics textbooks. How can you compute the expected value of X (the interview expense) using the simple law of probability 1.5? Let $E(X)$ be the expected value of X. It is obtained as follows:

$$E(X) = \$50 \times 0.7 + \$5 \times 0.3 = \$36.5. \qquad (1.6)$$

That is the expected value of X is the sum of its possible values, weighted by their respective probabilities of occurrence. On the other hand, if you want to know how far you should expect either dollar amount ($5 or $50) to be from the expected value, you could compute the *standard deviation* as follows:

$$SD(X) = \sqrt{(50 - 36.5)^2 \times 0.7 + (5 - 36.5)^2 \times 0.3,}$$
$$= \sqrt{425.25} = 20.62.$$

A standard deviation of 20.62 for an expected value of 36.5 indicates that any given value of the random variable X can be expected to deviate from the overall mean by as much 56 percent[8]. This substantial variation in the data was calculated using the possible values of X along with their respective probabilities of being observed.

But when I use the expression "probability of response" to a survey, what number am I exactly referring to? In this section, I am going to clarify the concept of probability. Because different experiments will lead to different estimations of the same probability, you need to have a good grasp of this notion to have an adequate interpretation of a statistical result. This is a good place to have a thorough discussion about the probability concept since in chapter 2, I will introduce a number of probability distributions that will be used throughout this book. I will also take this opportunity to mention a few fundamental concepts underlying the basic mathematical theory of probability.

1.7.1 What is a Probability?

The ordinary law of nature describes the conditions under which an event will occur or will not. For example *every*

---

[8]Note that 56% is obtained as the ratio of the standard deviation to the expected value. This ratio is often called the *Coefficient of Variation*

stone placed in water will sink to the bottom if no external force is exercised on it. Once you describe the conditions under which the experiment will be conducted, you will know before the experiment even takes place, what the outcome will be. However, you will be led to statistical methods in those situations where the prediction of the course of events is impossible. When conducting a telephone interview survey, no matter how careful the planning, sometimes a solicited individual will agree to participate, sometimes that person will decline. It is the *Law of probability* that can describe the conditions under which an individual will agree to participate with a specified level of chance called the probability of response. How do I know what is exactly the magnitude of that probability of response? Intuitively one may think that if I attempt to contact similar individuals under the same conditions a large number of times, the relative number of times someone will agree to participate will tend to stabilize around a certain number, giving me a sense of the probability magnitude.

Many statistics textbooks indeed define probability as a limiting value obtained following an unlimited series of experiments executed under the exact same conditions. This definition, although widespread, can only give you a broad intuitive sense of what probability actually is, and does not have a solid mathematical foundation. A formal, rigorous, and concrete definition of the notion of probability is sometimes possible in specific situations. For example, consider an experiment that consists of selecting 5 individuals randomly out of a predetermined group of 20. The total number of samples of 5 out of 20 is 15,504. Therefore, the probability for a particular event occurring is defined as the relative number of samples of 5 individuals for which that event occurs.

For experiments involving a predetermined group of units or individuals in their design, it will generally be possible to have a concrete definition of the notion of probability. Probabilities based on experiments for which the experimental conditions are not fully specified will generally not be defined in a way that is tied to reality. Suppose for example that you want to compute the probability the a light bulb manufactured under certain conditions will burn after 2,000 hours. The concrete meaning of such a number in this context is unclear, although you and I have some sense of what it represents. You should note however that, whether the real meaning of probability is clarified in a given context or not, the existence of the law of probability will be a reality that can and should be studied.

The notion of probability is intimately tied to that of experiments, primarily because it is with a repetition of experiments that one often quantifies the probability. Here are few definitions that will be needed later in this section:

Definition. 1.1.

> The *Experiment* is seen as a process that produces a single of several possible observations.

In the conduct of an telephone interview survey, you may define your experiment in a number of ways. The simplest of the experiments consists of selecting randomly one potential participant from a specified pool of individuals, and determining whether that person agrees to participate or not. Only two observations are possible. Either the selected individual agrees to participate or she declines.

Definition. 1.2.

> An *Outcome* is a particular result of an experiment. And the set of all possible outcomes of an experiment is called the *Sample space*

If your experiment consists of selecting 5 potential survey participants for the purpose of determining the number X of those who are willing to participate, then one possible outcome is $X = 4$. All the possible outcomes are $\{0,1,2,3,4,5\}$.

Definition. 1.3.

> An (experimental) *Event* is a collection of one or several outcomes from an experiment.

It is common practice to label an event with capital letters such as A, or B for easy reference. Using the previous experiment of a survey of 5 randomly selected individuals, I can define an event as *E= "More than 3 individuals agreed to participate in the survey"*. This event is the collection of the following 3 events: *A= "Exactly 3 persons agreed to participate"*, *B= "Exactly 4 persons agreed to participate"*, and *C= "Exactly 4 persons agreed to participate."* I will then say that event E is a *Compound event*, which the union[9] of the 3 *Simple events* A, B, and C. A simple event is one that matches one of the possible outcomes, while a compound event is derived from two or more simple events.

When an event E is the union of three events A, B, or C, one will generally express this as,

$$E = A \cup B \cup C.$$

A second compound event could be defined as $F = A \cup B$, which

---

[9]The union of 3 events occurs when any one of the 3 events occurs.

means that *"3 or 4 individuals agreed to participate"*. Additional operations are possible with events. For example, you could define event G that occurs only when both events E and F occur at the same time. Event G will be called "E *intersection* F" and denoted by,

$$G = E \cap F.$$

Definition. 1.4.

> Two events A and B are said to be *mutually exclusive* or *disjoint* when they have no outcome in common.

In practice, saying that two events are mutually exclusive amounts to saying that when one of the two occurs, the other cannot. For example the number of respondents to a survey cannot exceed 100 and still be below 55 at the same time.

Definition. 1.5.

> A *Random Event* is an event which has the property that, following the execution of a long sequence of experiments repeated under the same fixed conditions, their frequencies tend to become stable and concentrated around a certain level called its *Probability*.

In the next section, I will present some elementary mathematical tools that will enable you to manipulate probabilities under various circumstances. A mathematical framework for studying laws of probabilities is necessary because researchers often start by establishing certain laws of probability by experiment or in a subjective manner, then proceed to derive new laws of probability by logical means under general assumptions. The mathematical framework of the next section will provide guidelines for an effective use of probabilities.

1.7.2 The Axioms of the Mathematical Theory of Probability

The whole mathematical theory of probability is based on the following 3 axioms[10]:

(1) The probability $P(A)$ (read P of A) of any given event A is assumed to lie between 0 and 1. That is,

$$0 \leq P(A) \leq 1, \textit{ for any given event A.} \qquad (1.7)$$

(2) The probability of the union of two mutually exclusive events is the sum of the probabilities of the individual events. That is, for any two mutually exclusive events A and B,

$$P(A \cup B) = P(A) + P(B). \qquad (1.8)$$

(3) Any event that is certain will have the maximum probability of 1. That is, if event U is certain, then

$$P(U) = 1. \qquad (1.9)$$

Axiom 1.7 is merely a mathematical formulation of the fact that the probability is often represented in practice in the form of a proportion. Therefore, it can only take values between 0 and 1. As for the second axiom 1.8, it represents the fact that you will observe the union of 2 disjoint events by observing one after the other, and the frequency of the union will be the sum of the frequencies of the individual components. The last axiom 1.9 reflects the fact that any event, which systematically occurs following the execution of an experiment should have the maximum probability of 1.

---

[10]An axiom is a statement considered to be true, and which is used in the foundation of a system of thoughts

The 3 axioms 1.7, 1.8, and 1.9 can be used to deduce several other probabilities once a few initial fundamental probabilities have been provided through experimentation or prior knowledge. I now present some of the best known and most useful properties of the law of probability that are deduced from the 3 foundational axioms. In order to help understand how probabilities must be evaluated here is what the great Russian mathematician A. N. Kolmogorov said (see Aleksandrov et al. (1999), Vol. Two, Part 3, Page 252):

> *The basic concepts of the theory of probability, namely random events and their probabilities, are completely analogous in their properties to plane figures and their areas. It is sufficient to understand by $A \cap B$ the intersection (i.e. common part) of two figures , by $A \cup B$ their union, by $\emptyset$ the conventional "empty" figure, and by $P(A)$ the area of figure A, whereupon the analogy is complete.*

I found this analogy very useful for conceptualizing the notion of probability when dealing with complex probabilistic issues. It will be further illustrated in chapter 2. Here are a few properties that are derived from the 3 foundational axioms:

▶ For any event A, $P(A) = 1 - P(A')$ where A' is the complement of A (i.e. any outcome that is not included in A).

▶ If events A and B are disjoint, then $P(A \cap B) = 0$.

▶ For any two events A and B, $P(A \cup B) = P(A) + P(B) - P(A \cap B)$.

Here is an example that illustrates how the probability can be calculated in practice.

Example 1.3

A random sample of 2,000 licensed drivers revealed the follo-wing numbers of speeding violations.

Table 1.5: Distribution of 2,000 Drivers by Number of Speeding Violations

Number of Violations	Number of Drivers
0	1,910
1	46
2	18
3	12
4	9
5 or more	5
Total	2,000

a *What is the experiment ?*

The experiment in this example consists of selecting a group of 2,000 drivers randomly and to obtain the number of speeding violations of each of them. You cannot be more specific than that because a particular sampling scheme has not been described. As for the sample space, depending on the timeframe used for observing the number of speeding violations, this one may vary from 0 to a large number.

b *List one possible event.*

One possible event is A={*A selected driver has 3 speeding violations or more*}

c What is the probability that a particular driver had exactly two speeding violations ?

Let me define a generic event A_k ={*A selected driver has exactly k speeding violations*} for $k = 0$ through 4, and A_5 ={*A selected driver has 5 violations or more*}. The problem is to compute the probability $P(A_2)$, which is the ratio of the number of drivers with 2 violations to the total number of drivers. That is, $P(A_2) = 18/2000 = 0.009..$

A alternative, and perhaps more convenient way to approach this is to define the random variable $X = \{$*The number of speeding violations of a randomly selected driver*$\}$, and compute $P(X = 2) = 18/2000 = 0.009$.

▶ The method that I used to respond to question (c) is the *Empirical approach* for assigning probabilities. Empirical probability is calculated as the ratio of the number of times the event occurs to the total number of observations. In other experiments, you may have to use the *Classical approach* for assigning probabilities. The *Classical probability* is calculated as the ratio of the number of favorable outcomes to the total number of possible outcomes.

CHAPTER $\boxed{2}$

Statistics and the Law of Probability

OBJECTIVE

The purpose of chapter 2 is to review some of the most popular discrete and continuous probability models. The practical motivations of these models will be particularly emphasized.

CONTENTS

2.1   Overview

The exact law of probability associated with many natural phenomena that researchers study in practice is generally unknown due to complexity, or limited information, since observed data simply gives you a snapshot of the phenomenon at a given point in time. Moreover, the experiment that led to these observations may even have been contaminated by external factors beyond the experimenter's control, distorting the very nature of the subject under investigation. What statisticians have done overtime to get around this difficulty was to develop a large collection of probability models that were thoroughly investigated. You would then identify one specific model that appears to match your data reasonably well, and use that data to fully specify the model during a model-fitting process. The fully specified model will then be available for conducting a more sophisticated analysis of the variation underlying the variable of interest.

Model-fitting is a critical step for going from observed data towards exploring the larger universe that produced these observations. The purpose of this chapter is to review the most useful of these theoretical probability models that you will need in the later chapters of this book, devoted to statistical inference. The discussion on model-fitting strategies is delayed until chapter 3, devoted to point and interval estimation.

In the next few sections, I am going to consider the probability model to be a two-element set (X, f) where X is a random variable, and f a family of probability distributions of which the distribution of X is assumed to be a possible member. In section 2.2, the probability models are associated with discrete random variables. A random variable is discrete when its set of possible values is either finite or countable (i.e. all possible values can be explicitly listed). Section 2.3 will be devoted to continuous random variables. These are random variables whose possible values lie in a continuum[1] such as an interval. Examples of continuous variables include the weight, the height.

2.2. Probability Models for Discrete Variables

The 5 discrete laws of probability that I am considering in this section are Bernoulli, Binomial, Hypergeometric, Poisson, and Multinomial. These are among the probability laws that I have found most useful in practice.

2.2.1 The Bernoulli Distribution - $Ber(p)$

Suppose that a researcher wants to study the incidence of a disease in a given region. The experiment will consist of selecting an individual and diagnosing the disease. Two outcomes are possible from such an experiment: the presence or the absence of the disease. One may consider a random variable X defined as follows:

$$X = \begin{cases} 1 & \textit{if a patient is diagnosed with the disease,} \\ 0 & \textit{otherwise.} \end{cases}$$

The probability law of X assigns a probability of occurrence to each of the 2 possible values 0 and 1 that X can take. The probability that the random variable X takes value 1 for example, is denoted by $P(X = 1)$ or by $p(1)$. Likewise, you will have $p(0)$ or $P(X = 0)$. The general form of the Bernoulli family of proba-

---

[1]One characteristic of a continuum is that one cannot list 2 consecutive numbers from a continuum without omitting more numbers in between

bility distributions is defined as follows :

$$p(x) = \begin{cases} \alpha & \textit{if } x = 1, \\ 1 - \alpha & \textit{if } x = 0, \\ 0 & \textit{otherwise,} \end{cases} \tag{2.1}$$

where α is a number greater than 0 and smaller than 1 ($0 < \alpha < 1$). It represents the parameter of the Bernoulli family of probability distributions, that must be estimated from observed data. The parameter α may represent the prevalence of a disease in a particular region for example. A value such as $\alpha = 0.05$ will define a specific member of the Bernoulli family of probability distributions. $p(x)$ is often called the *Probability mass function* (p.m.f.), or the *Probability distribution function* of X.

To conceptualize the notion of probability, it is convenient to see the law of probability as a mass system, where a total probability mass of 1 must be distributed across a number of mass points. In the case of a Bernoulli probability model, there are two mass points[2], which are 0 and 1, the portion of the unitary probability mass going to 1 is α, and the portion going to 0 is $1 - \alpha$.

The Swiss mathematician Jacob Bernoulli (1654-1705) was first to formally study the properties of this distribution. The expectation $E(X)$ and variance $V(X)$ of a Bernoulli variable X are given by:

$$E(X) = \alpha \textit{ and } V(X) = \alpha(1 - \alpha). \tag{2.2}$$

The Bernoulli probability model labeled as $\mathcal{B}er(p)$ has found an important application in the study of logistic regression (see chapter 11) used to predict the outcome of dichotomous variables.

---

[2]A mass point is simply a point in the sample space that is to receive a positive probability mass.

2.2.2 Binomial Distribution - $\mathcal{B}(n,p)$ -

A situation of practical interest often occurs when a Bernoulli trial is performed n times, and the experimenter observes the number of successes[3] obtained. Let X be a random variable representing the number of times the "Success" outcome is observed. A classical hypothetical example is where the same coin is tossed n times with heads representing success (S) and tails representing failure (F). X for this experiment will be the number of tails observed. The problem to be resolved is to determine the probability mass function associated with X.

The statistical model applicable to the experiment described above requires the use of $n+1$ mass points. Why? Because there is a possibility to have 0, 1, or up to n successes. The sample space of X is given by:

$$\mathcal{A}(X) = \{0, 1, \cdots, n\}. \tag{2.3}$$

In the model to be constructed, we require the probability of success to be the same on each Bernoulli trial. A sufficient condition for this to be achieved is to require the $n = 20$ trials to be performed independently, and under the exact same conditions. The *independence* condition guarantees that the outcome of one trial does not affect that of the next one, while *homogeneity* of experimental conditions guarantees uniqueness of the success probability, denoted as p.

The p.m.f. of X denoted by $b(x; n, p)$ is the probability that

---

[3]Success in this context is achieved when the Bernoulli variable takes value 1, the value 0 representing failure.

X equals x, and is formulated as follows:

$$b(x; n, p) = \begin{cases} \binom{n}{x} p^x (1-p)^{n-x} & \text{if } x = 0, 1, \cdots, n, \\ 0 & \text{otherwise.} \end{cases} \qquad (2.4)$$

with n and p being the two parameters of the statistical model. To understand why the p.m.f of the Binomial model is formulated as shown in equation 2.4, note that you would observe x successes following n trials if you observe x successes and $n - x$ failures. This will occur with a probability of $p^x (1-p)^{n-x}$. However, the x successes may occur on the first x trials or on any other grouping of x trials. It can be established that the number of ways that x successes and $n - x$ failures can be observed is given by the number of combinations of x among n. That is,

$$\binom{n}{x} = \frac{n!}{x!(n-x)!},$$

where $n!$ (read n factorial) is defined as $n! = 1 \times 2 \times 3 \times \cdots \times (n-1) \times n$. The symbol $\binom{n}{x}$ is read n choose x. To summarize,

A Binomial Model is characterized by a random variable X that represents the number of successes observed during an experiment, which conforms the following list of requirements :

(a) The experimenter performs a predetermined number n of trials, resulting each to a success (S) or to failure (F).

(b) All trials are identical (i.e. performed under the same conditions) so that the probability of success p stays the same from trial to trial.

(c) The trials are independent in the sense that the outcome of one trial does not affect that of another trial.

*The family of probability distributions corresponding
to this model is the p.m.f. associated to X and given
by equation 2.4*

The Binomial probability is used later in this book in the statistical test of hypotheses regarding proportions based on small sample sizes.

Example 2.1

When a car comes to a dead-end intersection, it may either turn left or turn right. If the cars arrive at the intersection under similar traffic conditions, then one may consider that they will be turning left or right independently of one another. Suppose that it is known from past experience that about 60% of the cars coming at that intersection under current traffic conditions turn left. If the experimenter decides to observe the next 5 cars, he may apply the binomial model to the number of cars turning left. Figure 2.1 shows the 6 mass points that are part of the Binomial model, as well as an example of what their respective probability masses could be.

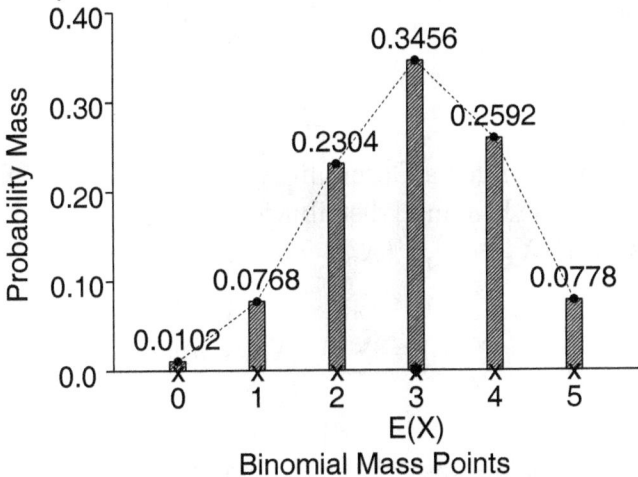

Figure 2.1. Graphical representation of Binomial statistical model

The unitary probability mass is distributed among the 6 mass points of the sample space, according to the Binomial law of probability $B(5, 0.6)$ formulated in equation 2.4. The mass point 0 for example is assigned a probability mass of 0.0102 (i.e. $p(0) = 0.0102$), while mass point 4 is assigned a probability mass of 0.2592. All 6 probability masses sum to the total probability mass of 1.

Expectation and Variance

The mathematical expectation of X denoted by $E(X)$, which represents the position on the horizontal axis of the center of mass is calculated as the sum of the mass points weighted by their respective probability masses. This definition is not convenient for computational purposes. In fact, the center of mass is more conveniently calculated using the fact that X is the sum of n independent Bernoulli variables $X_1, X_2, \cdots, X_n$. Because of independence, the center of mass of the Binomial variable equals the sum of the centers of mass of all n Bernoulli variables $E(X_i)$. That is,

$$E(X) = \sum_{i=1}^{n} E(X_i) = np, \qquad (2.5)$$

since $E(X_i) = p$ for a Bernoulli X_i as seen before. Likewise, the variance of the Binomial distribution is the sum of the n Bernoulli variances $V(X_1), \cdots, V(X_n)$. That is,

$$V(X) = \sum_{i=1}^{n} V(X_i) = np(1 - p). \qquad (2.6)$$

The center of mass of the Binomial model of example 2.1 is given by $E(X) = 5 \times 0.6 = 3$, and its variance given by $V(X) = 5 \times 0.6 \times 0.4 = 1.2$.

The Cumulative Distribution Function

In addition to the variance and the expectation, other quantities related to the Binomial distribution are often of interest. One of them is the cumulative distribution function (c.d.f.) $B(x; n, p)$, which associates each mass point x with the cumulative probability masses of x and all mass points below x. That is, $B(x; n, p) = P(X \leq x)$ and is defined as,

$$B(x; n, p) = \sum_{t=0}^{x} b(t; n, p) \quad \text{for } x = 0, 1, \cdots, n. \tag{2.7}$$

In other words, the cumulative distribution function represents the probability mass of the entire collection of mass points defined by the event $\{X \leq x\}$.

The Software Solution

Excel 2010 has two useful functions for dealing with the Binomial probability model. The first function is BINOM.DIST for computing the probability mass (equation 2.4) as well as the cumulative probability (equation 2.7) of a given mass point x. The second function is BINOM.INV for computing the percentiles of the Binomial distribution.

▶ BINOM.DIST(x, n, p, FALSE) will calculate the p.m.f $b(x; n, p)$ of equation 2.4.

▶ BINOM.DIST(x, n, p, TRUE) will calculate the c.d.f $B(x; n, p)$ of equation 2.7.

▶ BINOM.INV(n, p, α) will produce the αth percentile of the Binomial distribution $\mathcal{B}(n, p)$.

Excel 2007 offers a single function BINOMDIST that is similar to BINOM.DIST. Calc too has the same function BINOMDIST with the same functionality.

As for the R package, see section F.2.1 of Appendix F for a detailed description of what is offered. The R package offers more options than Excel and Calc. In addition to the p.m.f and the c.d.f, you can simulate the Binomial distribution with R by generating a data series where each mass point has a frequency of occurrence that is close to the probability mass predicted by equation 2.4.

In real life, trials are rarely performed independently under the exact same conditions. Therefore the conditions of the Binomial model will often be met only approximately as shown in example 2.2.

Example 2.2 _____

A small city has 10,000 workers, of whom 4,500 have a high school (HS) diploma. A sample of 20 ($n = 20$) workers is selected *without replacement* from the list of all workers. Suppose that the ith trial is a success (S) if the selected worker has a high school diploma, and is a failure (F) otherwise. Define the random variable X = *number of workers in the sample of 20 who have a high school diploma*.

The experiment is the random selection of 20 workers without replacement and the question is whether the Binomial model is applicable. This experiment has been performed 20 times. Thus, condition *(a)* is satisfied. However, the 20 trials are neither identical nor independent. If the probability for the very first worker selected to have a HS diploma is known to be $p = 4,500/10,000 = 0.45$, the probability for the second worker selected to have a high school diploma depends on the outcome of the first trial. If the first worker happens to have a high school diploma, then there will be 4,499 workers left with HS diploma out of 9,999 yet to be selected. Therefore the second worker will have HS diploma with probability $4,499/9,999$. This probability becomes $4,500/9,999$ if the first selected worker does not have HS diploma. Because the probability of success changes from trial to trial, conditions (b) and (c) of the Binomial model are not satisfied. However, if the total number of workers is reasonably large (e.g. $10,000$ or more), then the success probability will always remain close to 0.45 and it will still be safe to assume the Binomial model.

Example 2.2 shows an experiment that is approximately described by the Binomial statistical model. An interesting question is whether there is a statistical model that can describe with precision the experiment of Example 2.2. The answer is yes, and that model is referred to as the *hypergeometric distribution*.

2.2.3 The Hypergeometric Distribution - $\mathcal{H}(n, N_s, N)$ -

The hypergeometric model is suitable for any problem that amounts to selecting successively n individuals without replacement from a population made up of N individuals that are of 2 types. The characteristic of interest X is the number of selected individuals of the type labeled as "success". This experiment is called the *hypergeometric process*. Each selection of an individual is a trial, which yields either a success or a failure and the problem is to evaluate the probability to obtain x successes after n trials.

Although the hypergeometric process is very similar to the binomial, there are two main differences:

▶ The maximum number of trials is limited to N. It is not the case in a binomial model where there is no maximum set for the number of trials that can be performed.

▶ In the hypergeometric model, the trials are performed under different conditions as the population from which individuals are selected shrinks by 1 after each individual is selected.

What is the number of mass points in a hypergeometric model? The answer depends on how the number of trials compares to the number of "failures" in the population. For example, in a population of 10 individuals of whom 3 are labeled as F (for "failures") and 7 as S (for "success"), 5 trials will necessarily lead to a minimum of 2 S's and a maximum of 5 (although there are 7 S's in the population), excluding thereby 0, 1, 6, and 7 as possible mass points in the hypergeometric system. Let N_s and $N_f = N - N_s$ be the number of successes and failures respectively in the population of N individuals. All mass points in the system are between $\max(0, n - N_f)$ and $\min(n, N_s)$. An integer value k is a mass point (i.e. belongs to the sample space[4] $\mathcal{A}(n, N_s, N)$) of a Hypergeometric distribution $\mathcal{H}(n, N_s, N)$ if,

$$\max(0, n - N_f) \le k \le \min(n, N_s). \tag{2.8}$$

The probability mass function of the hypergeometric process is defined as follows :

$$h(k; n, N_s, N) = \begin{cases} \dfrac{\dbinom{N_s}{k}\dbinom{N - N_s}{n - k}}{\dbinom{N}{n}} & \textit{if } k \in \mathcal{A}, \\[2em] 0 & \textit{otherwise.} \end{cases} \tag{2.9}$$

Excel, R, and Calc provide functions for computing this p.m.f. However, only Excel 2010, and R provide functions for computing the c.d.f. I will describe these functions later.

---

[4]For the sake of simplicity, I will occasionally use $\mathcal{A}$ instead of $\mathcal{A}(n, N_s, N)$

Example 2.3

Suppose that in a small company of 40 employees, only 5 have an annual salary over $90,000.00. If 15 employees are selected without replacement, then the number X of employees in the sample with an annual salary over $90,000.00 follows a hypergeometric model. The associated probability mass function is given by :

$$
h(k; 15, 5, 40) = \begin{cases} \dfrac{1}{658,008} \dbinom{5}{k} \dbinom{35}{15-k} & \text{if } 0 \le k \le 5, \\[2mm] 0 & \text{otherwise.} \end{cases}
$$

$$(2.10)$$

Expectation and Variance

The expectation $E(X)$ and the variance $V(X)$ of a hypergeometric variable X having a probability mass function $h(k; n, N_s, N)$ are given by :

$$
E(X) = np \quad \text{and} \quad V(X) = \frac{N-n}{N-1} \cdot n \cdot p(1-p), \qquad (2.11)
$$

where the success probability p is defined as $p = N_s/N$.

The hypergeometric and the binomial distributions can both be seen as the sum of n Bernoulli variables. That is, $X = X_1 + \cdots + X_n$, where X_i is a Bernoulli variable with a success probability p. However, there are some differences and Table 2.1 shows a comparison between these two distributions with respect to some of their characteristics.

Table 2.1: Comparison of hypergeometric and binomial distributions

Characteristic	Binomial	Hypergeometric
Success probability p	p (unknown)	$p = N_s/N$ (Known)
Number of Bernoulli trials	n	n
Maximum number of trials	Unlimited	N
Dependency of the trials	Independent	Dependent
Expectation $E(X)$	np	np
Variance $V(X)$	$np(1-p)$	$\left(\dfrac{N-n}{N-1}\right)np(1-p)$

It follows from Table 2.1 that the differences between the two distributions do not affect the center of mass $E(X)$ of the distributions. However, the variance of the hypergeometric distribution is different from that of the binomial. This is essentially due to the fact that the hypergeometric distribution is characterized by two related conditions, which are the dependency of the bernoulli trials and the existence of a maximum N on the number of trials that can be performed. The variance of the hypergeometric variable can be rewritten as follows :

$$V(X) = \left(\frac{N}{N-1}\right)(1-f)np(1-p), \text{ where } f = n/N.$$

If the population size N is reasonably large and the *sampling fraction* f reasonably small (i.e. smaller than 5%), then the variance of the hypergeometric variable gets very close to that of the binomial. Thus, when the parameters n, N_s, N of the hypergeometric distribution are very large, computing the probability mass for the hypergeometric can be tedious and one can use the binomial approximation to solve the problem.

The Software Solution

Excel 2007 and Calc 3.2.0 only allow you to compute the probability masses (see equation 2.9) at the mass points with their

HYPGEOMDIST function. You will call = HYPGEOMDIST(k, n, N_s, N) with Excel 2007, or = HYPGEOMDIST($k; n; Ns; N$) with Calc.

However, you will need Excel 2010 or R to compute cumulative probabilities. Excel 2010 offers the **HYPGEOM.DIST** function (see section E.9 of Appendix E for a detailed description), and R offers the **qhyper** function (see section F.2 of Appendix F for a detailed description). Only R will allow you to compute the percentiles of the Hypergeometric distribution.

I have discussed the binomial distribution in section 2.2.2. I also indicated in section 2.2.3 that when the number of trials has an upper limit that is not very large, it may be necessary to use the hypergeometric distribution rather than the binomial. Now, I want to know what would happen if in a binomial experiment, the number of trials increases indefinitely. The interest of this inquiry lies in the practical desire to know about the chance of occurrence of an event of a particular type overtime.

2.2.4   **The Poisson Distribution - $\mathcal{P}(\lambda)$ -**

In a book published in 1837 and entitled *Recherches sur la probabilité des jugements em matière criminelle et en matière civile* (i.e. Researches on the probability of criminal and civil verdicts), the French mathematician Siméon Denis Poisson (1781-1840) stated that if the number of trials n is very large and the success probability p very small, the binomial probability mass function can be approximated by :

$$P(X = k) = \frac{\lambda^k}{k!}e^{-\lambda}, \text{ for } \lambda > 0 \text{ and } k = 0, 1, \cdots \qquad (2.12)$$

Although Poisson presented equation 2.12 (actually a variant of it) simply as an approximation to the binomial p.m.f., this expression turned out to fulfill itself the mathematical conditions of a probability mass function. The problem was that for a longtime no known real-life experiment could be associated with Poisson's equation, making it impossible to fully describe a Poisson statistical model. It took several decades before an English mathematician named William Sealy Gosset (1876-1937) found a clear example of a practical experiment related to yeast fermentation where the Poisson law was applicable. This happenned when he was working as a chemist in the Dublin brewery Arthur Guinness and Son.

Example 2.4 _____

Yeasts are organisms that are cultivated in jars of fluid and used in the manufacturing of beer. The amount of yeast used in the mash affects the beer quality and must be known. In jars, yeast cells multiply and divide continuously (at a unknown speed that decreases with time) so that at a given point in time the concentration of yeast cells is difficult to determine. Since the speed with which yeast cells multiply is neither known nor constant, the concentration of cells in the jar at time t is a random variable. If its probability distribution is known, then yeast cells concentration can be predicted with better accuracy. Gosset examined the data available and concluded that the Poisson distribution was suitable.

William Sealy Gosset is better known in the statistical community under the pseudonym of Student that he used to publish his work secretly as publishing was a forbidden activity at Arthur Guinness and Son to protect manufacturing secrets.

There are a few fancy mathematical assumptions of little statistical interest that are necessary to justify the derivation of the Poisson p.m.f (see Hogg, Craig, and McKean (2004)). What should be retained about a typical Poisson process is that it takes place overtime and,

▶ The relative number of times it generates a single new event of interest (i.e. "success") during a unit time is constant.

▶ The relative number of times it generates 2 successes or more during a unit time is negligible.

▶ The number of times success is observed in a given time interval does not affect the number of times that success will be observed in another nonoverlapping time interval. This is a form of independence.

Other examples of Poisson processes include the number of defects on a manufactured article, the number of car accidents at a given intersection, or the number of insurance claims, within a certain time period.

Example 2.5 _____

Let X denote the number of automobile insurance claims during a specified time interval Δt from t_0 to t_1 (i.e. $\Delta t = t_1 - t_0$). Suppose that X has a Poisson distribution with $\lambda = 5$. This indicates that during a time interval Δt, there will be 5 claims on average. The probability mass function associated with X is depicted in figure 2.2.

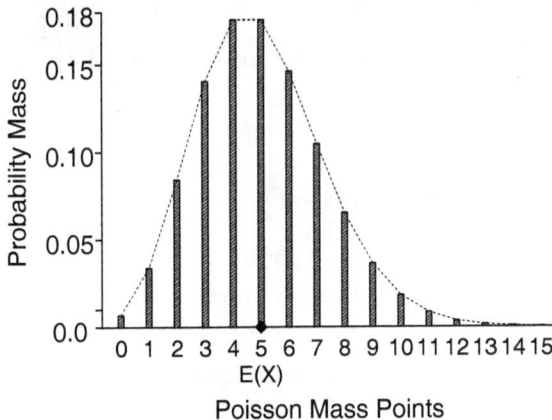

Figure 2.2. Graphical representation of a Poisson statistical model

The probability mass associated with having exactly two claims during Δt is given by:

$$p(X = 2) = \frac{5^2}{2!}e^{-5} = 0.084.$$

The probability of having at least two insurance claims during time period Δt is

$$P(X \le 2) \;=\; \sum_{k=0}^{2} P(X = k),$$

$$= \sum_{k=0}^{2} \frac{5^k}{k!}e^{-5} = \left(1 + 5 + \frac{25}{2}\right)e^{-5} = 0.125.$$

The Poisson process in example 2.4 can be seen as an approximation of the binomial process where a trial is performed by verifying the filling of a insurance claim every second for instance. A success is observed if one finds a claim, otherwise it is seen as a failure. The number of trials in a time interval of 1hour is 3,600. Since the average number of claims is 5, one concludes that the probability of success (i.e. observing a claim) is about $p \approx 5/3600 = 0.0014$, which is very small.

The Poisson distribution can in general be used in practice to approximate the law of probability of a Binomial experiment with a large number of trials and a small success probability.

Expectation and Variance

If X follows a Poisson distribution $\mathcal{P}(\lambda)$, what is the center of mass $E(X)$ of the distribution, and what is the expected distance from the mass points to the center of mass measured by the variance $V(X)$? These two quantities are given by,

$$E(X) = \lambda, \;\; and \; V(X) = \lambda. \tag{2.13}$$

This result is derived using basic results from introductory calculus.

You may also want to know that the Poisson distribution has found an important application in the development of log-linear regression also known as Poisson regression. Poisson regression is briefly introduced in chapter 11.

Software Solution

Excel (2007 & 2010), Calc 3.2.0, as well as R provide functions for computing the probability mass and the cumulative probability of the Poisson distribution. Only R however, will allow you calculate the percentiles of Poisson.

Excel 2007 and Calc 3.2.0 offer the **POISSON** function. In Excel 2007, **POISSON(k,λ,FALSE)**, and **POISSON(k,λ,TRUE)** produce the probability mass and the cumulative probability respectively at mass point k. In Calc 3.2.0, **POISSON(k;λ;0)**, and **POISSON(k;λ;1)** produce the probability mass and the cumulative probability respectively at mass point k

For Excel 2010, see section E.9 of Appendix E, and section F.2 of Appendix F for R, to obtain more information on the functions available.

2.2.5 The Multinomial Distribution

In section 2.2.2, I presented the Binomial distribution characterized by the replication of n independent Bernoulli trials. This can be seen as the replication of n independent and identical trials, each of which has only 2 possible outcomes, 0 and 1. One can then define 2 random variables X_0 (*the number of failures*) and X_1 (*the number of successes*) representing the number of times the experiment produces 0 and 1 respectively. X_1 will be the Binomial variable as defined in section 2.2.2. You can also rephrase this by saying the pair (X_0, X_1) follows the Binomial distribution. This is not often done because $P(X_0 = x_0, X_1 = x_1)$

is identical to $P(X_1 = x_1)$ as long as $x_0 + x_1 = n$, which justifies using the simpler expression $P(X_1 = x_1)$.

The Multinomial probability model represents the replication of n independent and identical trials, each of which has r possible outcomes labeled as[5] $1, 2, \cdots, r$. You can then define r random variables $X_1, X_2, \cdots, X_r$ where X_2 for example represents the number of times the n trials produced outcome "2". In the Multinomial model, the r variables can only take r values $x_1, x_2, \cdots, x_r$ that satisfy the following 2 conditions:

$$\begin{cases} x_i = 0, 2, \cdots, n, \;\; for \; i = 1, 2 \cdots, r \\ x_1 + x_2 + \cdots + x_r = n. \end{cases} \quad (2.14)$$

Assume that on each of the n trials, any particular outcome i can be observed with probability p_i. The probability mass function of the set of variables $(X_1, X_2, \cdots, X_r)$ is defined as follows:

$$p(x_1, \cdots, x_n) = \begin{cases} \dfrac{n!}{x_1! x_2! \cdots x_r!} p_1^{x_1} p_2^{x_2} \cdots p_r^{x_r}, & if \; (2.14) \; met, \\[2mm] 0 & otherwise. \end{cases}$$

$$(2.15)$$

The Multinomial probability model is used in hypothesis testing, when comparing equality of several population proportions when the number of observations per population is very small. I will address this issue in the chapter that discusses the chi-square test.

---

[5] The labeling of outcomes is subjective. Whether it ranges from 0 to $r-1$ or from 1 to r is irrelevant as long as you use one choice consistently.

2.3. Probability Models for Continuous Variables

This section deals with the quantification of chance for measurement variables. I indicated in the previous section that the calculus of probability for categorical variables is based on a system of mass where a probability mass is associated with each value that the categorical variable can take. Since the system has a total probability mass of 1, computing the probability of an event essentially comes down to summing the probability mass of all the mass points that can be associated with the particular event. For measurement variables, the calculus of probability is based on an entirely different computation model. The model is now based on a surface of a certain form[6] with a surface area of unity. Computing a probability comes down to assigning a portion of the baseline surface to an event and evaluating its surface area, which will necessarily be smaller than 1, a condition that all probability values must satisfy.

I want to illustrate this approach for quantifying probabilities by applying it to a practical problem. Consider the rectangle shown in figure 2.3. The experiment consists of selecting randomly, a point P from the rectangle's surface, and measuring the distance from P to the AC side. The random variable I am interested in is $X = $ *Distance to Side AC*, which may take any value from 0 to 2. However, only those points P situated in the shaded area on the surface, will lead to a distance that is below 0.5yd. In other words by choosing point P randomly, the probability $P(X \leq 0.5yd)$ that the distance from the selected location to the AC side is less that 0.5yd, corresponds to the ratio of the shaded area to the total area of the rectangle. The area of the shaded surface of figure 2.3 equals $0.5 \times 1.5 = 0.75$ while the total rectangle

---

[6]The particular form of the surface will be specific to each probability model.

surface area is $2 \times 1.5 = 3$. Thus, $P(X \leq 0.5yd) = 0.75/3 = 0.25$. More generally, for any value d, the probability $P(X \leq d)$ can be expressed as follows :

$$P(X \leq d) = \begin{cases} 0, & \text{if } d < 0, \\ d/2, & \text{for } 0 \leq d \leq 2, \\ 1, & \text{if } d \geq 2 \end{cases} \qquad (2.16)$$

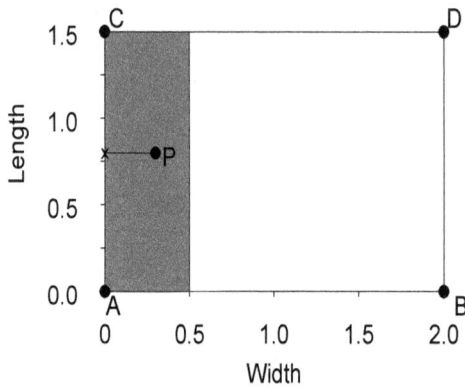

Figure 2.3. Rectangle's Surface

Although equation 2.16 describes the cumulative distribution function of X, deriving this function in practice will often be difficult for many complex real-life problems. It may even be impossible to derive the cumulative distribution function as a close expression. What is really needed, is a standard graphical display of the probability distribution of X that would be a more refined version of the histogram often used to depict the empirical probability of a measurement variable. Such a graph will help match your data with an existing and known theoretical probability model using histograms.

Figure 2.4 represents a rectangle whose surface area is 1 (i.e. $2 \times 0.5 = 1$). This surface area represents the overall probability mass of 1, which is uniformly distributed across all values

0 through 2 that the random variable X can take[7]. Now, the probability $P(X \leq 0.5)$ is the dashed-shaded surface area (the probability mass associated with the particular event $X \leq 0.5$), which is $0.5 \times 0.5 = 0.25$ (no need to divide by the overall probability mass, since it is 1).

The narrow solid-shaded strip on the other hand, represents the probability that X falls between $d = 0.6$ and $d + 0.1 = 0.7$, and is equal to $0.1 \times 0.5 = 0.05$. That is, if the random distance X increases by $\Delta = 0.1$ from an initial value d, the event probability mass will change by 0.05 for a change rate of $0.05/0.1 = 0.5$. You can then say that the density of probability or the *Probability Density* over the interval 0.6 to 0.7 is 0.5. Actually no matter how small the change Δ is, the probability density within the interval 0.6 to $0.6 + \Delta$ remains identical to 0.5. I can then conclude that the probability density at point 0.6 is 0.05. The function that links each point to the associated density is called the *Probability density function (pdf)*. In this case, the pdf is defined by,

$$f(x) = \begin{cases} 0.5, & \textit{if } 0 \leq x \leq 2, \\ 0, & \textit{otherwise.} \end{cases} \qquad (2.17)$$

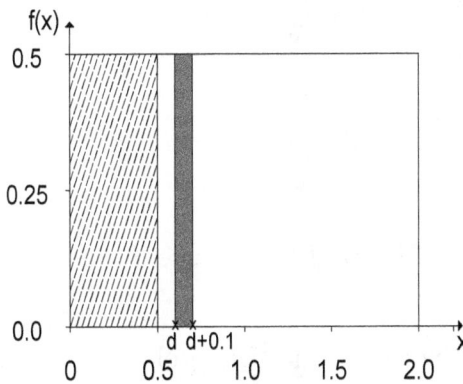

Figure 2.4. Probability Distribution of X

---

[7]The interval 0 to 2 is called the *Support* the probability distribution

The rectangle shown in Figure 2.4, is nothing else than the graphical representation of the probability density function. When this function is constant over the support of the probability distribution as is the case in equation 2.17, you may conclude that the probability mass has a uniform density across the support. The associated random variable X is said to belong to the family of the uniform distributions. The pdf completely specifies the probability distribution of a random variable, and most theoretical probability distributions discussed later in this chapter will be described by their pdf only.

Consider another experiment where a point P must be selected randomly from the triangle shown in Figure 2.5, and the distance from the selected point to the AC side of the triangle measured. $X = Distance\ from\ P\ to\ side\ AC$ is the random variable I am interested in.

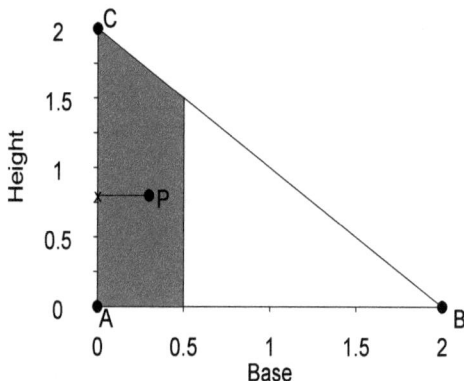

Figure 2.5. Triangle Surface

It follows from Figure 2.5 that the probability $P(X \leq 0.5)$ is calculated as the ration of the solid-shaded area to the overall triangle surface area. That is, $P(X \leq 0.5) = (2 \times 0.5 - 0.5^2/2)/2 = (1 - 0.125)/2 = 0.875$. More generally, for any value d, the probability $P(X \leq d)$ is calculated as follows:

$$P(X \leq d) = \begin{cases} d - d^2/4, & \text{if } 0 \leq d \leq 2, \\ 0, & \text{otherwise.} \end{cases} \tag{2.18}$$

The standard representation of the law of probability of X in the form of the distribution of the overall unitary probability mass is shown in Figure 2.6. The surface area of this triangle is 1, and you may notice that this probability mass is distributed differently from that of Figure 2.4. The dashed-shaded area represents the probability for X to remain below 0.5. As for the solid-shaded strip, it represents the change in probability mass as a result of a small change in X from d to $d + 0.1$. This shaded area can be evaluated at $0.1(2 - d) - 0.1^2/2$. That is from d to $d + 0.1$, the probability mass changes at the rate of $0.1 \times (2 - d - 0.1/2)/0.1 = 2 - d - 0.1/2$, which represents the probability mass density in the interval d to $d + 0.1$.

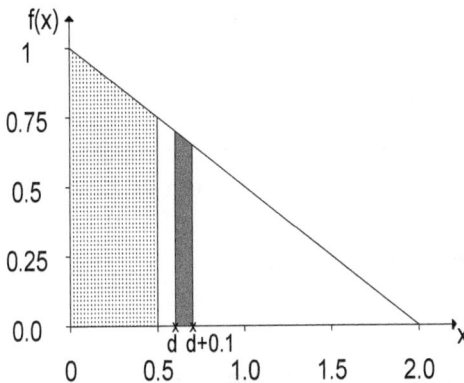

Figure 2.6. Graphical Representation of the Probability Distribution of Equation 2.20

Consequently, the area around 0 is one of high density of probability mass. Such areas are generally of great interest in practice as they are much more likely than any other area, to be the location where any given value of X will be found.

More generally, the probability mass density in an interval d to $d + \Delta$ for an arbitrary value Δ is given by $2 - d - \Delta/2$. If you set $\Delta = 0$, you will get the probability mass density at point d, which is given by $2 - d$. Consequently, to each value d you will associate the density function $f(d)$ defined as,

$$f(d) = \left\{ \begin{array}{cc} 2 - d, & \textit{if } 0 \leq d \leq 2, \\ 0, & \textit{otherwise.} \end{array} \right. \tag{2.19}$$

In case you have been wondering how graphs such as that of Figure 2.6 are constructed, you may have already realized that they are obtained by graphing the probability density function. Histograms with very narrow intervals can give you a reasonably good approximation of the pdf associated with your variable of interest. Since the probability distributions in subsequent sections are characterized with their probability density function, it is interesting to know when a given function can be called a probability density function.

Definition. 2.1.

If X is a measurement variable then its probability mass distribution is fully specified by a probability density function f that satisfies the following conditions :

▶ $f(x) \geq 0,$

▶ The area under the curve representing f is 1.

Readers with some background in Calculus know that for any two numbers a and b, the probability that X takes on a value in the interval $[a, b]$ equals the integral of the p.d.f f over that interval. But the area under the curve and within the interval $[a, b]$ can generally be well calculated with a series of approximations using an Excel spreadsheet if necessary.

2.3.1 Center of Mass and Variance of a Measurement Variable

In section 2.2, I introduced the location of the center of mass as well as the variance (or the dispersion parameter) of a discrete random variable. The random variable was then seen as a system of indivisible and enumerable mass points, each of which having a predetermined probability mass. I found out that the location of the center of mass is determined by taking the sum of the locations of each mass point, weighted by their respective probability mass. The variance on the other hand was determined by calculating the differences from the mass points to their center of mass and by taking the sum of these squared differences weighted by the probability masses

When the random variable is continuous, its support is a continuum and can no longer be seen as an enumerable set of indivisible mass points. Instead, it is the overall unitary probability mass that is distributed across the support according to a certain law to be determined. An interesting question that arises is how to determine the location of the center of mass and its variance. I will describe the well-known method known as "the method of limits" to resolve this problem.

To fix ideas, consider a continuous random variable X whose probability density function f is given by :

$$f(x) = \begin{cases} (-3x^2 + 4x + 4)/8 & \text{if } 0 \le x \le 2, \\ 0 & \text{otherwise.} \end{cases} \qquad (2.20)$$

The graph of this density function is the smooth curve shown in figure 2.7(a). Since I already know from section 2.2 how to handle categorical variables, I am going to "categorize" the continuous variable X temporarily, so that I can use existing results derived in section 2.2. Let h be any number satisfying the condition $0 <$

$h < 1$ and let n be the smallest integer greater than $2/h$. The categorized version of X denoted by X' (say X prime) is defined as follows:

▶ X' can take any of the n values in the set $\{x_1, \cdots, x_n\}$ where $x_i = (i-1)h$.

▶ The probability mass point $p(x_i)$ of x_i is the surface area of the rectangle of length $f(x)$ and width h. The probability mass function associated with X' is defined as,

$$p(x_i) = \frac{D_{ih}}{D_h}, \text{ where } \begin{cases} D_{ih} &= hf(x_i), \\ D_h &= \sum_{j=1}^{n} D_{jh} \end{cases} \qquad (2.21)$$

As an example, let us assume that $h = 0.25$. This creates the categorical variable X' that takes one of the $n = 8$ values in the set $\{0.25, 0.50, 0.75, 1.00, 1.25, 1.50, 1.75\}$. The probability mass function calculated according to equation 2.21 and given in Table 2.2, is depicted in Figure 2.7(a) by the vertical bars. The mass points are identified on Figure 2.7(a) by the small ticks of the x-axis. The area of each vertical bar in Figure 2.7(a) represents the probability mass $p(x_i)$ of the associated points x_i as shown in Table 2.2.

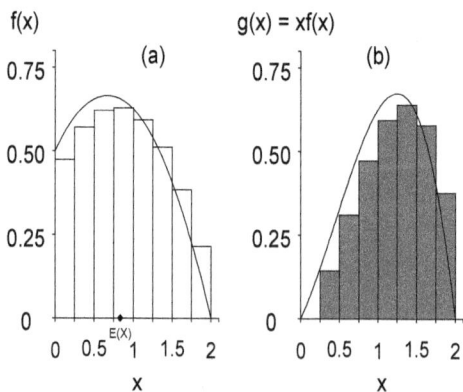

Figure 2.7. Graphical representation of the probability density function of equation 2.20

The bar area is obtained as the product $h \times l_i$ of its width h (the same for all bars) by its height l_i that varies for each mass points i and that is given by:

$$l_i = \frac{f(x_i)}{D_h}, \tag{2.22}$$

Table 2.2 : Probability distribution of the discretized random variable X'

x_i	l_i	$p(x_i)$
0	0.4741	0.1185
0.25	0.5704	0.1426
0.50	0.6222	0.1556
0.75	0.6296	0.1574
1.00	0.5926	0.1481
1.25	0.5111	0.1278
1.50	0.3852	0.0963
1.75	0.2148	0.0537
Total	4	1

Using the experience gained in section 2.2 with categorical variables, I can evaluate the center of mass of X' as follows:

$$E(X') = \sum_{i=1}^{n} x_i p(x_i),$$

$$= \sum_{i=1}^{n} h \left[x_i f(x_i)/D_h \right].$$

Note that $E(X')$ represents the shaded area in Figure 2.7(b), since $x_i f(x_i)/D_h$ is the height of ith vertical bar.

If I gradually decrease the width h of the vertical bars towards 0, I will begin the inverse process of moving from the discrete variable X' back to the original continuous variable X. During that process, the number of vertical bars under the curve will grow as their width decreases (see Figure 2.8(a) and 2.8(b)). The areas under these curves will be covered with more accuracy. In Figure 2.8(a), the area under the curve will be the overall probability mass of 1, and in Figure 2.8(b), the area under the curve will be the center of mass, also called the expectation of X. It is denoted by $E(X)$, and reported in Figure 2.8(a). In the jargon of Calculus, one would say that the expected value of X is the integral of the function $xf(x)$ over the support of X (i.e. the interval 0 to 2), and denoted as follows:

$$\mu_X = E(X) = \int_0^2 xf(x)dx. \tag{2.23}$$

The variance of a measurement variable X is defined as the expectation of the variable $(X - \mu_X)^2$, and is a measure of how far you would expect a value taken by X to be with respect to its center of mass. It can be calculated using the method of limits used to evaluate the center of mass. Again, in the jargon of calculus, this variance is said to be the integral of the function $(x - \mu_X)f(x)$ over the support of the measurement variable X, which is the interval 0 to 2. That is,

$$\sigma_X^2 = V(X) = \int_0^2 (x - \mu_X)^2 f(x)dx. \tag{2.24}$$

The *Standard Deviation* is defined as the square root of the variance.

I like to say a word about the "method of limits". The first step is to determine not the exact magnitude of the quantity of interest (such as the center of mass or the variance), but rather a rough approximation of it using simple arithmetic and algebra based upon enumerable objects. However, it is not just one approximation that should be made, but a whole series of them, the next one providing a more accurate result than the previous. The exact result is obtained by examining closely the series of approximation to detect what the limiting value looks like. That limited value will be a fixed constant that matches the exact value sought.

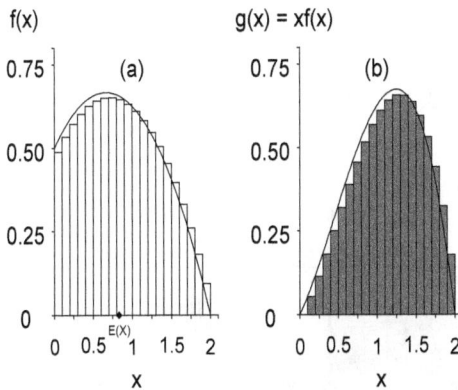

Figure 2.8. More accurate graphical representation of the probability density function of equation 2.20

There are several useful probability density functions that have been discovered during the course of the development of statistical science. Their usefulness is determined by the wide range of practical (random) phenomena they describe. In the next few sections, I will review the most popular of these density functions, and will discuss the conditions of their applicability.

During a statistical investigation, the researcher is often required to postulate a family of theoretical probability laws and to use collected data to determine the specific member of the family that can be used for statistical inference. Therefore, knowing the probability laws that have been successfully used in practice is essential for a practitioner.

2.3.2 Uniform Distribution

One of the simplest continuous probability distribution is the *uniform distribution*. This distribution is characterized by its probability mass density being constant across the distribution support. Let X be a measurement variable that can take any value within an interval (a, b) with $a < b$. Strictly speaking, by saying that X can take any value within the interval (a, b), what I mean is that *there is a real chance that a value of X may be observed in any subinterval included in (a, b) following a random experiment.* The probability distribution of X is said to be *uniform*, if the associated density function f is defined as follows :

$$f(x) = \begin{cases} \dfrac{1}{b-a} & \text{if } a < x < b, \\ 0 & \text{otherwise.} \end{cases} \qquad (2.25)$$

Equation 2.25 depends upon the 2 parameters a and b and therefore defines not one uniform probability distribution, instead it defines a family $\mathcal{U}(a, b)$ of uniform probability distributions. Each pair of values for (a, b) will specify one probability distribution.

Example 2.6

Suppose that my daily commute to work by train is expected to take one hour. However, because of a variety of circumstances the time that I actually spend in the train varies (randomly) from 45 minutes to 75 minutes. If X represents the commute time, then the set of possible values of X is the interval $(45, 75)$. If I believe that any subinterval included in $(45, 75)$ has the same chance to include the actual commute time of the next train, then I may say that the probability distribution of X is uniform and its probability density function given by:

$$f(t) = \begin{cases} \dfrac{1}{30} & \text{if } 45 < t < 75, \\ 0 & \text{otherwise.} \end{cases}$$

In Example 2.5, I treated commute time as a continuous variable. Do I have to? The answer is no, I do not. As discussed previously, continuity is a mathematical concept with no formal concrete representation. Commute time can be looked at as a discrete set of the 31 integers between 45 and 75, representing the number of minutes spent in the train. If this discretized version of commute time is uniformly distributed then the probability of observing any of the 31 values would be 1/31, which is close to the p.d.f shown in Example 2.5. Commute time may also be seen (like in example 2.5) as an entity not divided in parts (although divisible), that is a continuum allowing for fractional parts of one minute. The most important thing to retain here is that our model of continuity can sometimes provide a good approximation of a phenomenon that is discrete in nature. In Example 2.5, both discrete and continuous approaches can be used and will yield similar answers. However, there are situations where a discrete reality is difficult to describe and can usefully be approximated by a continuous model.

Going from a discrete structure to a continuous one has an important practical consequence. By taking commute T time as a continuous variable, the probability that it takes a specific value in the interval $(45, 75)$ is 0. This may a priori sound strange. The fact of the matter is that, if you believe there should be a "real chance" for the commute time to be exactly 52 minutes, then you have a discrete view of the problem that is incompatible with the continuity abstraction. A discrete model may then be more appropriate.

A continuum such as $(45, 75)$ is not made up of clearly distinguishable points such as 52. By allowing for unlimited accuracy in the determination of time, the concept of continuum takes away any meaning from the concept of exactness. If the observed commute time is $t_0 = 52$ minutes when time is seen as a continuous variable, then one could well assume that the actual time is $t_0 = 52.0000000001$ and not 52. One known thing is that the observed time lies between 51.5 and 52.5. Only intervals can be observed with a certain accuracy when the phenomenon under investigation is being explored with a continuous model in mind.

Cumulative Distribution, Expectation and Variance

▶ The cumulative distribution function (i.e. the probability $P(X \leq x)$ for an arbitrary x) of the uniform law of probability $U(a, b)$ is defined as follows:

$$F(x) = \begin{cases} 0, & \text{if } x \leq a, \\ (x-a)/(b-a), & \text{if } a \leq x \leq b, \\ 1, & \text{if } x \geq b \end{cases} \qquad (2.26)$$

▶ The expected value $E(X)$ and the variance $V(X)$ of the uniform distribution are given by:

$$E(X) = (a+b)/2, \quad and \quad V(X) = (b-a)^2/12. \qquad (2.27)$$

The Software Solution

The Uniform distribution is a very simple one to manipulate, even manually. This may be the reason why neither Excel nor Calc offer any command for manipulating this distribution. The R package however, offers some functions for the uniform distribution.

▶ dunif(x, min=a, max=b). This function calculates the probability density of the uniform distribution $\mathcal{U}(a, b)$ at point x. The default values for the parameters a and b are 0 and 1 respectively.

▶ punif(q, min=a, max=b, lower.tail = TRUE). This function calculates the cumulative probability distribution function of the uniform distribution $\mathcal{U}(a, b)$. If lower.tail = TRUE then the function calculates $P(U \leq q)$, and if lower.tail = FALSE then the function computes $P(U > q)$, which equals $1 - P(U \leq q)$.

▶ qunif(p, min=a, max=b, lower.tail = TRUE). This function calculates the pth quantile of the uniform distribution $\mathcal{U}(a, b)$. It represents the point x_p that satisfies the condition $P(U \leq x_p) = p$.

▶ runif(n, min=a, max=b). This function generates a random number following the Uniform distribution $\mathcal{U}(a, b)$.

2.3.3   Normal Distribution/Bell-Shaped Curve

There is one probability law that proved useful in providing good approximations to a wide variety of probabilities involving averages of large collections of numbers. In fact, that probability law approximates very well the probability distribution of the mean of any large set of numbers, regardless of what probability law governs the initial raw data. The fact that the process that generates your data is unknown does not preclude you from having a reasonably good knowledge about the probability distribution of the average of these numbers. And as the collection of observed numbers grows the probability distribution of the mean tends to get closer and closer to a well-known limiting distribution that was understandably called the *Normal distribution*. The behavior of individual observations may be unpredictable, but their average tends to have a "normal" behavior as the number of observations grows.

The Normal distribution has a long and glorious history. The French mathematician Abraham de Moivre (1667-1754) suggested a useful approximation of the Binomial distribution by the normal distribution. It is even thought that perhaps the Swiss mathematician Daniel Bernoulli (1700-1782) may have come across the normal distribution earlier. Nevertheless, the normal distribution is often referred to as the Gaussian distribution as the german mathematician Carl Friedrich Gauss (1777-1855) was once believed to be the first man to formally discussed this distribution. The normal probability distribution is by far the most popular of all probability distributions. Its popularity is not simply justified by its ability to describe the large-sample probability distribution of the mean, but also stems from the fact that it simplifies the mathematics of probability calculus considerably. Once random variables are assumed to follow the normal distribution, the study of their statistical properties becomes much more tractable, although this simplification is not as important in this computer age as it once was.

The convergence of the mean's probability distribution to the normal probability distribution is a very powerful result that is usually referred to as the *Central Limit Theorem* (CLT). The CLT will be further examined later in this section.

Normal Distribution: Definition and Properties

Figure 2.9 depicts the probability mass function (represented by the vertical bars) of the Binomial distribution $\mathcal{B}(20, 0.25)$ with a total of n trials and a success probability $p = 0.25$. On the same graph is a bell-shaped curve that represents the continuous function that the French Mathematician Abraham de Moivre (1667-1754) used to approximate Binomial probabilities. This function turned out to be the probability density function of what was later known as the normal distribution.

Today, the normal distribution is a well-defined and polished probability law that is usually introduced as a family of distributions, whose members are determined by a location parameter μ and a dispersion parameter σ^2.

Definition. 2.2.

Let X be a measurement variable that takes values in an interval (a, b) $(a < b)$. X is said to have a *normal probability distribution* with a location parameter μ and a dispersion parameter σ^2 $(\sigma > 0)$ if its probability density function is given by:

$$f(x; \mu, \sigma) = \frac{g(x; \mu, \sigma)}{I(a, b)} \text{ where } g(x; \mu, \sigma) = \frac{1}{\sigma\sqrt{2\pi}} e^{-(x-\mu)^2/2\sigma^2}$$

and $I(a, b)$ is a normalizing factor to ensure a mass probability of 1 over the interval (a, b).

In many statistical textbooks, measurement variables are typically assumed to take all possible values. Consequently, the interval (a, b) in the above definition is often replaced with $(-\infty, \infty)$, which should be read as "from negative infinity to infinity", and is a pure mathematical representation of all numbers. The value of the normalizing factor[8] $I(a, b)$ then becomes 1. Although such a presentation offers some mathematical elegance, it has the disadvantage of introducing the concept of infinity even when it is unnecessary, its value being essentially theoretical.

---

[8]The normalizing factor has only one purpose, which is to ensure that the surface area under the density curve, and limited to the interval (a, b) will be 1.

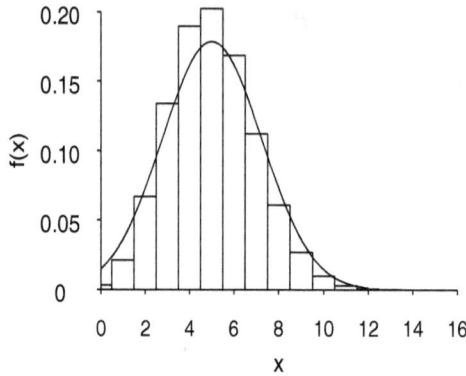

Figure 2.9. P.M.F. of a Binomial $\mathcal{B}(20, 0.25)$ with a normal approximation curve

Figure 2.10 shows the effect of the dispersion parameter σ on the form of the normal curve. This graph contains four normal curves $\mathcal{N}(\mu, \sigma^2)$ with the same location parameter $\mu = 12$ and a dispersion parameter σ that varies form 1 to 4. As it appears in Figure 2.10, the location parameter represents the center of symmetry of the normal distribution, while the dispersion parameter determines the narrowness of the bell under the curve.

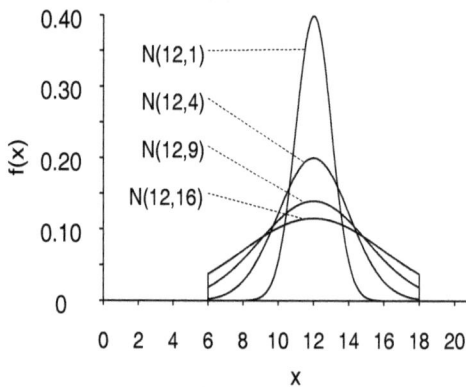

Figure 2.10. Graphs of normal density functions $\mathcal{N}(12, \sigma^2)$ for $\sigma = 1, 2, 3, 4$

The normal probability is used so often in statistical science that the practitioner is expected to be able to manipulate it with ease. This goal is achieved by mastering some of its basic and most important properties that will now be presented. The first thing to be noticed is that the density function of the normal distribution (see definition 2.2) is a complicated function. To evaluate cumulative probabilities or probability densities at various mass points, a natural solution could be to prepare numerical tables where various probabilities of the normal distribution are readily obtainable. But there is still a problem. The family of normal distributions is so broad (each value of μ or σ yields a different normal distribution) that the number of tables needed to cover them all will be excessive. Fortunately, the normal distribution has a nice property that helped resolve this problem.

Result 2.1.

Let us denote by Z a random variable that follows the normal distribution $\mathcal{N}(0,1)$ with a 0 location parameter and a dispersion parameter of 1. Let μ be an arbitrary number and σ a positive number. Then the random variable X defined by:

$$X = \sigma Z + \mu$$

follows the normal distribution $\mathcal{N}(\mu, \sigma^2)$ with location and dispersion parameters given by μ and σ^2 respectively.

Why is this proposition important? Suppose I like to compute the probability $P(X \leq x)$ that a random random variable X following a normal distribution $\mathcal{N}(\mu, \sigma^2)$ take a value that is less than x. This probability can be expressed as,

$$P(X \leq x) = P\left(Z \leq \frac{x - \mu}{\sigma} \right), \qquad (2.28)$$

where $Z = (X - \mu)/\sigma$ follows the normal distribution $\mathcal{N}(0,1)$.

Therefore, if we can tabulate the probabilities $P(Z \leq z)$ for various values of z, they can be used along with equation 2.28 to obtain the probabilities related to any normally-distributed random variable X. The normal distribution $\mathcal{N}(0, 1)$ is commonly referred to as the *Standard Normal Distribution* and plays a pivotal role in the family of normal distributions.

Cumulative Distribution, Expectation and Variance

The cumulative distribution function (often denoted by $F(x)$) of a random variable X is defined as the probability $P(X \leq x)$ that X takes a value smaller than or equal x. It is a widely-accepted convention to denote the c.d.f of the standard normal distribution as $\Phi(z)$ for any real number z. Equation 2.28 essentially stipulates that the c.d.f. F of any variable X that following the normal distribution $\mathcal{N}(\mu, \sigma^2)$ is related to the c.d.f of the standard normal distribution by the following relation:

$$F(x) = \Phi\left(\frac{x - \mu}{\sigma}\right). \tag{2.29}$$

Similarly, because we have that $P(\alpha \leq X \leq \beta) = F(\beta) - F(\alpha)$, the following relation is valid :

$$F(x) = \Phi\left(\frac{\beta - \mu}{\sigma}\right) - \Phi\left(\frac{\alpha - \mu}{\sigma}\right). \tag{2.30}$$

The Software Solution

Although many statistics textbooks offer statistical tables that help compute various probabilities related to the normal distribution, I found it more convenient and more accurate to compute these probabilities using Excel, Calc, or R. The equations 2.28, 2.29, and 2.30 have gained importance overtime because they allow you to use the probabilities associated with the standard Normal distribution in order to compute the probabilities of other non-standard normal distributions. Although it does not hurt to know about these equations, they are in my opinion only relevant to young mathematical statisticians who may need to study the math behind the normal distribution. For practitioners, these equations have become downright irrelevant in the computer age.

▶ The Excel 2010 function NORM.DIST($x,\mu,\sigma,cumulative$) applies to a random variable X that follows the Normal probability distribution $\mathcal{N}(\mu, \sigma^2)$. If cumulative = TRUE then NORM.DIST is the Normal cumulative distribution function, which calculates the cumulative probability $P(X \leq x)$. For cumulative=FALSE, it represents the Normal probability density function, which calculates the density of probability at x.

→ If X follows the Normal distribution $\mathcal{N}(12, 9)$ (i.e. $\sigma^2 = 9$) for example, then you may compute some probabilities using Excel 2010 as follows:

$$P(X \leq 13) = \text{NORM.DIST}(13,12,3,\text{TRUE}) = 0.63,$$
$$P(X > 13) = 1 - P(X \leq 13),$$
$$= 1\text{-NORM.DIST}(13,12,3,\text{TRUE}) = 0.37,$$
$$P(10 < X \leq 13) = P(X \leq 13) - P(X \leq 10),$$
$$= \text{NORM.DIST}(13,12,3,\text{TRUE})$$
$$-\text{NORM.DIST}(10,12,3,\text{TRUE})$$

▶ Appendix C contains tables where you may find R, Excel 2007, or Calc functions that are equivalent to this Excel 2010 function.

▶ You may look at section E.9 of Appendix E for more Excel 2010 functions related to the Normal distribution.

Expected Value of the Normal Distribution

I have discussed in section 2.3.1 how the expected value and the variance of a measurement variable with probability density function f can be calculated. For the Normal probability distribution, these two parameters have already been calculated and are given in the following result:

Result 2.2.

If X follows the normal distribution $\mathcal{N}(\mu, \sigma^2)$ then its expectation $E(X)$ and variance $V(X)$ are given by :

$$E(X) = \mu \text{ and } V(X) = \sigma^2. \qquad (2.31)$$

2.3.4 The Central Limit Theorem

If there was one single most important result in statistical science, it would probably be the *Central Limit Theorem* (**CLT**). It is because of the **CLT** that the normal distribution plays a central role in statistical inference today. I briefly mentioned in the beginning of the current section what the Central Limit Theorem was about. It states that the average of a large collection of observations is a random variable whose law of probability is well approximated by the Normal distribution, even if the probability distribution of original variable is non-normal.

More formally, the **CLT** is stated as follows:

Result 2.3.

> Let $X_1, \cdots, X_n$ be n (random) observations of a random variable X with expectation $\mu = E(X)$ and variance $\sigma^2 = V(X)$. When the number n of observations is *sufficiently large*, the probability distribution of the average $\overline{X}$ of the X_i's, is *approximately* normal $\mathcal{N}(\mu, \sigma^2/n)$. Similarly, the sum $S = X_1 + \cdots + X_n$ of the X_i's is *approximately* normal $\mathcal{N}(n\mu, n\sigma^2)$.

The sample mean that is based on a fixed number of observations will follow a certain unknown law of probability. But this law of probability will change as the number of observations at your disposal increases. However, this law of probability will not change indefinitely. After the body of observations grows to a certain level, the law of probability associated with the sample mean will take its final shape, its *limit* form, which is the Normal distribution. In this sense, Result 2.3 is a limit theorem. And it is because of its central role in statistical inference, it is referred to as the central limit theorem.

It follows from the **CLT** that if you want to calculate the probability that the average is smaller than a quantity x, you may accomplish that as follows:

$$P(\overline{X} \leq x) = \Phi\left(\frac{x - \mu}{\sigma/\sqrt{n}}\right). \qquad (2.32)$$

where Φ is the cumulative distribution function of the standard Normal distribution. Equation 2.32 can be used as long as the two parameters μ and σ are known. In chapter 3, I will discuss some practical methods for approximating these parameters based on the observations if necessary.

Example 2.7

The human body temperature varies from one individual to another, and its distribution among individuals is generally unknown. Suppose you know from past experience that such a temperature can be expected to be around $\mu = 98.20^oF$ and has a standard deviation $\sigma = 0.62$. Your problem is to determine how often the average body temperature falls between the two values 98.15^oF and 98.6^oF.

With no new data and using the CLT, the researcher can determine the probability $P(98.15 \leq \overline{X}_{100} \leq 98.6)$ that the average of 100 randomly collected body temperatures be in the specified interval. You know from the CLT that $\overline{X}_{100}$ follows approximately the normal distribution $\mathcal{N}(98.20, \sigma_{\overline{X}}^2)$, where $\sigma_{\overline{X}} = 0.62/\sqrt{100} = 0.062$. Thus,

$$P(98.15 \leq \overline{X}_{100} \leq 98.60) = P(\overline{X}_{100} \leq 98.6) - P(\overline{X}_{100} \leq 98.15),$$
$$= 1 - 0.209991 = 0.79$$

You may compute $P(\overline{X}_{100} \leq 98.15)$ for example, using Excel 2010 as =NORM.DIST(98.15,98.2,0.062,TRUE).

Example 2.6 shows how the CLT can be used to approximate certain probabilities without having to conduct an experiment. The CLT can be used to approximate probabilities whenever the sum or average of variables is being used. I discussed in section 2.2.2 how the Binomial variable $\mathcal{B}(n,p)$ can be seen as a sum of n Bernoulli variables. Thus, the CLT can be used to approximate probabilities related to the Binomial distribution provided n is large and p is neither too close to 0 nor too close to 1.

Although the CLT is a very powerful result, its routine use still poses two problems. The practitioner should normally be able to answer the following questions before deciding to apply the CLT with confidence.

(*a*) The **CLT** requires the number n of observations to be large. *How large should it be ?*

(*b*) The **CLT** provides an approximation of the law of probability of the mean. *How good is that approximation ?*

A rule of thumb often used in practice is to achieve a sample size of at least 30 before using the **CLT**. In reality, $n = 30$ will be unnecessarily high for some applications and dramatically low for others. The optimal n depends on the (usually unknown) probability distribution of the original data. Although the **CLT** is appealing for letting us ignore the distribution of our data, its performance still depends on it. Very skewed original data will require a bigger sample than non-skewed data. Although averaging normalizes the original probability distribution, the more distorted the original distribution, the harder and the longer the normalization.

It should clearly be understood that the normal distribution does not describe a concrete fact that can be observed from experience. It is an abstract concept that precisely represents what all averages $\overline{X}$ (all concrete variables and their associated actual probability distributions confounded) have in common. Although the **CLT** guarantees a wide applicability of the normal distribution, it does not really provide an accurate description of any particular situation. Therefore, the **CLT** should be expected to provide at best a partial answer to any real-life problem, a start from which further investigation can be conducted on the basis of our knowledge about the problem under investigation.

2.3.5 The Exponential distribution

In section 2.2.4, I indicated that the Poisson distribution was useful for modeling the number of occurrences of an event within a predetermined time period of length t. In such an experiment, the time period length t is fixed and it is the number X of occurrences of the event within that period that is random. I indicated that in this case X follows a Poisson distribution $P(\lambda t)$ where λ is the expected number of occurrences of the event in one time unit.

In this section, I am considering the reverse problem where the number of observed events of interest is fixed at 1. However, the time period length T necessary to observe that first event is random. Consider for example the number of patients calling a health insurance company call center to request a specific medical service. The time length T from the opening of the call center to the reception of the first call will vary from day to day. I am interested in the probability distribution of T. Since I do not want to impose a limitation on the precision with which the length of time T can be measured, I will take it to be a continuous variable (you as practitioner will want to verify that this assumption does not distort the nature of the problem under study).

The probability that T is smaller than a specified time t is given by:
$$F(t) = P(T \leq t) = 1 - P(T > t).$$

Remember that all I want is to observe the first event and to record the elapsed time T until that occurrence. For that elapsed time to be greater than t, there should be no event observed until time t. That is, the number of events X observed until time t, which follows the Poisson distribution $P(\lambda t)$, must be equal to 0. We have that,

$$F(t) = 1 - P(X = 0) = 1 - \frac{(\lambda t)^0 e^{-\lambda t}}{0!} = 1 - e^{-\lambda t},$$

which corresponds to the *cumulative distribution function (C.D.F.)* of T. This leads to the following probability density function:

$$f_\lambda(t) = \lambda e^{-\lambda t}.$$

Definition. 2.3.

A random variable T has a probability distribution in the family of *exponential distributions* $\mathcal{E}(\lambda)$ if it has a probability density function, which is of the following form:

$$f_\lambda(t) = \begin{cases} \lambda e^{-\lambda t} & \text{if } t \geq 0, \\ 0 & \text{otherwise.} \end{cases}$$

where λ is a positive parameter to be determined from each particular application.

Figure 2.12 shows the graph of the exponential density function when the λ parameter takes the values 0.5, 1.5 and 2.5.

The exponential distribution was derived as the distribution of the time length until the occurrence of the first event. One may also be interested in the distribution of the time length T_k until the occurrence of the kth event. This leads to a generalization of the exponential distribution discussed in the next sub-section.

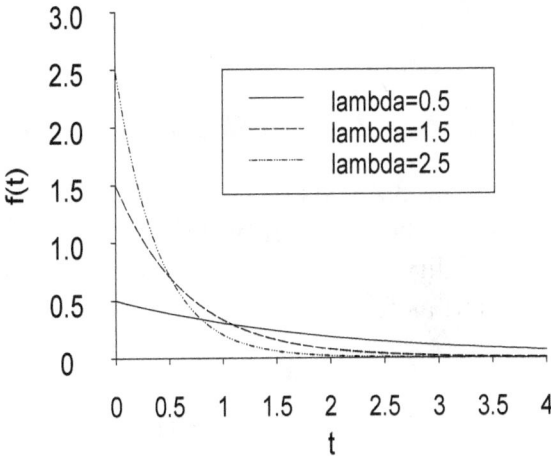

Figure 2.12. Exponential density functions for $\lambda = 0.5$, 1.5, 2.5

Cumulative Distribution, Expectation, and Variance

The cumulative distribution function of the Exponential probability distribution is defined as follows:

$$F(t) = \begin{cases} 1 - e^{-\lambda t} & \textit{if } t > 0, \\ 0 & \textit{otherwise.} \end{cases} \tag{2.33}$$

If a random variable X follows the exponential distribution $\mathcal{E}(\lambda)$, then its expected value $E(X)$, and variance $V(X)$ are given by:

$$E(X) = 1/\lambda \ \textit{and} \ V(X) = 1/\lambda^2. \tag{2.34}$$

The Software Solution

▶ Both versions of MS Excel 2010 & 2007 offer functions for calculating the probability density $f(t)$ as well as the cumulative probability $F(t)$. These calculations are done as follows:

Excel 2007 $f(t)$ =EXPONDIST(t,λ,FALSE) ,

$F(t)$ =EXPONDIST(t,λ,TRUE) .

Excel 2010 $f(t)$ =EXPON.DIST(t,λ,FALSE) ,

$F(t)$ =EXPON.DIST(t,λ,TRUE) .

▶ OpenOffice Calc 3.2.0 also offers the function EXPONDIST for calculating the probability density $f(t)$ and the cumulative probability $F(t)$ at a given point t. These quantities are calculated as follows:

$f(t)$ =EXPONDIST($t;\lambda;0$) ,

$F(t)$ =EXPONDIST($t;\lambda;1$) ,

▶ The R Package offers the dexp and pexp functions for calculating the probability density $f(t)$ and the cumulative
probability $F(t)$ respectively at a given point t. These quantities are calculated as follows:

$$f(t) = \text{dexp}(t, \text{rate} = \lambda),$$

$$F(t) = \text{pexp}(t, \text{rate} = \lambda).$$

*You may want to look at section F.2 of Appendix F for
more R functions related to the exponential distribution*

2.3.6   The Gamma Distribution

Suppose you want to obtain the probability distribu-
tion of the time length T_k until the occurrence of the kth event.
Starting with the smaller values of k and using the same approach I used to derive the p.d.f. of the exponential distribution,
you can see that the probability density function of T_k is given
by:

$$f_\lambda(t|k) = \begin{cases} \dfrac{\lambda^k t^{k-1} e^{-\lambda t}}{(k-1)!} & \text{if } t \geq 0, \\ 0 & \text{otherwise.} \end{cases} \tag{2.35}$$

Equation 2.35 is the probability density of a generalized exponential distribution where the number of occurrences of an event
is an arbitrary integer k.

Let me consider an even more general situation where a ho-
meowner spends a time duration T_α to mow an area of α yard
square of his lawn. In this particular case, the phenomenon being
observed during the time interval T_α is no longer discrete. It is
continuous (the surface area of the lawn). Therefore, equation
2.35 cannot be used to describe the probability density of T_α. This
problem is resolved by replacing k with α and $(k-1)!$ in the denominator with the surface area under the curve $g(t) = \lambda^\alpha t^{\alpha-1} e^{-\lambda t}$

for $t \geq 0$. This surface area is generally denoted by $\Gamma(\alpha)$ and referred to as the *Gamma Function*. This leads to the following definition:

Definition. 2.4.

A random variable T_α has a *gamma probability distribution* $\mathcal{G}(\alpha, \lambda)$ if its probability density function is given by :

$$f_\lambda(t|\alpha) = \begin{cases} \dfrac{\lambda^\alpha}{\Gamma(\alpha)} t^{\alpha-1} e^{-\lambda t} & \text{if } t \geq 0, \\ 0 & \text{otherwise.} \end{cases}$$

The Software Solution

▶ Both versions of MS Excel 2010 & 2007 offer functions for calculating the probability density $f(t)$ as well as the cumulative probability $F(t)$ for the Gamma distribution. These calculations are done as follows:

Excel 2007 $f(t)$ =GAMMADIST($x_0,\alpha,1/\lambda$,FALSE) ,

$F(t)$ =GAMMADIST($t,\alpha,1/\lambda$,TRUE) .

Excel 2010 $f(t)$ =GAMMA.DIST($t,\alpha,1/\lambda$,FALSE) ,

$F(t)$ =GAMMA.DIST($t,\alpha,1/\lambda$,TRUE) .

▶ OpenOffice Calc 3.2.0 also offers the function GAMMADIST for calculating the probability density $f(t)$ and the cumulative probability $F(t)$ at a given point t. These quantities are calculated as follows:

$f(t)$ =GAMMADIST($t;\alpha;1/\lambda;0$) ,

$F(t)$ =GAMMADIST($t;\alpha;1/\lambda;1$) .

▶ The R Package offers the dgamma and pgamma functions for calculating the probability density $f(t)$ and the cumulative probability $F(t)$ respectively at a given point t. These quantities are calculated as follows:

$$f(t) = \text{dgamma}(t, \alpha, \text{rate} = \lambda),$$

$$F(t) = \text{pgamma}(t, \alpha, \text{rate} = \lambda).$$

You may want to look at section F.2 of Appendix F for more R functions related to the gamma distribution

2.3.7 The Chi-Square Distribution

There is a special gamma distribution that is worth mentioning because of its importance in statistical inference. Many statistical procedures on hypothesis testing discussed in chapters 4 through 10, use this special gamma distribution called the *chi-square distribution*, which is defined as follows:

Definition. 2.5.

A random variable X follows the *Chi-square distribution* with parameter r (where r is a *positive integer*), if its probability distribution is a gamma $\mathcal{G}(r/2, 1/2)$. The corresponding probability density function is given by:

$$f(t|r) = \begin{cases} \dfrac{1}{2^{r/2}\Gamma(r/2)} t^{r/2-1} e^{-t/2} & \text{if } t \geq 0, \\ 0 & \text{otherwise.} \end{cases}$$

This probability distribution is often denoted by χ_r^2.

The parameter r associated with the chi-square distribution is known in the statistical literature as the *Number of degrees of* _freedom._ The name "number of degrees of freedom" reflects the fact that the chi-square distribution with r degrees of freedom often describes the law of probability of statistical aggregates in which r observations have the "freedom" to take any value, the remaining observations being derived from the information already available.

The software solution presented for the Gamma probability distribution can be used for the chi-square distribution as well. The Gamma parameters in this case will be $\alpha = r/2$, and $\lambda = 1/2$.

2.3.8 The Chi-Square, F, and Student t-Distributions

There are a few important results on probability distributions that have proved very useful in practice. I have selected a few of them to be presented in this section.

(a) The chi-square distribution has found numerous important applications in statistical inference. This partly stems from the fact that if $Z_1, Z_2, \cdots, Z_n$ are n independent standard Normal random variables, then the sum of their squares U is known to follow the chi-square distribution with n degrees of freedom.

$$\text{If } Z_i \sim N(0, 1) \text{ then } U = \sum_{i=1}^{n} Z_i^2 \sim \chi_n^2 \qquad (2.36)$$

The Chi-square distribution is used in chapter 7 for comparing two proportions based on data from two dependent samples. It is also used extensively in chapter 9 to test the equality of multiple population proportions.

(b) Let X and U be two independent random variables such that X follows the standard Normal distribution, and U the chi-square distribution with n degrees of freedom. Then the random variable T derived as the ratio of X over $\sqrt{U/n}$ follows a probability distribution called the *Student's t Distribution* with n degrees of freedom and denoted by t_n. Mathematically, this is expressed as,

$$\text{If } X \sim N(0,1) \ \& \ U \sim \chi_n^2 \text{ then } T = \frac{X}{\sqrt{U/n}} \sim t_n \quad (2.37)$$

The t distribution has found numerous applications in statistical inference as will be seen in subsequent chapters. This probability distribution has been thoroughly investigated, and is well documented in the statistical literature.

Tables 2.3 and 2.4 provide Excel, Calc, and R functions for calculating probability densities and cumulative probabilities of the t distribution.

Table 2.3 : Excel Solutions for Student's t_n Distribution

Software	Probability Type	Commands
Excel 2007	Probability Density	N/A
	Cumulative Probability	1-TDIST(x,n,1)
Excel 2010	Probability Density	T.DIST(x,n, FALSE)
	Cumulative Probability	T.DIST(x,n, TRUE)

Table 2.4 : R and Calc Solutions for Student's t_n Distribution

Software	Probability Type	Commands
R	Probability Density	N/A
	Cumulative Probability	1-TDIST(x ;n ;1)
Calc	Probability Density	dt(x, n)
	Cumulative Probability	pt(x, n)

(c) If two random variables U_1 and U_2 are independent and follow the chi-square distribution with n_1 and n_2 degrees of freedom respectively, the derived random variable $F = (U_1/n_1)/(U_2/n_2)$ follows a distribution function called F. This is expressed mathematically as,

$$\text{If } U_1 \sim \chi^2_{n_1} \ \& \ U_2 \sim \chi^2_{n_2} \text{ then } F = \frac{U_1/n_1}{U_2/n_2} \sim F_{n_1,n_2} \quad (2.38)$$

The F distribution has been thoroughly studied, and is often used in practice (see chapters 10 and 11 in this book). However, its probability density function is very complex and is not often used in practice. Nevertheless practitioners need to be able to compute the percentiles and the cumulative probabilities of the F distribution. Tables 2.5 and 2.6 show some software options that you can used to handle the F distribution. Section E.9 of Appendix E describes more Excel functions related to F distributions. Section F.2 of Appendix F will show you R functions for the F distribution.

Table 2.5 : Excel Solutions for F_{n_1,n_2} Distribution

Software	*Probability Type*	*Commands*
Excel 2007	*Prob. Density*	N/A
	Cum. Probability	1-FDIST(x,n_1,n_2)
Excel 2010	*Prob. Density*	F.DIST$(x,n_1,n_2,$FALSE$)$
	Cum. Probability	F.DIST$(x,n_1,n_2,$TRUE$)$

Table 2.6 : R and Calc Solutions for F_{n_1,n_2} Distribution

Software	*Probability Type*	*Commands*
R	*Probability Density*	N/A
	Cumulative Probability	1-FDIST$(x\,;n_1\,;n_2)$
Calc	*Probability Density*	df(x,n_1,n_2)
	Cumulative Probability	pf(x,n_1,n_2)

(d) If $Z_1, Z_2, \cdots, Z_n$ are n independent variables that follow the normal distribution $N(\mu_1, 1), N(\mu_2, 1), \cdots, N(\mu_n, 1)$, the sum of their squares follows a distribution that is called the *Non-central chi-square* distribution with n degrees of freedom and the *non-centrality parameter* $\lambda = (\mu_1^2 + \cdots + \mu_n^2)/2$.

$$\text{If } Z_i \sim N(\mu_i, 1) \text{ then } Q \sim \chi_n^2(n, \lambda), \qquad (2.39)$$

where $Q = Z_1^2 + Z_2^2 + \cdots + Z_n^2$. Non-central chi-square distributions are often used to quantify the propensity for rejecting the assumption of equality among proportions when these are indeed not all equal.

The chi-square distribution described in (a) is often called the *Central Chi-Square* distribution to indicate that its centrality parameter is 0.

The mean and variance of the $\chi^2(n, \lambda)$ are $n+2\lambda$ and $2n+8\lambda$ respectively.

(e) Let X and U be two independent random variables such that X follows the Normal distribution $N(\mu, 1)$, and U the chi-square distribution with n degrees of freedom. Then the random variable T derived as the ratio of X over $\sqrt{U/n}$ follows the *Non-Central t Distribution* with n degrees of freedom, and non-centrality parameter μ.

$$\text{If } X \sim N(\mu, 1) \text{ and } U \sim \chi_n^2 \text{ then } T = \frac{X}{\sqrt{U/n}} \sim t(n, \mu).$$
$$(2.40)$$

The t distribution has found numerous applications in statistical inference as will be seen in subsequent chapters. Among other applications, the non-central t-distribution is used to study the quality of the t-test as shown in chapter 6.

(f) If U_1 and U_2 are 2 independent random variables such that U_1 follows the non-central chi-square distribution $\chi^2(n_1, \lambda)$

and U_2 follows the central chi-square $\chi^2_{n_2}$ then the random variable F derived as the ratio of U_1/n_1 to U_2/n_2 follows a probability distribution called the *Non-central F-Distribution* and denoted by $F(n_1, n_2, \lambda)$.

$$\text{If } \left\{ \begin{array}{l} U_1 \sim \chi^2(n_1, \lambda), \\ U_2 \sim \chi^2_{n_2} \end{array} \right\} \text{ then } F = \frac{U_1/n_1}{U_2/n_2} \sim F(n_1, n_2, \lambda),$$

(2.41)

Note that of all software options I am considering in this book, only the R package can handle non-central chi-square, non-central t, and non-central F distributions. See section F.2 for more details.

CHAPTER $\boxed{3}$

Point and Interval Estimation

OBJECTIVE

The purpose of chapter 3 is to discuss some key fundamental properties of a good estimation procedure. The focus is placed on procedures for estimating a population mean with a single number (point estimation), and with a range of values (interval estimation)

CONTENTS

3.1. Overview

Many research studies revolve around the desire to quantify population parameters of interest such as the probability of response to a credit card company's solicitation campaign, or the average lifetime of a certain type of light bulb. Oftentimes the goal is evaluate quality or efficiency.

The probability of response for example, is a theoretical concept that is approximated in practice by selecting a random sample of prospects during a survey and using the proportion of respondents as an estimation. Can you use such a proportion as a reliable estimation of the probability of response in all subsequent campaigns? How good is this approximation?

If you compute the expected lifetime of light bulbs by averaging the life times of a sample of light bulbs, one could ask the same questions about the quality of the sample mean as a reliable estimate of the expected lifetime. Proportions and means are widely used metrics. But are they reliable? Are there better alternatives available in some situations? What criteria should you use for evaluating the quality of an estimation? These are the types of questions that I want to address in this chapter.

Theoretical constructs are assigned concrete (i.e. estimated) values using information from random samples. These concrete values are called *Point Estimations*, because they characterize the parameters they approximate with a single value that can be conveniently used by practitioners. However, these estimations are subject to sampling errors induced by your decision to limit the investigation to a small collection of units. To account for these specific types of errors, some researchers prefer to approximate the parameters not with a single estimation, but with a range of values that is likely to include the true value of the parameter. This range of values is an *Interval Estimation* of the parameter. These interval estimations are also better known under the naming *Confidence Intervals*. Section 3.2 will focus on point estimation while section 3.3 is entirely devoted to interval estimation.

3.2. Point Estimation: Introduction

To study point estimations effectively, I need to start by clarifying the terminology that will be used throughout this section. When someone tells you she is looking for statistics on education for example, that person is generally looking for concrete numbers on educational attainment. In the study of point estimation, a concrete number is called an estimate or an estimation in order to stress out the fact that it represents an approximation of a reality (often expressed in the form of a parameter), which although elusive, still deserves your full attention and consideration. What is a *statistic* then?

Before you start collecting data, you should think about what you are going to measure. If you are going to measure the breaking strength of 10 wire ropes, you may name these 10 measurements as $x_1, x_2, \cdots, x_{10}$, and their mean value as $\bar{x} = (x_1 + x_2 + \cdots + x_{10})/10$. All the quantities $x_1, x_2, \cdots, x_{10}$ are random variables because their actual values will only be known after the sample has been selected and the measurements taken. In this case, $\bar{x}$, which is a function of the random variables $x_1, x_2, \cdots, x_{10}$ is called a *statistic*. More generally, any function of the random sample $x_1, x_2, \cdots, x_n$ of size n is called a *statistic* as long as it does not depend on unknown quantities.

If μ denote the expected breaking strength of a wire rope that you want to approximate using 10 measurements $x_1, x_2, \cdots, x_{10}$, then any statistic derived from the 10 measurements that you formulate as your best guess of the value of μ is called an *esti-*

mator of μ (generally denoted by $\widehat{\mu}$ (read mu hat)). Therefore an estimator of a parameter μ is essentially a special type of statistic that is somehow tied to the parameter. It describes a protocol for obtaining concrete estimations once actual data are gathered.

3.2.1   Concepts of Point Estimation

Let X be a random variable representing the breaking strength of a wire rope. You as researcher may want to quantify the expected value μ of X and its standard deviation σ. The general notation for an estimator of μ will be $\widehat{\mu}$, and the estimator of σ will be $\widehat{\sigma}$. The standard approach is to select a random sample $x_1, x_2, \cdots, x_n$ of n rope strengths and use $\widehat{\mu} = \overline{x}$ and $\widehat{\sigma} = s$ where $s = \sqrt{s^2}$ and s^2 defined as,

$$s^2 = \frac{1}{n-1} \sum_{i=1}^{n} (x_i - \overline{x})^2 \tag{3.1}$$

Unbiased Estimators

To be considered good, an estimator must be *unbiased* and has a small *standard error*. As function of random variables, estimators are random themselves. Therefore they have their own expected values and standard deviations. The standard error is simply the standard deviation of an estimator, and when the expected value of an estimator is identical to the parameter it estimates, that estimator is said to be unbiased.

Here are a few standard statistical results about the goodness of some basic statistics that will often be used throughout this book. These statistics include essentially the mean and the variance, and the results are about their unbiasedness and the magnitude of their standard errors.

Result 3.1.

> If $x_1, x_2, \cdots, x_n$ is a random sample from a theoretical distribution with mean μ, then the sample mean $\bar{x}$ is always an unbiased estimator of μ.

It follows from this result that the sample mean yields neither a systematic overestimation nor a systematic underestimation of the population mean. But you will not know how far any given sample mean strays from the population mean.

Result 3.2.

> If $x_1, x_2, \cdots, x_n$ is a random sample from a theoretical distribution with mean μ and variance σ^2, then the sample variance s^2 defined as,
>
> $$s^2 = \frac{1}{n-1} \sum_{i=1}^{n} (x_i - \bar{x})^2,$$
>
> is an unbiased estimator of σ^2.

Result 3.2 indicates that the sample variance s^2 too will not introduce systematic overestimation or underestimation into the estimation procedure.

Minimum Variance Unbiased Estimators

In addition to being unbiased, the sample mean has another advantage, which is to be a *linear estimator*[1]. And linear estimators are interesting given the convenience with which their variances can be obtained mathematically. However, the variance

---

[1]A linear estimator based on sample values $x_1, x_2, \cdots, x_n$ is defined as any linear combination of these values such as $a_1x_1 + a_2x_2 + \cdots + a_nx_n$, with coefficients a_i's that are independent of the sample

of the sample mean is known to be unduly high for skewed distributions such as those in the family of Gamma. The question then becomes, is it possible to have a Minimum Variance Unbiased Estimator (MVUE)? The answer depends on the context you are in. Here are two results that may help:

Result 3.3.

> If $x_1, x_2, \cdots, x_n$ is a random sample from a theoretical distribution with mean μ and variance σ^2, then the sample mean $\overline{x}$ is the minimum variance unbiased estimator of μ among all linear estimators.

Result 3.3 indicates that if linear estimators are the ones you are interested in due to the simple form they take, then the sample mean is your best bet for estimating the population. Do not look elsewhere.

Result 3.4.

> If $x_1, x_2, \cdots, x_n$ is a random sample from the Normal distribution with mean μ and variance σ^2, then the sample mean $\overline{x}$ is the minimum variance unbiased estimator of μ.

Result 3.4 stipulates that if the law of probability of your variable of interest is Normal, then the sample mean is the most precise estimator of all unbiased estimators that exist. All these results show that the sample mean has numerous qualities. It becomes problematic only when your data is skewed. In this case, alternative location measures such as the median are indicated.

3.2.2   The Maximum Likelihood Method

In section 3.2.1 I have mentioned the sample mean $\overline{x}$ and the sample variance s^2 as two estimators with good statistical properties for a wide variety of probability distributions. You may consider yourself lucky that the particular form taken by these two estimators was given to you. That will not always be case in practice. There are situations where you may need to estimated a population parameter without knowing what its general form looks like.

Suppose that you are running a towing service and want to estimate the probability that the number of requests for assistance per hour on any given Friday, exceeds 7. One option is to collect data every Friday for 20 Fridays in row, and consider the relative number of times the number of requests exceeded 7, as an estimator of your probability. There is one major drawback with this simplistic approach. If on most of the Fridays in your sample the number of requests for assistance happens to be smaller than usual, your experiment will result in an underestimation of the probability in question. Since the number of requests for assistance may vary from 0 to a large number, a better option is to model that number using the Poisson family of probability distributions.

If X is a random variable representing the number of requests for assistance on a randomly chosen Friday, then its law of probability will be defined as,

$$P(X = k) = \frac{\lambda^k}{k!} e^{-\lambda} \ for \ k = 0, 1, \cdots \qquad (3.2)$$

Before you can use this expression to compute the probability that the number of requests exceed 7, you need to know what the value of λ is. Here is where you will need your sample data $k_1, k_2, \cdots, k_{20}$ representing the number of requests you have collected over 20 consecutive Fridays. But how do you use that data to obtain $\widehat{\lambda}$, an estimator of λ? There is a general method called the *Maximum Likelihood Method* that can be used in most situations (not always easily) to resolve this problem. Let me briefly explain what this method is all about.

On any given Friday i, you will record k_i number of requests. Over a period of n Fridays (in the example above $n = 20$), you would have observed the n numbers $(k_1, k_2, \cdots, k_n)$. Since the number observed on any given Friday does not influence that of another Friday in any way, the joint probability $P(k_1, k_2, \cdots, k_n)$ of obtaining the specific set of numbers $(k_1, k_2, \cdots, k_n)$ during the observation period is the product of all n individual probabilities $P(X = k_i)$ for $i = 1, \cdots, n$. These individual probabilities are defined in equation 3.2. The maximum likelihood estimator of λ will be that specific value of λ that will yield the highest join probability possible.

The maximum likelihood principle is based upon the fact that you were able to observe the specific data you have, because it must have had a good chance of being observed. The probability of observing the set of numbers of requests $(k_1, k_2, \cdots, k_n)$ is given[2] by,

$$P(k_1, k_2, \cdots, k_n) = P(k_1, \lambda) \times \cdots \times P(k_n, \lambda), \qquad (3.3)$$

where $P(k_i, \lambda) = \lambda^{k_i} e^{-\lambda} / k_i!$. Note that the value of λ that maximizes the likelihood function is the exact same value that will maximize the logarithm of that function[3]. The *log likelihood* is then written as,

$$\ln[P(k_1, \cdots, k_n)] = \sum_{i=1}^{n} \ln[P(k_i, \lambda)]. \qquad (3.4)$$

---

[2] Note that this probability is called the *likelihood function*

[3] The interest in using the logarithm function lies on its property that says the log of a product equals the sum of the logs. Manipulating sums is generally much more convenient than manipulating products

Using Excel's Solver to Find λ

Once you collect your data points $k_1, k_2, \cdots, k_n$, the Excel Add-In solver can be used to compute the Maximum Likelihood Estimator of λ by minimizing the likelihood equation 3.4. To demonstrate how this is done, consider the Towing service example of the beginning of this section, and suppose that data on the number of requests for assistance has been collected for 20 consecutive Fridays along with the temperature (^0F) on those days as shown in Figure 3.1.

	D7			f_x	=LN(($E$3^B7/FACT(B7))*EXP(-$E$3))		
	A	B	C	D	E	F	G
1							
2	Day	Requests	Temperature	Log	Lambda		
3	1	10	39	-2.104	**9.30**		
4	2	10	39	-2.104			
5	3	12	36	-2.527			
6	4	11	35	-2.272			
7	5	8	50	-2.064			
8	6	10	40	-2.104			
9	7	15	30	-3.749			
10	8	8	41	-2.064			
11	9	12	31	-2.527			
12	10	5	55	-2.937			
13	11	9	43	-2.032			
14	12	6	52	-2.499			
15	13	7	54	-2.215			
16	14	9	43	-2.032			
17	15	14	32	-3.271			
18	16	0	60	-9.300			
19	17	7	55	-2.215			
20	18	9	40	-2.032			
21	19	13	32	-2.862			
22	20	11	36	-2.272			
23			Log Likelihood =	**-55.184**			

Figure 3.1. Setting up Excel to fit the Poisson Probability Model

Column D on Figure 3.1 contains the individual log probabilities $\ln[P(k_i, \lambda)]$, and cell D23 the sum of these log probabilities (i.e.

the log likelihood of equation 3.4. In figure 3.1, cell E3 contains the final maximum likelihood estimate of λ, you may start with an initial value such as 2, 3, or 4.

Launch the Excel Add-In Solver by selecting Data (from Excel's main menu) and by clicking the Solver icon. The Solver's parameters dialog form will be displayed as shown in figure 3.2. Fill out this form with the same information you see on that figure, and click on Solve. This will produce 9.30 in cell E3, and this will be the maximum likelihood estimation of λ. You may not have noticed that, by 9.30 is the mean number of requests. This can be proved using Calculus. But you do not need Calculus to compute the maximum likelihood estimate of the Poisson parameter, nor that of any other probability distribution. Solver is a very powerful tool in that regard.

Using Equation 3.2, I can now formulate the final fitted Poisson probability model as follows:

$$P(X = k) = \frac{9.3^k}{k!} e^{-9.3} \ for \ k = 0, 1, 2, \cdots \qquad (3.5)$$

Recall that the initial problem was to calculate the probability that the number of requests for assistance will exceed 7 (i.e. $P(X > 7)$). Note that $P(X > 7) = 1 - P(X \leq 7) = 1 - \{P(X = 0) + P(X = 1) + \cdots + P(X = 7)\}$. Using Equation 3. ? you could evaluate all these probabilities. But $P(X > 7)$ can conveniently be evaluated using Excel 2010 as follows[4]:

0.71 =1-POISSON.DIST(7,9.3,TRUE) .

---

[4]If you are using Excel 2007 you may use the POISSON function. R users may refer to Appendix C to find the equivalent R function.

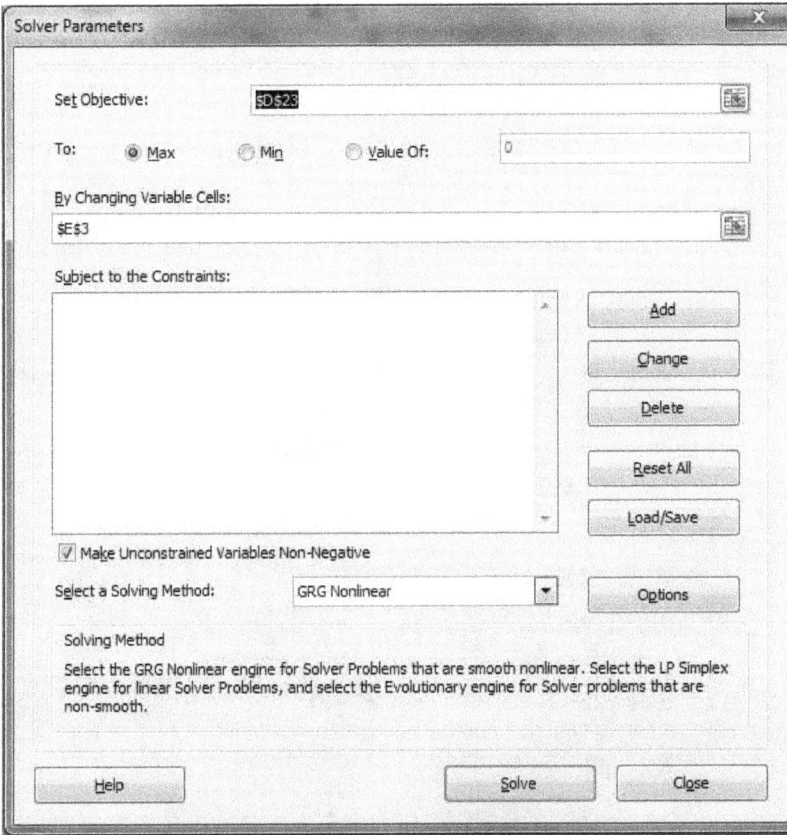

Figure 3.2. Specifying Solver's Parameters

A More Sophisticated Probability Model

In the previous example, I estimated the probability for the number of requests for assistance to exceed 7 using the fitted Poisson model of equation 3.5. However, there could be some suspicions that such a probability may well be higher in low temperatures than in higher ones. How can you factor temperature in? A simple solution to this problem is to consider that the average number of requests for assistance λ is a function of temperature t that can be formulated as $\lambda(t) = e^{\alpha + \beta t}$, and to use the same data to compute the maximum likelihood estimate of α and β. Why link $\lambda(t)$ to the exponent of linear function $\alpha + \beta t$ instead of linking it directly to the linear function? It is because I want to satisfy the constraint $\lambda(t) > 0$ of the Poisson probability model that must be met for all values of t.

The newer and more sophisticated Poisson model is formulated as follows:

$$p(x, t, \alpha, \beta) = \frac{[\lambda(t)]^k}{k!} e^{-\lambda(t)}, \ \ where \ \ln[\lambda(t)] = \alpha + \beta t. \quad (3.6)$$

The parameters α and β can be calculated using Excel's Solver by first setting up the spreadsheet as shown in Figure 3.3.

	E3			f_x	=EXP($F$2+$G$2*C3)		
	A	B	C	D	E	F	G
1	Day	Requests	Temperature	Log	Lambda	alpha	beta
2	1	10	39	-2.079	9.95	3.923	-0.042
3	2	10	39	-2.079	9.95		
4	3	12	36	-2.191	11.27		
5	4	11	35	-2.150	11.75		
6	5	8	50	-2.183	6.29		
7	6	10	40	-2.089	9.54		
8	7	15	30	-2.288	14.47		
9	8	8	41	-2.045	9.15		
10	9	12	31	-2.302	13.88		
11	10	5	55	-1.741	5.11		
12	11	9	43	-2.046	8.42		
13	12	6	52	-1.833	5.79		
14	13	7	54	-2.144	5.32		
15	14	9	43	-2.046	8.42		
16	15	14	32	-2.262	13.32		
17	16	0	60	-4.145	4.14		
18	17	7	55	-2.218	5.11		
19	18	9	40	-2.042	9.54		
20	19	13	32	-2.212	13.32		
21	20	11	36	-2.129	11.27		
22			Log Likelihood =	**-44.225**			

Figure 3.3. Setting up Excel to compute α and β of Model 3.6

I essentially used the same set up of figure 3.1, added the "Lambda" column since $\lambda(t)$ now varies by temperature t, also added cells F2 and G2 to contain the parameters α and β respectively.

The next step is to launch Excel solver and to fill out the parameters dialog form as shown in Figure 3.4. *Objective* still represents cell $D22$ containing the log likelihood function to be maximized.

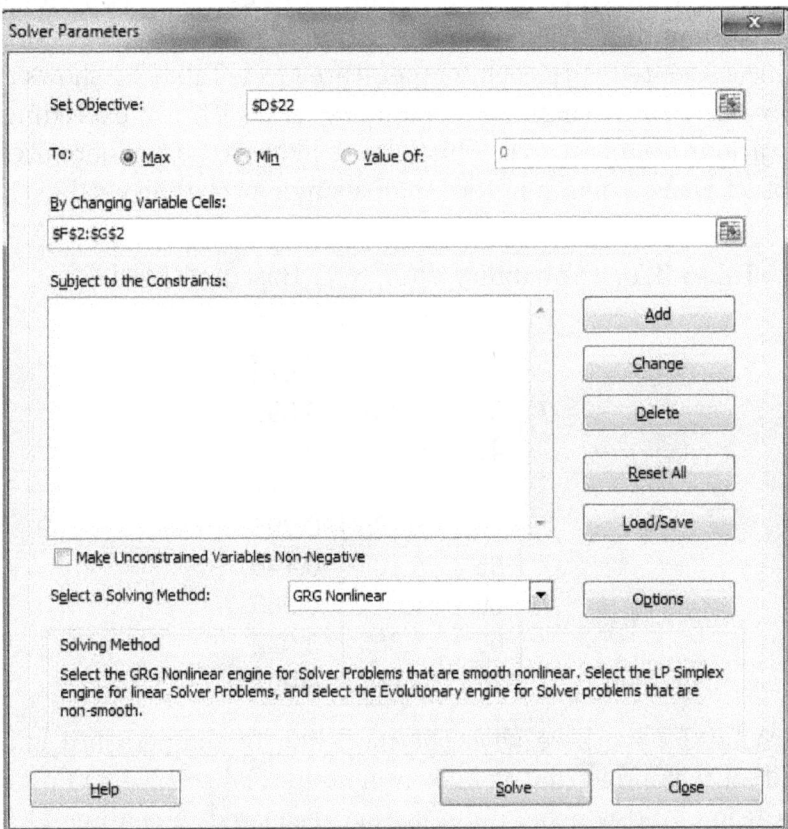

Figure 3.4. Specifying Solver's Parameters for Model 3.6

Changing Variable Cells represent the 2 cells F2 and G2 for α and β respectively. After you click **Solve** you should see the maximum likelihood estimations of α and β in cells F2 and G2 (see Figure 3.3). The new fitted model is formulated as,

$$p(k,t) = \frac{[\lambda(t)]^k}{k!} e^{-\lambda(t)}, \quad where \; \lambda(t) = e^{3.923-0.042t} \tag{3.7}$$

The main advantage of model 3.7 is that the probability that the number of requests for assistance will exceed 7 is no longer a single number. Instead, it is a series of numbers, one number being associated with each temperature level. Table 3.1 shows the probability of the number of requests for assistance exceeding 7 for various temperatures. This table is likely to be far more useful in practice as a planning tool than a single probability value that does not account for weather conditions.

Table 3.1: Probability $P(X > 7)$ Based on Model 3.7

k	t	λ	$p(k,t)$
7	30	14.474	0.976
7	35	11.751	0.899
7	40	9.540	0.735
7	45	7.745	0.511
7	50	6.288	0.297
7	55	5.105	0.145
7	60	4.145	0.060
7	65	3.365	0.022
7	70	2.732	0.007

What I have here with the modeling of temperature is actually a technique that is known in statistical science as Poisson regression, and which will be further discussed in chapter 11.

General Maximum Likelihood Method

What I have done in this section so far, only focused on the Poisson probability model. But in reality, the same estimation technique can be used with any probability distribution to compute any parameter or function of parameters that you are interested in. Let $f(x, \theta)$ be a probability distribution that depends on a parameter θ. In this context, θ could well represent several parameters. Moreover, if the random variable X is discrete, $f(x, \theta)$ will be a probability mass function, and for measurement variables it represents the probability density function.

To compute the maximum likelihood estimate of θ based on a sample of n data points $x_1, x_2, \cdots, x_n$, you will first form the log likelihood function as follows:

$$\ln\big[f(x_1, \cdots, x_n, \theta)\big] = \sum_{i=1}^{n} \ln\big[f(x_i, \theta)\big]. \qquad (3.8)$$

From this this point on, Excel solver can be used the same way I used it with the Poisson probability model.

3.3. Interval Estimation: Introduction

The 1994-1995 Canadian National Population Health Survey revealed that 30.5% of Canadians aged 20 to 64 are obese (see Birmingham et al., 1999). An individual is considered obese when his/her Body Mass Index (BMI) is 27 or greater. Because of its proven negative effects on people's health and the potential cost implications associated with it, obesity has become a major concern to public policy makers in many industrialized nations.

Birmingham et al. (1999) estimated the cost of obesity in Canada. One key component of their methodology was the use of the Population Attributable Fraction (PAF), which allowed them to determine the proportion of the total costs of a disease attributed to obesity. The PAF, which involves the probability of a person in a given segment of the population to be obese, is estimated from the probability-sample-based National Health Survey.

Consequently, the estimated total cost of obesity in Canada is subject to sampling variability and is only an approximation of the actual cost. It is natural for a policy maker in this situation to attempt to know how low the actual obesity cost is likely to go below its estimated value, and how high it may go above that estimated value. Aware of this problem, Birmingham et al. (1999) conducted a sensitivity analysis based on the use of *confidence intervals* for the estimated PAFs. The confidence interval is a range of values that are *likely* to contain the "true" PAF. The lower and upper bounds of these intervals allowed these researchers to determine a range of values likely to contain the "true" obesity cost.

It is desirable to obtain a narrow confidence interval for the obesity cost. For, if the interval is too wide then the two confidence bounds may lead policymakers to different intervention policies. The lower confidence bound may indicate an obesity cost lower than that of other diseases suggesting that the healthcare budget be primarily allocated to the prevention of these other diseases. However, if the upper confidence bound suggests an obesity cost higher than that of many other diseases, the impact on the budget allocation may be different from what was suggested by the lower confidence bound.

The type of problems discussed in the past few paragraphs demonstrates the importance of interval estimation in research. Point estimates are important for obtaining our best guess of the magnitude of a social phenomenon. As seen in section 3.2, point estimates are subject to sampling and non-sampling errors. To increase the chance of success of a given policy, policymakers often need to devise policies that are flexible enough to account for possible discrepancies between estimates and reality.

Although the problem of interval estimation has been known for a longtime, it is only in 1934 that a Polish mathematician in the name of Jerzy Neyman (1894-1981) proposed a simple procedure to obtain an interval estimate. He was first to use the expression "Confidence Interval" and to call the lower and upper bounds of the interval, "Confidence bounds". In spite of the difficulties posed by the interpretation and meaning of confidence intervals, Neyman's method remains today the most widely-used approach for obtaining confidence intervals and will be discussed in sections 3.4, and 3.5.

The availability of cheap and powerful computers these days has made possible the use of computer-intensive methods such as the *bootstrap* for creating nonparametric confidence intervals. The bootstrap belongs to the family of resampling methods, and was introduced by Efron and Tibshirani (1994) as a general method that can be applied to virtually any situation to create confidence intervals. Section 3.6 is devoted to a discussion of some of the most widely-used bootstrap methods for interval estimation.

One of the simplest interval estimation problem consists of constructing a confidence interval for a proportion. Therefore, it is convenient to expose the main ideas of interval estimation for this simple case.

3.4. Interval Estimation of a Proportion

Let us consider a small survey where 15 randomly selected students aged 12 to 18 are asked questions about their use of computers in school. The survey reveals that 80% (i.e. 12) of them use computers in school. Because, 80% is your best guess of the actual and unknown proportion of students in the 12-18 age range who use computers in school, you may want to know how low and how high the actual proportion is likely to be. That is you want to have the likely lower bound, and the likely upper bound for your parameter of interest. The whole range of values between these two bounds is an interval estimation of the "true" proportion, since any value within that range could possibly be that "true" proportion.

Since students who participated in the survey were randomly selected, the number of computer users follows a Binomial distribution $\mathcal{B}(n,p)$ with a number of trials $n = 15$ and an unknown probability of "success" p, which is the actual proportion of students using computers. The survey provided an estimate $\hat{p} = 0.80$ of proportion p, which corresponds to $m = 12$ students out of 15. Our problem is to find two statistics p_L (lower bound) and p_U (upper bound) such that the interval (p_L, p_U) contains the "true" proportion p with a certain (high) probability called the *Confidence level* or the *Coverage probability*. In the next two sections, I will present two methods for finding the two bounds. When the number of successes and the number of failures in your experiment both exceed or equal 5, then you should use the simpler *large-sample approach*[5], while the more complex *small-sample approach* is reserved for the situation where either the number of successes or the number of failures is below 5, and is based on the direct use of the Binomial distribution.

3.4.1 Small-Sample Intervals for Proportions

In general, the confidence level is denoted by $1 - \alpha$ where α is the (small) probability of the undesirable situation where our interval does not contain the "true" proportion. The most commonly-used confidence level is 95% (i.e. $\alpha = 5\%$). Constructing a confidence interval at the confidence level $1 - \alpha$ amounts to finding two statistics p_L and p_U with $p_L < p_U$, such

---

[5]This is the only approach presented many introductory statistics textbooks. Because the sample is large in this case, the Binomial distribution is well approximated by the Normal distribution, which is generally simpler to manipulate.

that,

$$P\left(p_L < p < p_U\right) = 1 - \alpha. \qquad (3.9)$$

When dealing with this problem, your input data will typically include the number of trials n, and the number of successes. Let $B(x; n, p)$ be the cumulative probability at point x of a binomial variable $\mathcal{B}(n, p)$. The lower bound p_L is calculated using equation 3.10, while p_U obtained from equation 3.11.

$$\begin{cases} B(m - 1; n, p_L) = 1 - \alpha/2, & \text{if } m \geq 1, \quad (a) \\ p_L = 1 - \sqrt{1 - \alpha/2}, & \text{if } m = 0. \quad (b) \end{cases} \qquad (3.10)$$

If you observe no success in your sample, then use equation 3.10(b) to determine the lower bound. Otherwise, you would use equation 3.10(a) that implicity defines that lower bound.

$$\begin{cases} B(m; n, p_U) = \alpha/2, & \text{if } 0 \leq m < n, \quad (a) \\ p_U = 1, & \text{if } m = n. \quad (b) \end{cases} \qquad (3.11)$$

If all of your trials result in successes (i.e. $m = n$), then use equation 3.11(b) to determine the upper bound. Otherwise, you would use equation 3.11(a) that implicity defines that upper bound.

Example 3.1 _____

Consider the survey on student computer usage in school, that I discussed in the beginning of section 3.4. The sample size is $n = 15$ students, and the number of successes $m = 12$ represents the number of computer users among the 15 students selected in the sample. The "true" proportion p of computer users was estimated as $\hat{p} = 0.80$.

I now want to construct a confidence interval around the "true" parameter p at the 95% confidence level (i.e. $1 - \alpha = 0.95$ or $\alpha = 0.05$). This problem must be addressed using the small-sample approach since the number of failures (i.e. non-users) is $n - m = 15 - 12 = 3$, which is below 5.

▶ Since $m = 12$ exceed 0, I need to use equation 3.10a to obtain p_L. This equation can be re-expressed as,

$$B(11, 15, p_L) = 0.975. \tag{3.12}$$

The binomial cumulative probability of equation 3.12 can be calculated using Excel 2010 as, =BINOM.DIST(11,15,p_L,TRUE). To resolve equation 3.12, I created Table 3.2 in Excel, which resulted in 0.52 being the solution I was looking for since $B(11, 15, 0.52) = 0.9746$, which is close enough to 0.975.

Table 3.2: Calculation of the lower bound p_L

p	0.5	0.51	**0.52**	0.53	0.54	0.55
$B(11, 15, p)$	0.9824	0.9788	**0.9746**	0.9697	0.9641	0.9576

▶ To compute p_U, I need to use equation 3.11(a), since $m = 12 \neq 15$. This equation can be re-expressed as,

$$B(12, 15, p_U) = 0.025. \tag{3.13}$$

To resolve equation 3.13, I proceed by trying several values in an Excel spreadsheet as shown in Table 3.3. The cumulative probabilities $B(12, 15, p_U)$ were computed again using the Excel 2010 function BINOM.DIST as before. This led to 0.957 being the approximate solution for p_U.

Table 3.3: Calculation of the lower bound p_U

p	0.955	0.956	**0.957**	0.958	0.959	0.96
$B(12, 15, p)$	0.0276	0.0260	**0.0245**	0.0231	0.0217	0.0203

The 95% confidence interval of the proportion of computer users in school is given by:

$$\text{CI}(p) = (0.52, 0.957)$$

In Example 3.1, I used Excel to compute the cumulative binomial probabilities. However, any other software can be used for that purpose including OpenOffice Calc, or R. See Table C.2 of Appendix C or section F.2 of Appendix F for alternative software options.

How were Equations 3.10 and 3.11 Developed ?

I like to briefly discussed the logic behind the small-sample method for constructing confidence intervals as formulated by equations 3.10 and 3.11. Note that $P(p_L < p < p_U) = P(p_L < p) - P(p_U \leq p)$. Therefore, for equation 3.9 to be satisfied, it is sufficient to have $P(p_L < p) = 1 - \alpha/2$ and $P(p_U \leq p) = \alpha/2$). How do you determine p_L so that $P(p_L < p) = 1 - \alpha/2$? This can be done by realizing that the expected number of successes in a Binomial experiment is proportional to its nominal probability of success p.

Now let m_L be the observed number of successes in a Binomial trial $\mathcal{B}(n, p_L)$, and $E(m_L)$ its expected value. Likewise, m is the observed number of successes in a Binomial trial $\mathcal{B}(n, p)$, and $E(m)$ its expected value. If $p_L < p$ then $E(m_L)$ will be smaller than $E(m)$. And vice-versa, if $E(m_L)$ is smaller than $E(m)$, you will conclude that $p_L < p$. I can then conclude that $p_L < p$ is equivalent to $E(m_L) < E(m)$. This naturally suggests that a good candidate for p_L will satisfy the relation,

$$P(p_L < p) = P(m_L < m),$$
$$= P(m_L \leq m - 1) = B(m - 1; n, p_L) = 1 - \alpha/2.$$

This equation stems from the fact that m_L follows the Binomial distribution $\mathcal{B}(n, p_L)$, and is identical to equation 3.10. Equation 3.11 can also be justified using the same argument.

On the Validity of the Procedure

The small-sample procedure for constructing confidence intervals for proportions discussed in this section is often referred to in the statistical literature as the "Exact" procedure. I like to discuss a few issues related to its validity, and to some possible alternative approaches.

(1) *Do the two confidence bounds p_L and p_U prescribed by equations 3.10 and 3.11 satisfy equation 3.9? Is this method really exact?*

(2) In the process of constructing the confidence interval for p, equations 3.10 and 3.11 were obtained by setting $P(p_L \le p) = 1 - \alpha/2$ and $P(p_U \le p) = \alpha/2$. The question here is the following: since the objective was to have $P(p_L \le p) - P(p_U \le p) = 1 - \alpha$, *instead of using $1 - \alpha/2$ and $\alpha/2$, could we have used any two numbers α_L and α_U such that $\alpha_L + \alpha_U = 1 - \alpha$?*

(3) Computing the cumulative binomial probabilities may become complex because of the necessity to evaluate $\binom{n}{k}$. The question is *how should the confidence interval be calculated when the sample size n is very large?*

Issue (1)

The term "exact" is used in the statistical literature to refer to the small-sample method for constructing confidence intervals. What is exact in this method is certainly not the coverage probability of the resulting confidence interval. Instead, the term exact refers to the fact that the procedure is based on the exact probability distribution of the observed number of successes, which is Binomial. The confidence interval proposed in equations 3.10 and 3.11 is not quite exact with respect to its coverage rate, since it sometimes contains the "true" proportion with a probability that is slightly higher than its nominal value $1 - \alpha$.

Issue 2.

In the process of constructing confidence intervals, it is indeed not mandatory to use $1 - \alpha/2$ and $\alpha/2$ to determine the confidence bounds. Using different probabilities such as $1 - \alpha/3$ and $2\alpha/3$ or $1 - 2\alpha/3$ and $\alpha/3$ will lead to confidence intervals more to left or to right side of the 0-1 interval. In fact, after using $1 - \alpha/3$ and $2\alpha/3$ with $\alpha = 0.05$ in example 3.1, I obtained the two confidence bounds $p_L = 0.497$ and $p_U = 0.951$ and the new confidence interval,

$$\mathsf{CI}(p) = (0.497, 0.951). \tag{3.14}$$

Note that a broad family of confidence intervals can be found at the 95% confidence level by resolving the following system of equations:

$$\begin{cases} B(m; n, p) = \alpha_L, \\ B(m; n, p) = \alpha_U, \end{cases} \text{ where } \begin{cases} \alpha_L - \alpha_U = 0.95, \\ 0.5 \leq \alpha_L < 1, \\ 0 < \alpha_U < 0.5. \end{cases} \tag{3.15}$$

Issue 3.

Computing the cumulative binomial probabilities in order to derive the confidence bounds can indeed become less convenient as the sample size n and possible number of successes increase. The solution to this problem uses the normal approximation to the binomial distribution discussed earlier in section 2.2.2 of chapter 2. This approach is discussed in section 3.4.2.

3.4.2   Large-Sample Intervals for Proportions

When the size of the sample gets large, computing "true" confidence bounds becomes time-consuming, and may not even be feasible based on a direct evaluation of the Binomial distribution. The solution is to use the normal approximation to the Binomial distribution based on the Central Limit Theorem

discussed in section 2.3.4 of chapter 2 (Result 2.3). Many software products will automatically implement this approximation
as soon as the sample size exceeds a certain threshold.

The method presented in this section was recommended by
Agresti and Coull (1998) also by Brown, Cai and DasGupta
(2001) . The objective remains to find two confidence bounds p_L
and p_U such that equation 3.9 is satisfied. If $\widehat{p}$ is the estimated
value of the population proportion p for which the confidence
interval is sought, then the solution based on the Normal approximation, which is found in most statistical textbooks, is the
following:

$$p_L = \widehat{p} - z_{\alpha/2}\sqrt{\widehat{p}(1 - \widehat{p})/n}, \qquad (3.16)$$

$$p_U = \widehat{p} + z_{\alpha/2}\sqrt{\widehat{p}(1 - \widehat{p})/n}, \qquad (3.17)$$

where $z_{\alpha/2}$ is the $100(1-\alpha/2)^{th}$ percentile of the standard Normal
distribution.

Example 3.2 _____

Suppose that you are running for mayor of your city of residence. A sample of 400 voters reveals that 300 would support
you in the upcoming election. Here are the questions:

    a. Estimate the value of the population proportion.

    b. Develop a 99 percent confidence interval for the popula-
        tion proportion.

    c. Interpret your findings.

Solution:

    a. Estimate the value of the population proportion. *The po-*
        pulation proportion p in this case, is the relative number

of eligible voters in your city willing to support you. What you have at your disposal, is not the entire population of voters. Instead, it is a sample of 400 voters. Since 300 of them will support you, the sample proportion is evaluated at $\widehat{p} = 300/400 = 0.75$, and represents the estimatde value of p.

b. Develop a 99 percent confidence interval for the population proportion. *This confidence interval will be developed using the 2 equation 3.16 and 3.17. The confidence level is $1 - \alpha = 0.99$, which leads to $\alpha = 0.01$. However, $100(1 - \alpha/2) = 99.5$, and the 99.5^{th} quantile of the standard Normal can be calculated with Excel 2010 as $z_{0.005}$ =NORM.S.INV(0.995) = 2.576 (see Tables C.3 and C.4 of appendix C for alternative methods for obtaining quantiles). The two confidence bounds are then calculated as follows:*

$$p_L = 0.75 - 2.576 \times \sqrt{0.75(1 - 0.75)/400} = 0.694,$$
$$p_L = 0.75 + 2.576 \times \sqrt{0.75(1 - 0.75)/400} = 0.806.$$

The 99% confidence interval for p is $(0.694; 0.806)$.

c. Interpret your findings. *Our best approximation of the population proportion of supporters is $\widehat{p} = 0.75$. However, the population proportion is included in $(0.694; 0.806)$ with 99% certainty.*

In practice, you will often do what I did in example 3.2, which to produce both the point and the interval estimations of the population proportion. Do not make the mistake of considering the interval $(0.694; 0.806)$ to be an attribute of the point estimate 0.75. The confidence interval is instead, an attribute of the method you used for obtaining the point estimate 0.75.

Do p_L and p_U of equations 3.16 and 3.17 work?

One can claim that the two confidence bounds work, if they satisfy equation 3.9. A direct consequence of the Central Limit Theorem is that when the sample size is reasonably large, the sample proportion $\hat{p}$ follows approximately the normal distribution with mean $\mu = p$ and variance $\sigma^2 = p(1-p)/n$. Let Z be defined as follows:

$$Z = \frac{\hat{p} - p}{\sqrt{p(1-p)/n}}. \tag{3.18}$$

The statistic Z of equation 3.18 follows the standard normal distribution (with mean 0 and variance 1).

I know from chapter 2 that the Normal distribution is bell-shaped, its center of symmetry being its mean (see figure 2.10), which is also the point with the highest probability density. Because the ultimate goal is to find the smallest interval possible that covers the true proportion p with probability $1 - \alpha$, it is natural to seek an interval that is centered at p. This is how you will obtain the shortest interval for a given confidence level.

Let Φ denote the cumulative distribution function of the standard normal distribution, and $z_{\alpha/2}$ be such that $\Phi(z_{\alpha/2}) = 1 - \alpha/2$. Because of the symmetric nature of the normal distribution, $\Phi(-z_{\alpha/2}) = \alpha/2$. Therefore,

$$P(-z_{\alpha/2} \leq Z \leq z_{\alpha/2}) = \Phi(z_{\alpha/2}) - \Phi(-z_{\alpha/2}),$$
$$= 1 - \alpha. \tag{3.19}$$

It follows from equations 3.18 and 3.19 that you will want the following event,

$$-z_{\alpha/2}\sqrt{p(1-p)/n} \leq \hat{p} - p \leq z_{\alpha/2}\sqrt{p(1-p)/n}$$

to occur with probability $1 - \alpha$. If the difference $\hat{p} - p$ reaches either bounds (lower or upper bound), then,

$$(\hat{p} - p)^2 = z_{\alpha/2}^2 p(1-p)/n.$$

Resolving this quadratic equation equation in p yields the two confidence bounds p_L and p_U sought, which are,

$$
\begin{aligned}
p_L &= \frac{\widehat{p} + \dfrac{z_{\alpha/2}^2}{2n} - z_{\alpha/2}\sqrt{\dfrac{\widehat{p}(1-\widehat{p})}{n} + \dfrac{z_{\alpha/2}^2}{4n^2}}}{1 + z_{\alpha/2}^2/n}, \\[2em]
p_U &= \frac{\widehat{p} + \dfrac{z_{\alpha/2}^2}{2n} + z_{\alpha/2}\sqrt{\dfrac{\widehat{p}(1-\widehat{p})}{n} + \dfrac{z_{\alpha/2}^2}{4n^2}}}{1 + z_{\alpha/2}^2/n},
\end{aligned}
\tag{3.20}
$$

Note that the two confidence bounds of equation 3.20 will become equivalent to the two bounds of equation 3.16 and 3.17 only if the ratio $z_{\alpha/2}^2/(2n)$ is neglected, which is often the case in practice when the sample size n is reasonably large. Therefore, the confidence bounds of equation 3.16 and 3.17 work fine even for sample sizes as small as 20.

Sample Size Determination

I indicated previously that a desired feature of any confidence interval is to be as narrow as possible. The shorter the interval length, the more information you have about the location of the population proportion. A confidence interval such as $(0.02; 0.98)$ does not carry any useful information regarding the location of the population proportion, since it can literally be anywhere from 0 to 1. On the other hand, a confidence interval such as $(0.58; 0.62)$ gives you a very good appreciation of the magnitude of the population proportion even if it is unknown. The problem I want to resolve is to be able to determine the sample size that is required in order to achieve a confidence interval with a prescribed length.

It follows from equations 3.16 and 3.17 that the length L of the confidence interval is $p_U - p_L$, which is,

$$\mathsf{L} = 2 \times z_{\alpha/2} \sqrt{\widehat{p}(1 - \widehat{p})/n} = 2E, \tag{3.21}$$

where $E = z_{\alpha/2}\sqrt{\widehat{p}(1 - \widehat{p})/n}$ is referred to as the *Error Margin* or *Margin of Error* associated with the sample proportion. Equation 3.21 leads to the following expression for the required sample size n given the desired confidence level $1 - \alpha$.

$$n = \hat{p}(1 - \hat{p})\left(\frac{z_{\alpha/2}}{E}\right)^2. \tag{3.22}$$

Note that the knowledge of the required sample size is usually needed at the study planning stage when no estimation $\hat{p}$ of p is available. To get around this difficulty, a conservative approach is often adopted that consists of replacing $\hat{p}$ with 0.5, a value that maximizes the sample size n. The resulting value for n will always yield a confidence interval length that is a little shorter than the desired one L. The expression commonly used in practice to determine the sample size is the following:

$$n = \left(\frac{z_{\alpha/2}}{2E}\right)^2 \tag{3.23}$$

3.5. Interval Estimation for the Mean of a Measurement Variable

A proportion can be seen the mean of a 0-1 dichotomous variable and requires specific methods (exact or large-sample) for constructing the associated confidence intervals. These methods have been discussed in the last section. The present section deals with the important problem of interval estimation for the mean of a measurement variable such as the height, the weight, the income, or the length.

3.5.1 A Confidence Interval For Large Samples

To construct a confidence interval for the mean of a measurement variable when the sample size is large (i.e. the sample size n is 30 or more), I will use the Central Limit Theorem again in the same way I used it in section 3.5.2 to construct confidence intervals for proportions. Let the measurement variable be denoted by X with μ being its population mean (or expected value), and σ its population standard deviation. The population mean is always assumed unknown, otherwise there would not be any need for statistical inference. As for the standard deviation σ, the practical situations where it is known are very rare, particularly in business and social statistics. But when it is unknown, it will generally be replaced with its sample-based estimate s of equation 3.1. Many textbooks separately treat the situation where σ is known and when it is unknown. For large samples, this distinction is unnecessary because σ and s will be sufficiently close, and the Central Limit theorem will still be applicable. In this sub-section, I will systematically use s in place of σ. If σ happens to be available, it should be used instead.

Following a random experiment, the experimenter usually collects a sample $\mathcal{S} = \{x_1, \cdots, x_n\}$ of n data points from which an interval estimate of the mean μ of X must be constructed. The point estimates of the mean μ and variance σ^2 are respectively given by:

$$\overline{x} = \frac{1}{n}\sum_{i=1}^{n} x_i \ and \ s^2 = \frac{1}{n-1}\sum_{i=1}^{n}(x_i - \overline{x})^2. \tag{3.24}$$

The $100(1-\alpha)\%$ confidence interval for the population μ is given by:

$$\text{C.I}(\mu) = \left(\overline{x} - z_{\alpha/2}\frac{s}{\sqrt{n}} \ ; \ \overline{x} + z_{\alpha/2}\frac{s}{\sqrt{n}}\right) \tag{3.25}$$

Example 3.3

A sample of 53 cigarette smokers revealed that they spend on average $26.50 each week on cigarettes with a sample standard deviation of $5.2. What is the 90% confidence interval for the population mean dollar amount μ spent each week?

The sample size, sample mean, and sample standard deviation are respectively given by, $n = 53$, $\bar{x} = 26.50$, and $s = 5.2$. The confidence level is $1 - \alpha = 0.90$ or $\alpha = 0.10$. That is, $1 - \alpha/2 = 1 - 0.05 = 0.95$. The 95th percentile of the standard Normal distribution is 1.645 (I calculated this using Excel 2010 as =NORM.S.INV(0.95)). Feel free to use other software options proposed in Appendix C.

Using Equation 3.25, I can compute the lower bound (LB) and the upper bound (UB) of the interval as,

$$LB = 26.50 - 1.645 \times 5.2/\sqrt{53} = 25.32,$$
$$UB = 26.50 + 1.645 \times 5.2/\sqrt{53} = 27.68.$$

Consequently the 90% confidence of μ is $(25.32; 27.68)$. That is, 26.50 is your estimation of the actual mean weekly cigarette expenditures μ. This estimation is subject to sampling error, and the actual (but unknown number μ) will lie between $25.32 and $27.68 with 90% certainty.

How is the CI of Equation 3.25 Derived?

Here is what the Central Limit Theorem says,

If the sample size n is 30 or more, then the estimated variance s^2 is deemed sufficiently close to the population variance σ^2 for the Central Limit Theorem (CLT) to apply. That is, the standardized variable z

(also called the pivot) given by,

$$z = \frac{\overline{x} - \mu}{s/\sqrt{n}}, \qquad (3.26)$$

follows the standard Normal distribution

The statistic z has the advantage that it involves only one unknown, which is μ, the parameter you want to obtain an interval estimate for. It is this feature of z that makes it possible to construct a confidence interval. It follows from the CLT that,

$$P\left(-z_{\alpha/2} \le \frac{\overline{x} - \mu}{s/\sqrt{n}} \le z_{\alpha/2}\right) = 1 - \alpha,$$

where $z_{\alpha/2}$ is the $100(1-\alpha/2)^{th}$ percentile of the standard Normal distribution. The above equation can be rewritten as,

$$P\left(\overline{x} - z_{\alpha/2}\frac{s}{\sqrt{n}} \le \mu \le \overline{x} + z_{\alpha/2}\frac{s}{\sqrt{n}}\right) = 1 - \alpha,$$

leading to a $100(1 - \alpha)\%$ confidence interval for μ of equation 3.25.

If you have a desired length L for your confidence interval, you may use expression 3.25 to determine the required sample size n that can achieve that length.

$$n = \left(2z_{\alpha/2}\frac{s}{L}\right)^2, \qquad (3.27)$$

where $z_{\alpha/2}$ is the $100(1-\alpha/2)^{th}$ percentile of the standard Normal distribution. Note that the interval length L is twice the error margin $E = z_{\alpha/2}s/\sqrt{n}$. Therefore, the required sample size may be re-expressed as,

$$n = \left(z_{\alpha/2}\frac{s}{E}\right)^2, \qquad (3.28)$$

Both equations 3.27 and 3.28 require that you know the standard deviation. This quantity will generally come from an external source or a pilot study.

3.5.2   A Confidence Interval For Small Samples

The confidence interval of expression 3.25 is based upon the assumption that the sample at hand is sufficiently large to ensure the validity of the CLT. It is the statistician William Sealy Gosset (1876-1937), best known by his pseudonym "Student," who first asked what would happen if the sample is small as is often the case in many experiments. His remarkable discovery (Student (1908a, 1908b)) was that, the ratio of $\overline{x}$ to $s/\sqrt{n}$ follows a distribution that can be tabulated. This distribution is now known as Student's non-central t-distribution $t(n-1, \mu)$ with $n-1$ degrees of freedom and a non-centrality parameter μ.

Moreover, the standardized statistic $t = (\overline{x} - \mu)/(s/\sqrt{n})$ follows the Student (central) t-distribution $t(n-1)$. Although Student's work is based upon the assumption that the measurement variable x follows the normal distribution, experience suggests that the t distribution still holds even when the observations are only approximately normal.

Student's discovery provides a blueprint for constructing confidence intervals when the sample is small and the law of probability of the random variable is *approximately normal*. The $100(1-\alpha)^{th}$ confidence interval in this case is given by:

$$\text{C.I}(\mu) = \left(\overline{x} - t_{\alpha/2, n-1} \frac{s}{\sqrt{n}} \; ; \; \overline{x} + t_{\alpha/2, n-1} \frac{s}{\sqrt{n}} \right), \qquad (3.29)$$

where n is the sample size and $t_{\alpha/2, n-1}$ is the $100(1-\alpha/2)^{th}$ percentile of the t-distribution with $n-1$ degrees of freedom.

For small samples, I have only considered the situation where the measurement variable is normally distributed at least approximately. If the population standard deviation σ is known in this case, then $z_{\alpha/2}$ should replace $t_{\alpha/2,n-1}$ in equation 3.29, and σ should replace s. It is because when x is Normal, and σ known, then the sample mean $\overline{x}$ remains Normal with a standard deviation of $\sigma/\sqrt{n}$.

Example 3.4 _____

A sample of 11 employees who use child care reveals the following amounts spent last week: {$107, $92, $97, $95, $105, $101, $91, $99, $95, $104, $100}. The problem is to develop a 90 percent confidence interval for the population mean, and to interpret the result.

The basic statistics needed to construct this confidence interval are the sample size $n = 11$, the sample mean $\overline{x} = 98.73$, and the sample standard deviation $s = 5.274$. Since the confidence level is $1-\alpha = 0.90$, it follows that $1-\alpha/2 = 0.95$. The 95th percentile of the t-distribution with 10 degrees of freedom (i.e. $n-1 = 11-1 = 10$) is $t_{0.05,10} = 1.8125$. This percentile is obtained from Excel 2010 as T.INV(0.95,10). The two 90% confidence bounds are defined as LB = $91.79, and LB = $105.67.

A key question now, is what can you do if the sample size is small, and the law of probability underlying the data is not even close to being Normal? I will describe in the next section, one possible solution based on the bootstrap approach, which belongs to the larger family of resampling methods.

3.6. Bootstrap Confidence Intervals

The confidence intervals discussed in section 3.5 are all based on the pivotal method, where the pivot was either the z statistic of equation 3.26, or Gosset's t statistic (i.e. $t = (\overline{x}-\mu)/(s/\sqrt{n})$). However, the use of the pivotal method requires that one can find a pivot statistic with a probability distribution that is known (at least approximately). In section 3.5, when the sample size was small and the probability distribution underlying the data not Normal, I could not find a suitable pivot to construct a confidence interval. The difficulty of finding a suitable pivot statistic under specific circumstances has been problematic with statistics other than the mean, which created the need for a general approach to obtain reasonable confidence intervals without having to worry about the distributional properties of a pivot statistic.

Efron (1979) invented such a method. A computationally intensive method called the *Bootstrap*. The general idea behind the *Bootstrap* methodology is simple to understand. It is based upon the fact that the mean (actually many other statistics as well) can be expressed as a function of the probability distribution that underlies the data. If that probability distribution is known, so will be the parameter of interest. If on the other hand that probability distribution is unknown, it can be replaced with a reasonable surrogate distribution called the *Bootstrap distribution*, which will then be used for statistical inference.

Your statistical effort generally revolves around the need to have a good grasp of the cumulative probability distribution F created by the sampling of a variable of interest X. You will start by selecting a sample $\mathcal{S} = \{x_1, x_2, \cdots, x_n\}$ to obtain a snapshot of the universe $\mathcal{U}$ of all values the variable X can take. The sample values will lead to the sample mean $\bar{x}$ the distribution of which (denoted by $F_{\bar{x}}$) may be unknown. Only a repeated sampling of the universe $\mathcal{U}$ can give us a glimpse into the distribution of $\bar{x}$. Such a repeated sampling is impractical due to the cost of collecting sample data. The Bootstrap solution to this difficulty consists of using the sample $\mathcal{S}$ as your universe and to sample it with replacement[6]. A bootstrap sample will be something like $S^* = \{x_1^*, x_2^*, \cdots, x_n^*\}$, where x_i^* can be any value in the initial sample S. The bootstrap sample will naturally contains duplicates.

What will be associated with the bootstrap experiment are the bootstrap variable X^* and its bootstrap distribution F^*, which can be used as surrogate to the unknown distribution F. If $E_*(X^*)$ represents the expected value of characteristic X^* based on the bootstrap distribution, one can prove that $E_*(X^*) = \bar{x}$. Likewise, you will have $E_*(\bar{x}^*) = \bar{x}$, where $\bar{x}^*$ is the mean of the bootstrap sample. The problem that remains to be dealt with is that even the bootstrap distribution F^* is unknown[7]. However, you can approximate it by resampling the initial sample S (i.e. the bootstrap universe) a large number of times (e.g. hundreds of times) to obtain the Monte-Carlo approximation $\widehat{F}^*$ to the bootstrap distribution F^*.

Remember that, the inferential problem I am interested in is about the mean $\bar{x}$, which requires the probability distribution $F_{\bar{x}}$. But the only probability distribution the bootstrap approach is offering is $\widehat{F}_{\bar{x}^*}^*$, which is the bootstrap distribution of the bootstrap sample mean $\bar{x}^*$. For simplicity sake, I will often use the notation $\widehat{F}_{\bar{x}}^*$ instead of $\widehat{F}_{\bar{x}^*}^*$. Figure 3.5 shows the successive ap-

---

[6]With replacement sampling amounts to randomly selecting n numbers one at a time, from the set of values $\{x_1, \cdots, x_n\}$. Each new number is selected only after replacing the previously selected one into the pot of data

[7]You will need to select all conceivable samples with replacement from S to be able to completely specify the bootstrap distribution F^*.

proximations to the distribution of $\overline{x}$ that are of interest.

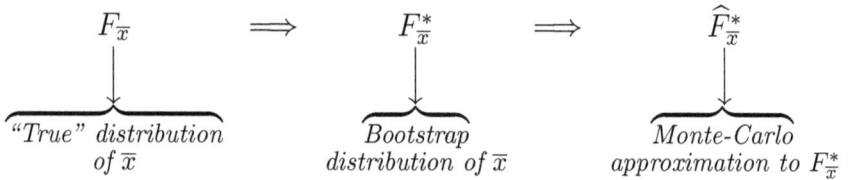

$$F_{\overline{x}} \quad \Longrightarrow \quad F_{\overline{x}}^* \quad \Longrightarrow \quad \widehat{F}_{\overline{x}}^*$$

| "True" distribution of $\overline{x}$ | Bootstrap distribution of $\overline{x}$ | Monte-Carlo approximation to $F_{\overline{x}}^*$ |

Figure 3.5. Three probability distributions of the mean

The Bootstrap Confidence Interval

Let F be any probability distribution function. The pth percentile of this distribution denoted by $F^{-1}(p)$ is defined as the smallest number x for which $F(x)$ equals or exceeds p. The $100(1 - \alpha)\%$ confidence interval is defined as follows:

$$\left(\widehat{F}_{\overline{x}}^{*-1}(\alpha/2); \widehat{F}_{\overline{x}}^{*-1}(1 - \alpha/2) \right) \tag{3.30}$$

The Bootstrap approach to confidence intervals only requires the ability to generate with a computer, a large number of bootstrap samples of size n, and to compute the sample mean $\overline{x}^*$ for each bootstrap sample. You will end up for example with 1000 bootstrap sample means $\{\overline{x}_1^*, \overline{x}_2^*, \cdots, \overline{x}_{1000}^*\}$, which is all you need to have the Monte-Carlo approximation $\widehat{F}_{\overline{x}}^*$ to the bootstrap distribution, necessary to construct the confidence interval 3.30. *Note that the 2.5^{th} and the 97.5^{th} of the 1,000 numbers $\{\overline{x}_1^*, \overline{x}_2^*, \cdots, \overline{x}_{1000}^*\}$ are the lower and upper bounds of the 95% bootstrap confidence interval.* That is why this method of constructing confidence intervals is known in the literature as the *Percentile Bootstrap*.

An alternative Bootstrap confidence interval known as the *Basic Bootstrap Confidence Interval* has been proposed and is given by:

$$\left(2\overline{x} - \widehat{F}_{\overline{x}}^{*-1}(1 - \alpha/2); 2\overline{x} - \widehat{F}_{\overline{x}}^{*-1}(\alpha/2)\right) \tag{3.31}$$

The basic confidence interval is a variant of the percentile interval and is supposed to make the latter look a little more symmetric around the estimated mean. Other bootstrap methods including the "Normal", and "BCA" (i.e. Bias-Corrected and Accelerated) have been suggested in the literature. Interested readers may read Efron (1982), Efron (1987), Schenker (1985), or Efron and Tibshirani (1994).

Bootstrap Confidence Interval & the R Package

If you are ever going to use the bootstrap method, my advice would be to use the R package. It is the most convenient tool out there for implementing the bootstrap methodology that I have seen. It is flexible and has a great deal of capabilities. I am only showing here how you can construct a confidence interval for the population mean, when the sample size is small, and the data is non-normal.

Suppose that you have selected a small sample of hospitals and collected the following data on the number of caesareans performed within a certain timeframe: {8, 16, 15, 23, 5, 13, 4, 19, 33, 19, 10}. You want to estimate the mean number of c-sections and the associated 95% confidence interval. Since the sample size is small, and you may not have any indication that the data is Normal, computing the bootstrap confidence interval should be a good option.

To resolve the above problem with the R package, you would need to submit the R commands listed in Example 3.5.

Example 3.5 _____

```
> library(boot)
> csections <- c(8,16,15,23,5,13,4,19,33,19,10)
> samplemean <- function(x, d) {return(mean(x[d]))}
> results <- boot(csections,statistic=samplemean,
+ R=1000)
> boot.ci(results, conf = 0.95, type="all")
```

In example 3.5, library(boot) loads the bootstrap package that provides the two functions boot and boot.ci needed to construct the bootstrap confidence interval. boot generates the bootstrap sample means, while boot.ci generates the confidence intervals. The variable csections contains input data on the number of c-sections per hospital, samplemean defines the R function for computing the sample mean from the input data contained in csections, results contains the 1,000 bootstrap sample means generated by the boot function, and boot.ci produces the bootstrap confidence intervals.

The results obtained from the R commands of Example 3.5 are the following:

Intervals :

Level	Normal	Basic
95%	(10.60, 20.09)	(9.62, 20.00)

Level	Percentile	BCa
95%	(10.00, 20.38)	(11.22, 20.82)

The R command boot.ci(results,conf = 0.95,type="all") requested all 4 bootstrap confidence intervals offered by the R package. I suspect you may want recommendation regarding the best confidence interval to use. Such a recommendation would assume that "the best" exists. It does not not. I personally use the Basic bootstrap interval more often. If you have good data, any of these options will be acceptable, even if there will always be some differences.

For further information on the boot package, you may want to read *Resampling Methods in R : The boot package*, by Angelo J. Canty, which appeared in the December 2002 issue of the R journal, and is free. You may download it at,

`http://journal.r-project.org/`

CHAPTER $\boxed{4}$

Statistical Tests of Hypothesis

OBJECTIVE

The primary objective of chapter 4 is limited to a high-level discussion about the logic behind the statistical test of hypothesis. The basic concepts and ideas underlying hypothesis testing are introduced within the context of a concrete real-world example. The two schools of thought due to Fisher, and to Neyman and Pearson are described heuristically. A more formal presentation of these tests is found in subsequent chapters.

CONTENTS

4.1. Overview

A simple test of hypothesis may be considered in a situation where you as a business manager want to benchmark the performance of your business on a known standard in your industry. To be more concrete, suppose you are running an auto insurance company, and want to know if the cost to process a claim is below the industry standard of $55.00 for instance. You may feel confident about your company's cost being below $55; but how do you prove it? One option would be to monitor the next 10 claims so that you can get a sense of how well your agents process them, and decide whether or you are below the industry standard. While this crude procedure may be all you need to be satisfied, you will find it difficult to convince somebody other than yourself of its merits.

The statistical (and supposedly more rigorous) solution to this problem exists and is not unique. There are be many possible and valid solutions. All these solutions will need to comply with a number of guiding statistical principles to be considered as statistical solution. I am going to review some of these principles now.

The Universe

You may need to define a timeframe within which you are going to limit your investigation. If you decide to concentrate your investigation on the period from January 1st of 2010 through April 30th of 2010, then all claims filed during that reference period will constitute your *Study Population* or *Study Universe* or your *Population of Interest*. The validity of your study will be restricted to that universe. The term population in this context refers to the population of claims, and the claim is seen as the population unit.

The definition of the study population will essentially be dictated by business and policy considerations. It may also happen that you will want to leave this study population open. The decision to leave open the definition of the study population will make perfect sense if you believe that the limited number of claims you study will use represent reasonably well all types of claims your business can expect to deal with. Many statistics textbooks refer to an open population as an "infinite population." As I previously indicated, the statistical consultant I am does not really like the term infinite very much, because it does not correspond to anything concrete in the mind of my clients. I would rather use the term open population.

The Selection of Claims

The population of claims you are interested in may be too vast, and too expensive to study in its entirety. The recommendation in this case is to restrict your investigation to a smaller set of claims called the claim *Sample*, selected randomly from the population of claims. The decision to select a sample should not be seen as a shift of interest from the target universe to its subset. The primary goal remains to study the entire universe. The sample is selected for the sole purpose of reducing cost. The premium to be paid for obtaining this cost reduction is to have to tie your sample results back to the target universe, using the principles of statistical inference.

Note that the random selection of a few claims will not necessarily result in a good sample that is representative of the characteristics of the entire population of claims you are interested in. Randomization has mainly the advantage of removing any possible personal bias from the claim selection process, ensuring thereby the integrity of the scientific investigation. The sample must also have an adequate size to ensure representativeness.

Collection of Cost Data

The data collection phase will yield for example 10 cost values such as $48.50, $56.00, $58.50, $49.50, $52.50, $45.50, $40.00, $51.50, $55.50, $39.50. At this stage you will want to average all these numbers to obtain the mean cost. The mean is often known to have a more rationale behavior from which you can study a phenomenon. Individual measurements are subject to all sort of variations, some of which will certainly not carry any useful information. The cost to process one specific claim may go up or down depending on whether it was initially reported with accuracy or not, or whether a particular agent was disrupted by a technical problem. Using the mean tends to smooth out this random variation that carries no information about the claim processing system you want to improve.

Because of the need to use the mean to eliminate random variation from individual measurements, it is natural to rephrase the whole initial test problem in terms of a statement regarding the population mean. That is, investigating whether your claim processing costs are below the industry standard of $55.00 amounts to testing whether the mean cost of claim processing calculated using all claims included in the large population of interest is smaller than $55.00. Our initial *Research Hypothesis* expressed in broad terms is now rephrased with terms such as the population mean and specific numbers to become a *Statistical Hypothesis*. This phase of translating the research hypothesis into the language of statistics is a critical step for resolving it using statistical principles. You as researcher must take the time to ensure that the relationship between the research and the statistical hypotheses is sufficiently strong to make the statistical hypothesis relevant. It is the research hypothesis that matters, but it is the statistical hypothesis that will be tested. Therefore, you must feel confident that the conclusion drawn about the statistical hypothesis will also apply to the research hypothesis.

Deciding about the Truthfulness of your Hypothesis

The mean cost for processing the 10 claims considered earlier is
$49.70. Now this number is clearly below the industry standard
of $55.00 mentioned previously. Here is the questions:

> *Can you conclude with confidence that the cost to pro-*
> *cess a claim in your company is below the industry*
> *standard?*

The temptation to answer yes in great, especially if a no answer
will lead your supervisor to start questioning your effectiveness as
manager. But if you want your approach to be methodologically
sound, here are a few things to consider before deciding:

- ▶ If I repeat the study by selecting 10 claims other than the
  10 previously analyzed, will I still get a mean cost below
  the industry standard?
- ▶ If I decide based on the mean, how often will I be wrong?
- ▶ If I conclude that my costs are below those of the industry,
  how strong is the data evidence upon which my decision is
  based?
- ▶ What control do I have over the frequency of errors in my
  decisions?
- ▶ How should I formulate my decision to be fair to all parties
  with a stake in what I am doing?

There is no one single statistical procedure that will provide answers to all of these questions. The most commonly-used tests
of hypothesis these days rely on 2 main approaches, which are
the *Fisher Approach*, and the *Neyman-Pearson Approach*. These
two approaches are fundamentally different. For an unknown reason, many introductory statistics textbooks now present a hybrid Fisher/Neyman-Pearson formulation that blended concepts

from both approaches, making it even more difficult to explain what hypothesis testing actually does. The Fisher and Neyman-Pearson approaches follow different logics. Combining concepts from both approaches into a single hybrid approach has made it more difficult to justify what is being done as you will see later in this chapter.

In my opinion, for many problems encountered in practice, only one of the 2 approaches is really necessary. In some cases however, using both approaches can be instructive, provided you clearly explain what each is bringing to the investigation. In this book, I will always indicate when I am using the Fisher approach and why, and when I am using the Neyman-Pearson approach and why. That distinction is essential for those like me, who wants to have a complete understanding of what they do.

4.2. The Fisher Approach to (Null) Hypothesis Testing

If you are interested in a batch of 1000 claims, the only way you are going to prove with no doubt your *Research Hypothesis* that their mean processing cost is below $55.00 is to actually review them all, compute the mean (i.e. the population mean), and compare it to $55.00 to ensure that it is below that industry threshold. However, if you can only afford to review 50 of the 1000 claims for example, then it becomes impossible to prove your conjecture with certainty. With 50 out of 1000 claims to review, you will have at your disposal only a portion of all the information you will need. The Fisher's approach in this situation consists of using that partial information to disprove the opposite of your Research Hypothesis, which is called the *Null Hypothesis*

The *"Research Hypothesis"* is our context here is the following:

> *The Population Mean Cost of Claim Processing is below $55.00*

and represents your conjecture that you want to prove. This conjecture is not part of Fisher's general approach to hypothesis testing. It is replaced with the *"Null Hypothesis"* defined as the opposite of the research hypothesis as follows:

> *The Population Mean Cost of Claim Processing equals or exceeds $55.00.*

Fisher called it the null hypothesis because it is the hypothesis to be nullified, the hypothesis to be rejected, and the only hypothesis considered in the statistical procedure. In other words Fisher's hypothesis testing is all about the **Null Hypothesis** testing. The idea of using the null hypothesis is justified by the fact that disproving a statement is supposedly easier than proving it. To disprove a statement, you only need one counterexample that is inconsistent with it. The 50 claims to be reviewed will produced 50 cost values from which you can compute the *Sample Mean*. This sample mean will differ from the population mean. But if your 50 sample claims reveal a sample mean of $85 for example, this should certainly raise a red flag. And this red flag will make it difficult for you to reject the null hypothesis.

The outcome of Fisher's hypothesis testing procedure is either the rejection of the null hypothesis, or the suspension of any decision (i.e. not rejecting the null hypothesis). There is no provision for accepting a hypothesis. That is rejecting the null hypothesis is not in any way an endorsement of the research hypothesis. It simply means that the data available to you is consistent with the research hypothesis, since the null hypothesis has been repudiated. There is only one possibility of error, which consists of rejecting the null hypothesis when it is indeed true.

How is the Null Hypothesis Rejected?

You will always formulate your null hypothesis in terms of the (unknown) population mean. The only known quantity, the sample mean is used in the testing procedure as a surrogate for the population mean you are interested in. Using the sample mean to test the population mean is possible because of the organic relationship that exists between the sample and the larger population it is selected from. One aspect of this relationship is defined by one of the most beautiful findings in statistical science: the **Central Limit Theorem (CLT)**, which links the sample mean and the population mean. Basically the **CLT** in our context says the following:

> *There is a large number of samples of 50 insurance claims that you can select from a given population of 1,000 claims. These samples will yield an equally large number of sample means.*
>
> *If you know what the population mean μ and the population standard deviation σ are, then you can compute the relative number of sample means that are below any given number. This computation is possible even without knowing any of the sample means nor the 1,000 raw individual cost values. It is because the sample means are always distributed according to a known and well documented law of probability that belongs to the family of normal distributions.*

The null hypothesis of our problem can be formulated as follows:

$$\text{H: } \mu \geq \$55, \tag{4.1}$$

where μ (read mu) represents the unknown population mean being tested. Here is the logic behind Fisher's procedure[1]:

---

[1]This procedure supposes that you have already selected your sample of 50 insurance complaints for review

(a) Assume that the unknown population mean equals \$55 (i.e. $\mu = 55$)

(b) Compute the mean of your sample of 50 complaints. I will denote it as $\overline{x}_{obs}$, where the subscript *obs* stands for "observed" as opposed to abstract.

(c) Compute the probability for a sample mean selected from that same population of claims to be more extreme[2] than the current one $\overline{x}_{obs}$ calculated in step (b). Fisher named this frequency the **P-value** or the Probability Value. Note that this calculation is feasible under the CLT.

A small P-value (smaller than 5% for example) indicates that if the experience is repeated several times, and if $\mu = \$55$, then the sample mean will be more extreme than the currently observed $\overline{x}_{obs}$, only 5% of thee times or less. This will be an indication that $\overline{x}_{obs}$ is substantially smaller than \$55, causing the rejection of the null hypothesis. The test is then considered to be statistically significant. For that reason, the P-value is often referred to as the *Significance Level of the Test*. This significance level is the only value that some researchers will report, leaving it up to the user to decide whether to reject the null hypothesis or not.

If the p-value is large (how large ?), Fisher recommended not to decide. This simply indicates that the data evidence is not sufficiently strong to reject the null hypothesis. In order to decide whether to reject or not reject the null hypothesis, Fisher suggested the 0.05 threshold for the P-value. That is whenever P-value fell below 0.05, the test will be called significant, and the null hypothesis rejected. This will also be an implicit decision in favor of the research hypothesis. The threshold of 0.05 looked reasonable to Fisher, and is still the standard is statistics today. Neverthe-

---

[2] The "extreme" condition will have to be defined in a way that is consistent with $\overline{x}_{obs}$ being substantially smaller than \$55, so that the null hypothesis can be rejected

less, Fisher himself later recognized that this cut-off point cannot be expected to be reasonable under all circumstances.

Some researchers in my opinion tend to rely too much on the magnitude of the P-value, attempting to see in its value more than it can actually offer. The most important thing to understand about the p-value is that it is an attribute of the experiment that you have conducted, and is solely based on the quality and quantity of information you have been able to collect. The statistical analysis is not expected to dominate the decision-making process in science. It must support it. It investigates only one aspect or a limited number of aspects associated with the scientific investigation.

4.3. Neyman-Pearson Approach to Hypothesis Testing

The P-value approach of Fisher can be a useful tool for testing a research hypothesis albeit it is not overused. It is generally simple to compute, and convenient to report. The most fundamental problem with the P-value is that it is not part of formal system of statistical inference. In other words, once you compute the P-value, you do not have specific guidelines that you can follow to decide about the research hypothesis, and what are the implications and risks associated with your decision. The Neyman-Pearson approach is supposed to offer such a comprehensive system. However, this approach requires more work, and it is perhaps the reason why it is less popular among researchers.

In a nutshell, what Neyman and Pearson have discovered is that it is impossible to know how good a procedure for rejecting a hypothesis is, unless the assessment is carried out with respect to a competing alternative hypothesis. This Neyman-Pearson observation is understandable. You will use the observations generated by your experiment to decide in favor or against a hypothesis, much more effectively if a competing hypothesis is fully spelled out.

Consider for example the conjecture that a six-sided dice is fair. That will be the case when each of the 6 sides has an equal chance of being observed. One way of testing this hypothesis is to pick one of the sides such as 2 and test the hypothesis H that the probability to observe side 2 is $1/6$ (i.e. H : $p = 1/6$). After rolling the dice 100 times for example and tallying the number of times 2 turns up, you will be able to obtain the relative number of times side 2 is observed. This quantity will be p_{obs}. You need to decide whether to reject the test hypothesis H or not based on the value of the observed proportion p_{obs}. It means you will need a *Decision Rule* for rejecting the hypothesis. Suppose you decide to use the following decision rule:

Decision Rule: Reject H if $p_{obs} < 0.10$

Now, whether this decision rule is good or not, depends on the true value of the probability p to observe side 2 of the dice. If $p = 0.2$, then the likelihood of seeing p_{obs} less than 10% of the times, thus rejecting H is very slim (Actually, this probability can be evaluated at 0.57%), although under this scenario the dice would clearly be unfair. If on the other hand, $p = 0.12$, then this decision rule would be excellent as the observed number p_{obs} will often be smaller than 0.10 (this probability can be evaluated at 98.85%).

The dice example shows that to be able to determine how good a hypothesis testing procedure is, you need to be able to evaluate the probability of deciding in favor or against the test hypothesis. Unfortunately this probability depends on the true value of parameter p. Because p is unknown, Neyman and Pearson decided to consider various values of p. The strategy that they adopted is to introduce 2 hypotheses into the procedure:

▶ The *Test Hypothesis* denoted as H_0 (H zero) to be rejected or not,

▶ The *Alternative Hypothesis* denoted as H_a (some statistics textbooks use the notation H_1). This is the hypothesis in favor of which the test hypothesis will be rejected.

You will note that in the Neyman-Pearson approach, there is no such a thing as the null hypothesis (à la Fisher) supposed to represent the opposite of the research hypothesis. Here is a major point of confusion that appears very often in many statistical textbooks. The 0 subscript in H_0 (the test hypothesis) has no relationship with the "null" term used by Fisher.

What is a Good Test ?

Suppose that you want to test the hypothesis

$$H_0 : p = 1/6$$

against the alternative

$$H_a : p > 1/6.$$

With two hypotheses involved, there will be 2 possibilities of errors. The first possibility is to reject the test hypothesis H_0 when in fact that hypothesis is true. This error is called the *Type I Error*, and the probability of committing it often denoted by α (alpha), and also called the *Significance Level* or the *Type I risk*. The second error you can make is to fail to reject the test hypothesis H_0 when in reality it is false. This is the *Type II Error*, and the probability of committing it is often denoted by β (beta). A hypothesis testing procedure is considered good when the probabilities associated with both error types are reasonably small. A test is considered powerful when the test hypothesis is rejected very often when it is false, and the associated probability, which is $1 - \beta$ is known as the *Power of the test*. It is essential to note that both error probabilities α and β change in opposite direction, meaning that when one goes up the other goes down. Therefore, a compromise must be found, when developing the test.

To recapitulate, you will want your hypothesis testing procedure to meet the following 2 objectives:

(1) The probability of Type I error must not exceed the significance level α that you have chosen.

(2) The probability of Type II error β will be reasonably low, which is equivalent to saying the power of the test is good. In practice you will not always be able to control this power. However, you must at least know the conditions under which the test you are using will have a reasonable power (or the ability to reject the test hypothesis when it is the alternative hypothesis that is true).

How to Determine the Test Hypothesis?

I indicated earlier that there is no such a thing as the null hypothesis in the Neyman-Pearson approach to hypothesis testing. To determine what hypothesis should be considered the test hypothesis, you as researcher must first decide about the most important risk against which you want to be protected. That risk will become your Type I risk α, and the associated hypothesis will be the test hypothesis. Consider again the the insurance claim processing cost example. As manager, you certainly do not want to accept the hypothesis that cost of processing claims is low when reality it is high, because you will lose money. In this case you will define your test hypothesis as follows:

$$\mathsf{H}_0 : \mu > \$55$$

Rejecting H_0 means accepting that the cost is low. You certainly want the minimize the risk of rejecting this H_0 when it is true. You will then assign this risk a low value such as 0.05. But it is up to you. Your decision here has some practical implications that I will elucidate later.

How To Formulate the Decision Rule?

Depending on the parameter being tested (population mean or population proportion etc...), you will at some point summarize your observations in a quantity called the *Test Statistic*[3]. It is the magnitude of this test statistic that will lead you to favor one hypothesis over another one. Depending on how the test hypothesis is formulated, it may be favored by a large or a small value for the test statistic.

For the test statistic to be useful, its probability law will have to be known, at least for some specific population parameters. The reason this law of probability is needed is related to the decision rule that will be formulated. This decision rule will take a form such as:

Decision Rule: Reject the test hypothesis H_0 if $T \geq c$,

where the number c, called the *Critical Value* must be determined. This critical value is determined in such a way that the predetermined Type I risk α is never exceeded. This goal is achieved if the probability of $T \geq c$ evaluated under the conditions of the test hypothesis does not exceed α. And this critical value c can be determined only if the law of probability of the test statistic is known.

---

[3]Typically this test statistic will be a simple expression involving the sample mean and the sample standard deviation, and eventually other quantities.

4.4.   The Different Types of Hypotheses

In the next few chapters I will concentrate on the testing of two types of population parameters, which are the population proportion denoted by π (Greek character Pi)[4], and the population mean denoted by another Greek character μ (read mu)[5]. However, a hypothesis testing problem may involve one, two, three or more parameters of the same type. This will lead to different formulations of the hypotheses and slightly different methods.

When the test consists of comparing a population parameter to a predetermined value of interest, then this problem involves one population of interest from which a single sample will be selected for testing. The test hypothesis may be of the form, $H_0 : \pi \geq 0.25$. This problem is called the *One-Sample Test of Hypothesis*, and will be investigated in chapter 5.

There are also situations where you may want to compare 2 population parameters. For example you want to compare 2 US states such as Maryland (MD) and Virginia (VA) with respect to their respective proportions of students who study at least 26 hours a week. Your conjecture may be that the MD proportion exceeds that of VA. Your null hypothesis would be $\pi_{MD} \leq \pi_{VA}$. This problem involves two populations of interest, which are the MD student population, and the VA student population. A student sample will be selected from each population for a total of 2 samples. This problem is therefore called the *Two-Sample Test of Hypothesis*, and will be addressed in chapter 6.

---

[4] π may represent for example the proportion of American students who study at least 26 hours each week. You may want to test if $\pi \geq 0.30$. The population here is that of American students at a given point in time

[5] As an example, μ may represent the average height of men in a particular city.

A more general problem you may also face in practice is to compare parameters associated with three populations or more. For example comparing the average cost of a compact car from 3 manufacturers such as GM, TOYOTA, and HYUNDAI. The null hypothesis may take the form $H_0 : \mu_G = \mu_T = \mu_H$, where μ_G, μ_T, and μ_H represent respectively the average cost of a GM, TOYOTA, and HYUNDAI compact car. This special hypothesis testing problem is tackled with a special statistical technique called the *Analysis of Variance* (also known as ANOVA), and will be discussed in chapter 7.

CHAPTER $\boxed{5}$

One-Sample Test of Hypothesis for Proportions

OBJECTIVE

From this chapter's content, you will learn various statistical methods for testing the magnitude of a single population proportion, by comparing it to a predetermined hypothetical value.

CONTENTS

5.1. Testing a Population Proportion

An example of a situation where you will be concerned about the one-sample test of hypothesis for proportions is that of a restaurant manager who claims that 90 percent of orders are delivered within 10 minutes from the time they are received. You may then want to test whether this claim is accurate or not. Because the manager's claim is about percentages, you are dealing with the notion of proportion and not that of the mean of a quantitative measurement. The population parameter in this case is,

$$\pi = \textit{Proportion of all orders that are delivered}$$
$$\textit{within 10 minutes after being received.}$$

Note that the population of interest (or of inference) here is the open list of all orders received at the restaurant. No timeframe has been specified, although it may seem more appropriate to limit the period of investigation to all orders received in the past 6 months for instance, if you are testing the effectiveness of a new delivery system that was not in place 6 months ago.

The problem stated above also mentions the *Hypothetical Value* 90% as the reference mark for comparing the population proportion π. I will denote this hypothetical value as π_0, and for this specific example $\pi_0 = 0.90$.

Suppose that a review of 100 orders reveals 82 orders that were delivered within 10 minutes. The number of orders used in the study represents the sample size, and is generally denoted by n. In this problem in particular $n = 100$. Intuitively you can see that you will have to compute the proportion of orders delivered within 10 minutes from reception based on the sample of 100 orders. This proportion is called the *Sample Proportion*. It is different from the population proportion, and will be denoted by

p as opposed to π that designates the population proportion. The concrete value of p based on a specific sample will be denoted by p_{obs}. For the example above $p_{obs} = 0.82$.

Let me recapitulate what you have at your disposal for tackling this problem:

- ▶ π = population proportion, unknown and being tested
- ▶ π_0 = hypothetical value (sometimes called null value) used to benchmark π. This is always an actual number that comes from the problem at hand (in the order delivery problem $\pi_0 = 0.90$)
- ▶ n = the sample size, sometimes imposed to the researcher, but could also be determined by the researcher using methods to be discussed later.
- ▶ p, is the proportion of sample orders delivered within 10 minutes after being placed.
- ▶ The conjecture on the relationship between the population parameter π and the hypothetical value π_0. In the order delivery example, the conjecture is $\pi = 0.90$. Actually investigating the conjecture $\pi > 0.90$ may be more appropriate[1].

P-Value for the Order Delivery Example

If your objective is simply to determine whether the sample data collected is consistent with the claim, you should evaluate the strength of evidence in the sample concerning the following null hypothesis (the opposite[2] of the claim that $\pi > 0.90$):

$$H: \pi \leq 0.90.$$

---

[1]In reality the manager will want 90% or more of the orders to be delivered within 10 minutes. I do not know why a manager will want exactly 90% of the orders to be delivered within 10 minutes.

[2]Sample data that are inconsistent with the opposite of your claim tend to support that claim. Hence, the pivotal nature of the null hypothesis.

Intuitively, you can see that if the null hypothesis described above is true then you would expect p_{obs} to be substantially smaller than 0.90 the hypothetical value, since p_{obs} follows the direction of π. Therefore the null hypothesis should be rejected if p_{obs} exceeds 0.90 instead, of if the difference $p_{obs} - 0.90$ is large. The method used to evaluate the magnitude of $p_{obs} - 0.90$ is probabilistic. This difference is considered large if the probability for $p - 0.90$ to exceed $p_{obs} - 0.90$ is small[3]. Computing this probability is possible because π being the population proportion, the ratio Z defined by:

$$Z = \frac{p - \pi}{\sqrt{\pi \times (1 - \pi)/100}}, \tag{5.1}$$

follows the standard normal distribution as suggested by the Central Limit Theorem.

▶ Let us now assume that π is equal to the hypothetical value 0.90. Under this assumption, you can conclude that the quantity $Z = (p - 0.90)/\sqrt{0.90(1 - 0.90)/100}$ follows the standard normal distribution.

▶ The specific value of Z observed from the sample of 100 orders at hand is $z_{obs} = (0.84 - 0.90)/\sqrt{0.90(1 - 0.90)/100} = -2.0$

▶ The P-value in this case is defined as the probability for Z to exceed z_{obs}. That is,

$$\text{P-value} = P(Z > z_{obs}),$$
$$= P(Z > -2.0)$$

At this stage, most statistics textbooks will direct you to one of the statistical tables prepared for the normal distribution. I personally never use them. Because I found it

---

[3]Note that the value of the sample proportion p varies from sample to sample, and is therefore a random variable. The difference $p_{obs} - 0.90$ is large when it will be exceeded only rarely by the random variable $p_{obs} - 0.90$

more convenient to use Microsoft Excel to compute these probabilities. For example to compute the P-value defined above, open Excel 2007 or 2010, and in a cell of your choice, type in the following (including the equal sign):

$$= 1 - \text{NORMSDIST}(-2.0)$$

and press enter. The answer will be 0.9773. **NORMSDIST** is an Excel 2007 function, which is still implemented in Excel 2010 to ensure compatibility of the 2 versions. Appendix C gives you alternative functions in Excel 2010, Calc 3.2.0, and R.

The order delivery problem led to a P-value of 0.9773, which represents the strength of evidence in the sample concerning the null hypothesis. Should you conclude that the evidence is sufficiently strong to reject the null hypothesis? It is common practice to decide against the null hypothesis if the P-value is smaller than 0.05. That is not the case here. Therefore, you cannot reject the null hypothesis, and you must suspend your judgement and wait to gather better data.

One last remark about what I just did. Z is defined as a ratio of the difference $p - \pi$ to a square root. The difference in the numerator is what I care about. The denominator is used for the sole purpose of obtaining an expression with a known law of probability (the Standard Normality Distribution) that can be used to compute the P-value.

5.2. Fisher's Null Hypothesis Testing

In general, how you compute the P-value depends upon the form of the research hypothesis as well as on the size of the experimental sample. Let me first look at the different forms that the research hypothesis can take. Recall that the null hypothesis is defined as the opposite of the research hypothesis that the researcher believes to be true.

The P-value was initially developed in the context where the researcher wanted to prove a gain in efficiency of a new treatment, which meant a conjecture such as $\pi > \pi_0$, or $\pi < \pi_0$ or $\pi \neq \pi_0$. In other words the objective was to prove an improvement over the status quo, which was $\pi = \pi_0$ (i.e. no improvement). Consequently the typical null hypothesis has always been formulated as,

$$\text{H: } \pi = \pi_0, \tag{5.2}$$

which is what you will find in most introductory statistical textbooks treating the P-value problem. These textbooks also use the symbol H_0 (borrowed from the Neyman-Pearson theory) to designate the null hypothesis.

5.2.1 Research Hypothesis: $\pi \neq \pi_0$

If you want to prove that $\pi \neq \pi_0$ (the research hypothesis) then you would reject the null hypothesis if the difference $p_{obs} - \pi_0$ is large in absolute value. And that will be the case if the probability that $|p - \pi_0| > |p_{obs} - \pi_0|$ is small. How this probability is calculated depends on the sample size n.

▶ If $n\pi_0 \geq 5$ and $n(1 - \pi_0) \geq 5$ then the P-value is calculated as follows:

$$\text{P-value} = P(|Z| > |z_{obs}|), \tag{5.3}$$

where $z_{obs} = (p_{obs} - \pi_0)/\sqrt{\pi_0(1 - \pi_0)/n}$. Many introductory statistics textbooks suggest to compute this P-value using statistical tables often presented in appendix. My recommendation is to use Excel and to compute the P-value of equation (5.3) as follows:

$$= 2*(1\text{-NORMSDIST}(\text{ABS}(z_{obs}))) \tag{5.4}$$

where z_{obs} is the Z value observed from the experimental sample.

In the order delivery example discussed previously, I had $\pi_0 = 0.90$, and $n = 100$, which led to the products $n\pi_0 = 90$ and $n(1 - \pi_0) = 10$. Therefore, the sample size condition was satisfied.

▶ If $n\pi_0 < 5$ or $n(1-\pi_0) < 5$ (that would the case if $\pi_0 = 0.96$ and $n = 100$ for example) then the P-value can no longer be calculated using the Normal distribution. It will have to be calculated using the Binomial distribution. Let us see how this can be done.

Consider again the order delivery example discussed earlier. Let X designate the number of orders delivered within 10 minutes from the time they were placed, out of the 100 orders in the sample. Since X can also be seen as the number of successes out of n independent trials, it follows the Binomial distribution $\mathcal{B}(n, \pi)$, where π is the unknown probability of success, n the sample size.

You would reject the null hypothesis H: $\pi = \pi_0$ if the relative number of successes X_{obs}/n is distant from the hypothetical value π_0. That is if $|X_{obs}/n - \pi_0|$ is large. That will be the case if the probability for the random variable $|X/n - \pi_0|$ to exceed the observed value $|X_{obs}/n - \pi_0|$ is small. It follows that the P-value is calculated as,

$$
\begin{aligned}
\text{P-value } = &\ P\Big(X > n(\pi_0 + |p_{obs} - \pi_0|)\Big) \\
&+ P\Big(X < n(\pi_0 - |p_{obs} - \pi_0|)\Big)
\end{aligned}
\tag{5.5}
$$

This probability can be calculated because X follows the Binomial distribution $\mathcal{B}(n, \pi_0)$ under the assumption that the null hypothesis is true. This P-value is calculated using MS Excel as follows:

(1) Execute the Excel expression $=n*(\pi_0 + \text{ABS}(p_{obs} - \pi_0))$ and store it in a cell **B1**

(2) Execute the Excel expression $=n*(\pi_0 - \text{ABS}(p_{obs} - \pi_0))$ and store it in a cell **B2**

(3) The P-value is calculated as follows:

$$= 1 - \text{BINOMDIST}(\text{MIN}(n,\text{B1}),n,\pi_0,1) \\ + \text{BINOMDIST}(\text{MIN}(n,\text{B2}),n,\pi_0,1) \quad (5.6)$$

Example 5.1 _____

The proportion of misshelved books was 5% at the last annual shelf inventory of a university library. The head librarian believes that the new shelving rules has had an impact on that percentage (in one way or another), and tested it based on a random sample of 500 books only, before the next annual inventory. This sample revealed 15 misshelved books.

Let π be the "true" proportion of books misshelved. The only 2 options here are the status quo (i.e. $\pi = 0.05$), and a difference (i.e. the conjecture $\pi \neq 0.05$). Therefore the null hypothesis that you like to reject is,

$$\text{H: } \pi = 0.05$$

The parameters of this problem are the following:

▶ Hypothetical proportion: $\pi_0 = 0.05$.

▶ Sample size $n = 500$,

▶ Observed misshelved books: $X_{obs} = 15$

The sample size requirements for normality are met since $n\pi_0 = 500 \times 0.05 = 25$, and $n(1 - \pi_0) = 500 \times 0.95 = 475$. Therefore, expression 5.4 can be used to compute the P-value. z_{obs} is calculated as follows:

$$z_{obs} = (p_{obs} - \pi_0)/\sqrt{\pi_0(1-\pi_0)/n},$$
$$= (0.03 - 0.05)/\sqrt{0.05 \times 0.95/500} = -2.052.$$

Note that $p_{obs} = X_{obs}/n = 15/500 = 0.03$ (the proportion of misshelved books observed from the experimental sample). Now using the Excel expression (5.4), the P-value is obtained by typing the following in Excel:

=2*(1-NORMSDIST(ABS(-2.052)))

The answer obtained is 0.0402. It is common practice to consider a P-value smaller than 0.05 to be a strong enough evidence to reject the null hypothesis. In this case the librarian could conclude that the new shelving method is changing the status quo.

In example 5.1, you would not be in the position to use expression 5.4 if the librarian had based the testing on a sample of 80 books, because $n\pi_0 = 80 \times 0.05 = 4$, and the normal distribution would have been a poor approximation to the law of probability associated with the test statistic Z. In this situation, you should use the Binomial distribution-based expression 5.6 to compute the P-value. With $n = 80$, $\pi_0 = 0.05$, and $X_{obs} = 1$, the P-value would be calculated as follows:

▶ First compute $n(\pi_0 + |p_{obs} - \pi_0|) =$
  $= 80 * (0.05 + \text{ABS}(1/80 - 0.05)) = 7.0$
▶ Next compute $n(\pi_0 - |p_{obs} - \pi_0|) =$
  $= 80 * (0.05 - \text{ABS}(1/80 - 0.05)) = 1.0$
▶ Now, you can use expression 5.6 with B1=7 and B2=1. This leads to a P-value of 0.063. Because P-value exceeds 0.05, many practitioners will consider the degree of significance against the null hypothesis to be insufficient. That

is the test is not significant, and the null hypothesis is not rejected.

5.2.2 Research Hypothesis: $\pi > \pi_0$

Consider a package delivery service that delivers about 90% of the packages within 3 hours after they have been brought into office. After providing additional training to the delivery agents, the manager wants to know whether the training program has been effective. The training program can be deemed effective if the (new) percentage π of packages delivered within 3 hours of being received exceeds the current value (i.e. the status quo) of 90%. That is the conjecture, expectation, or research hypothesis in this situation is $\pi > \pi_0$.

This conjecture could be tested by selecting a sample of n packages and computing the sample proportion p. After selecting a specific sample of packages, you will compute the associated proportion p_{obs} of packages delivered within 3 hours. The evidence in that sample against the null hypothesis H: $\pi = \pi_0$ and in favor of the research hypothesis $\pi > \pi_0$ would be considered strong, if $p_{obs} - \pi_0$ is large. A statistical procedure based on a straight evaluation of the observed difference will be weak, and will lead to different findings for each sample of packages used. *A more statistically sound procedure is to assess the magnitude of $p_{obs} - \pi_0$ "probabilistically" by calculating how often the random difference $p - \pi_0$ will exceed its currently observed $p_{obs} - \pi_0$, a quantity that represents the P-value.* But this requires that the law of probability of the $p - \pi_0$ be known. Generally, this law of probability is either known or a good approximation of it is known. The law of probability appropriate for your situation depends on whether you have a small sample size n or a large sample size.

▶ *The sample size n is considered small if $n\pi_0 < 5$ or $n(1 -$

$\pi_0) < 5$. In this case, you look at the P-value as being the probability $P(X > X_{obs})$. This probability can be calculated because if the null hypothesis is true, then the law of probability associated with X is the Binomial distribution $\mathcal{B}(n, \pi_0)$. Using MS Excel, this probability is computed as follows:

$$= 1 \text{ - BINOMDIST}(X_{obs}, n, \pi_0, 1)$$

▶ *The sample size n is considered large if $n\pi_0 \geq 5$ and $n(1 - \pi_0) \geq 5$.* I previously indicated that the variable Z defined by,

$$Z = \frac{p - \pi}{\sqrt{\pi(1 - \pi)/n}}, \tag{5.7}$$

follows the standard normal distribution $\mathcal{N}(0, 1)$. If you assume the null hypothesis to be true (i.e. $\pi = \pi_0$) then Z becomes $Z = (p - \pi_0)/\sqrt{\pi_0(1 - \pi_0)}$ and its observed value from the specific sample selected in your study is $z_{obs} = (p_{obs} - \pi_0)/\sqrt{\pi_0(1 - \pi_0)}$. The P-value when the research hypothesis is in the form $\pi > \pi_0$ is given by:

$$\boxed{\text{P-value} = P(Z > z_{obs}),} \tag{5.8}$$

This P-value is easily calculated from Excel by typing the following expression in any Excel cell:

$$=1\text{-NORMSDIST}(z_{obs}) \tag{5.9}$$

This supposes that the observed value z_{obs} has previously been calculated. In Example 5.1, the value of z_{obs} was evaluated at -2.052. If the research hypothesis is $\pi > \pi_0$ then the corresponding P-value would be:

$$=1\text{-NORMSDIST}(-2.052) = 0.9799.$$

This P-value is very large, and therefore provides a very weak evidence against the null hypothesis. This is not surprising, since the negative value of z_{obs} shows that p_{obs} is smaller than the hypothetical value π_0. This cannot favor the research hypothesis that $\pi > \pi_0$.

5.2.3 Research Hypothesis: $\pi < \pi_0$

A research hypothesis in the form $\pi < \pi_0$ will be heavily favored if the observed difference $p_{obs} - \pi_0$ is very small[4]. Once again, the value of $p_{obs} - \pi_0$ is small if the random variable $p - \pi_0$ often exceeds it. In other words, $p_{obs} - \pi_0$ is small if the probability $P(p - \pi_0 > p_{obs} - \pi_0)$ is large, which is equivalent to saying that $P\big(p - \pi_0 < -(p_{obs} - \pi_0)\big)$ is small. Now, the question is why switch from $P(p - \pi_0 > p_{obs} - \pi_0)$ to $P\big(p - \pi_0 < -(p_{obs} - \pi_0)\big)$? The answer lies on the desire to have a unique interpretation of the magnitude of the P-value, which is valid for all forms that the research hypothesis may take. That interpretation is that *a small P-value is a strong evidence against the null hypothesis, and in favor of the research hypothesis.*

▶ *If the sample size is small (i.e. $n\pi_0 < 5$ or $n(1 - \pi_0) < 5$) then the P-value should be computed using the Binomial distribution $\mathcal{B}(n, \pi_0)$, which will be a good approximation of the law of probability associated with X, the number of subjects that bear the characteristic of interest.*

$$\text{P-value} = P\big(X > n\big[\pi_0 - (p_{obs} - \pi_0)\big]\big), \qquad (5.10)$$

This P-value is calculated with MS Excel as follows:

---

[4]Note that, a large negative value such as -1,356.0 for example is considered a very small value.

(a) Compute the quantity $n * (\pi_0 - (p_{obs} - \pi_0))$ in a cell (e.g. cell A1)

(b) Type the formula =1-BINOMDIST(MIN(n,A1),n,π_0,1) in any cell.

▶ *If the sample size n is large (i.e. $n\pi_0 > 5$ and $n(1-\pi_0) > 5$)* then the P-value, which I defined as $P(p - \pi_0 < -(p_{obs} - \pi_0))$ can rewritten as:

$$\boxed{\text{P-value} = P(Z > -Z_{obs})} \qquad (5.11)$$

where $Z = (p - \pi_0)/\sqrt{\pi_0(1 - \pi_0)/n}$ and $Z_{obs} = (p_{obs} - \pi_0)/\sqrt{\pi_0(1 - \pi_0)/n}$. This P-value is calculated with MS Excel as follows:

(a) Compute the Z_{obs} as shown above in a cell (e.g. cell A1)

(b) Type the formula =1-NORMSDIST(A1) in any cell, to obtain the P-value.

Example 5.2

Consider the problem of misshelved books in a university library discussed in Example 5.1. The percent of misshelved books at the last inventory was 5%, and now the head librarian wants to know if there has been improvement since, by testing a sample of 80 books, one of which happened to be misshelved. The hypothetical percentage is $\pi_0 = 0.05$, the sample size $n = 80$, and the observed percentage of misshelved books $p_{obs} = 1.25\%$ (obtained as 1/80). The null hypothesis of no improvement is H: $\pi = 0.05$. But unlike example 5.1, the librarian wants to reject the null hypothesis now only if there is an improvement. That is the rejection of the null hypothesis must be in favor of the research hypothesis $\pi < \pi_0$.

The sample of 80 books is considered small since $n\pi_0 = 80 \times 0.05 = 4$ is smaller than 5. Therefore the P-value must be calculated using equation 5.10. This number is obtained with Excel as follows:

$$= 1\text{-BINOMDIST(MIN(7,80),80,0.05,1)} = 0.047,$$

where 7 is obtained from expression $n \times (\pi_0 - (p_{obs} - \pi_0))$. The P-value is small and represents a strong evidence against the null hypothesis, and in favor of an improvement.

5.3. Neyman-Pearson Method

Fisher's approach to hypothesis testing discussed in the previous section supposes a problem where the researcher wants to demonstrate a change from a status quo. The status quo (the null hypothesis to be rejected) is expressed in one the forms an equality of proportions $\pi = \pi_0$, and the conjecture in the form of $\pi \neq \pi_0$, $\pi > \pi_0$, or $\pi < \pi_0$. The P-value measures the strength of evidence in the sample against the null hypothesis. The interpretation of this P-value is left in the hands of the practitioner, although a commonly-used rule of thumb is to consider the evidence against the null hypothesis sufficiently strong when the P-value is smaller than 0.05 (a subjective threshold). However, Fisher's procedure does not lead to a formal endorsement of one hypothesis over another one, although one often tends to endorse the researcher hypothesis whenever the null is rejected. Only the Neyman-Pearson approach will allow you to formally endorse a hypothesis.

Consider again the example of the restaurant manager who wants to verify that 90% of the orders are delivered within 10 minutes from the time they are received. To tackle this problem from the Neyman-Pearson perspective, you as manager need to first determine which mistake you do want to avoid the most in your decision to endorse one hypothesis. In the case of a restaurant manager, you do not want to claim that the proportion of orders delivered within 10 minutes exceeds 90% when in reality it is below 90%. Such an error could be costly in terms of customer dissatisfaction, and will eventually result in a loss of business. If π represents the proportion of orders delivered within 10 minutes, you want to minimize the risk of rejecting the hypothesis that $\pi \leq 90\%$ when it is true. Once you determine the risk you want to minimize, the corresponding hypothesis becomes your *Test Hypothesis*, and you will denote it as H_0 (read H zero[5]).

The opposite of the test hypothesis is denoted by H_a (read H a) and called the *Alternative Hypothesis*. To be more accurate, the alternative hypothesis is the hypothesis that will be true if the null hypothesis is not. By rejecting the test hypothesis, you will tacitly endorse the alternative hypothesis. With two competing hypotheses in the Neyman-Pearson approach, there are 2 possible errors you can mae:

▶ *Reject the test hypothesis H_0 when it is true.* This is the *Type I Error* that you want to control. You will set a threshold that the probability of committing the Type I Error must not exceed. That threshold is called the *Significance Level* of the test, and often denoted by α (read alpha).

▶ *Fail to reject the test hypothesis H_0 when it is false.* This is the *Type II Error*, often denoted by β (read beta). Note that $1 - \beta$ is the probability to reject the test hypothesis when it is false, and represents the *Test Power*.

The Neyman-Pearson's method is commonly implemented following a five-step procedure that I now describe.

---

[5]As mentioned before, many introductory statistics textbooks referred to this hypothesis as the "Null" hypothesis. I am unsure what the benefit is to blend Fisher's terminology into the Neyman-Pearson framework.

Five-Step Hypothesis Testing Procedure

Step 1. Definition of Hypotheses

The test and alternative hypotheses H_0 and H_a will be defined here. For the order delivery example, they take the following form:

$$(Hypotheses) \begin{cases} H_0: & \pi \leq \pi_0, \\ H_a: & \pi > \pi_0. \end{cases}$$

H_0 and H_a may have other formulations that will be discussed later in this section.

Step 2. The Significance Level

In practice, this significance level must be determined by the researcher. It represents the threshold that the probability of Type I error must not exceed. A value commonly used by practitioners is $\alpha = 0.05$.

Step 3. The Test Statistic

The **Test Statistic** is a summary statistical measure used to evaluate the discrepancy between the data and the test hypothesis. Small and large sample sizes will lead to 2 different forms for the test statistic. Small-sample tests are based on the test statistic $X = $ *Count of sample subjects that bear the characteristic of interest,* while large-sample tests use the test statistic $Z = (p - \pi_0)/\sqrt{\pi_0(1 - \pi_0)/n}$.

Step 4. Decision Rule

The decision rule is the condition that the test statistic must meet before the test hypothesis can be rejected. This condition will be determined in such a way that the Type I error probability does not exceed the significance α of step 2.

Step 5. Final Decision

> This step consists of actually computing the test statistics using the data gathered, and applying the decision rule of step 4 before deciding whether or not the null hypothesis must be rejected in favor of the alternative.

5.3.1 The Two-Sided Test of Hypothesis

The Hypotheses

A test of hypothesis is two-sided when the test and alternative hypotheses have the following form:

$$(Hypotheses) \quad \begin{cases} H_0\colon & \pi = \pi_0, \\ H_a\colon & \pi \neq \pi_0. \end{cases} \qquad (5.12)$$

This supposes that the risk that you want to control the most is that of rejecting equality between the "true" and hypothetical proportions when it is true. The definition of these 2 hypotheses will complete step 1 of the 5-step procedure.

The Significance Level

As indicated earlier, step 2 is up to you to decide. Researchers commonly decide that the probability of rejecting H_0 given that it is true should not exceed $\alpha = 0.05$. Other possible values used in practice are 0.01 and 0.10. Can you set this significance level to 0? After all you want the probability of Type I error the smallest possible. For a given sample size, the only way to achieve a 0 risk when rejecting H_0 is to never reject H_0. A 0 significance level will actually force you in step 4 to set the threshold for rejecting H_0 so high that it will never be reached. Even in our daily lives, people who do not want to take any risk do nothing.

The Test Statistic

► If the sample n is small then,

Test Statistic = X. This test statistic follows the Binomial distribution $\mathcal{B}(n, \pi)$.

► If the sample size is large then,

Test Statistic = Z, where Z is defined as follows:

$$Z = \frac{p - \pi}{\sqrt{\pi(1 - \pi)/n}}.$$

The standard Normal distribution is the law of probability associated with the Z test statistic.

The Decision Rule

► If the sample n is small then,

Determine 2 numbers c_1 and c_2 that satisfy the equation:

$$P(X \leq c_1) + P(X > c_2) = \alpha \tag{5.13}$$

The 2 numbers c_1 and c_2 that satisfy condition 5.13 are referred to as the *Critical Values*.

Decision Rule:
Reject H_0 if $X > c_1$ or $X \geq c_2$ $\tag{5.14}$

I will show in the next example how these critical values can be determined in practice.

► If the sample size is large then,

Determine a number c that satisfies the following equation:

$$P(|Z| > c) = \alpha \tag{5.15}$$

Since the test statistic Z follows the standard Normal distribution, the critical value c_α that satisfies equation 5.13 can be calculated using MS Excel 2007 or 2010 as follows:

$$=\text{NORMSINV}(1-\alpha/2)$$

Once the critical value is calculated, you can formulate the decision rule as follows:

Decision Rule:
Reject the test hypothesis H_0 if $|Z_{obs}| > c_\alpha$ (5.16)

Let us consider an example.

Example 5.3

Suppose you want to verify the claim of a package delivery service that 65% of packages received are delivered within 1 hour from reception. A sample of 100 packages reveals that 67 of them were delivered within 1 hour of reception[6]. Can you conclude that the claim of 65% packages delivered within 1 hour is false? Assume for the purpose of this exercise that the risk of rejecting the claim if it is true is the one you want keep under control with a significance level of 5%.

(1) The hypotheses are defined as follows:

$$\begin{cases} H_0: & \pi = 0.65, \\ H_a: & \pi \neq 0.65. \end{cases}$$

(2) The significance level has been set at $\alpha = 5\%$.

---

[6] The sample proportion of packages delivered within 1 hour of reception is 67%, which exceeds 65%. Because the sample proportion is subject to sampling error, you cannot yet celebrate the interesting story the sample is telling. You want to ensure that the observed difference between 67% and 65% is not caused by the sampling errors only. Hence the need of statistical test of hypothesis.

(3) Before deciding which test statistic to use, you need to evaluate the size of the sample. The sample of 100 packages is considered large because $n(1 - \pi_0) = 100 \times (1 - 0.65) = 35$ is greater than 5, $n\pi_0 = 100 \times 0.65 = 65$ also greater than 5. Consequently, the test statistic to use is the following:

$$Z = \frac{p - \pi_0}{\sqrt{\pi_0(1 - \pi_0)/n}}$$

If the test hypothesis H_0 is true, then the test statistic Z follows the standard normal distribution $\mathcal{N}(0, 1)$.

(4) If H_0 is true then you should expect the test statistic Z to have a small value. Therefore, you would reject H_0 if Z is large in absolute value. That is H_0 is rejected if $|Z|$ (i.e. the absolute value of Z) exceeds a certain threshold c called the *Critical Value*. Because the significance level must be $\alpha = 0.05$, the critical value c must be determined in such a way that,

$$P(|Z| > c) = 0.05.$$

This critical value is obtained using MS Excel as follows:

=NORMSINV(1-0.05/2)

The implementation of this formula in Excel produces the critical value $c = \mathbf{1.96}$.

The decision rule is formulated as follows:

Decision Rule: Reject H_0 if $|Z_{obs}| > 1.96$

where Z_{obs} is the observed value of the statistic Z calculated from the sample selected for the study. This is equivalent to rejecting H_0 if $Z_{obs} > 1.96$ or $Z_{obs} < -1.96$. The decision rule defines a range of values for the test statistic that will result in the rejection of the test hypothesis. This range of values ($x < -1.96$ and $x > 1.96$) is called the critical region.

(5) To reach a final decision, you must compute Z_{obs} and apply the decision rule of step 4. I know from step (3) that $Z_{obs} =$

$(p_{obs}-\pi_0)/\sqrt{\pi_0(1 - \pi_0)} = (0.67-0.65)/\sqrt{0.65 \times (1 - 0.65)/100}$
$= 0.419$. This number lies between -1.96 and 1.96. Consequently, it does not belong to the critical region, and the test hypothesis H_0 cannot be rejected.

Now that you have not rejected the test hypothesis H_0 that the proportion of packages delivered within 1 hour of reception is 65%, how do you interpret this result? The standard interpretation of this result is that there is not sufficient evidence about the sample at hand to warrant the rejection of the test hypothesis. Statisticians often recommend not to endorse H_0 as being true, because that is not what the test has proved. In practice however, people often have to decide whether to endorse or to reject. Recommendations must be made. Simply suspending a judgment, leaving the decision up in the air is often not an option in the real world. So what should you do when you cannot reject H_0?

My recommendation in this situation to do what is known as the *Power Analysis*. The Power Analysis helps answer the following 2 fundamental questions:

[**Q₁**] With the current sample size of $n = 100$, what are the values of π that could have led to the rejection of the test hypothesis? And what would the rejection probability be? *Note that the probability to reject H_0 when it is false is what I referred to earlier as the Power of the Test*

[**Q₂**] Since the current sample size could not reject H_0, what sample sizes could have done the job, and under what conditions?

The answer to question 1 will give you a range of values for π that could have led to the rejection of the test hypothesis. Based on the experience you have in your area of expertise, you may determine that none of these values is realistic. This is all you need to know before actually endorsing the test hypothesis.

Here is where your knowledge, experience and expertise come into play. Statistical analysis will support the decision-making process by rationalizing it. It cannot and should not replace the practical knowledge you have accumulated over time in relation to your main professional activity. Your judgement will remain essential. Always.

The answer to the second question is mainly useful for planning purposes. It will tell you whether by simply increasing the sample size, you could have rejected the test hypothesis with the current procedure. If a larger sample size would still not have led to the rejection of the test hypothesis, then you could decide to endorse it. The power analysis is in my opinion the most useful, and yet neglected component of the Neyman-Pearson theory to hypothesis testing.

Power Curves Pertaining to Example 5.3

The Power Curve depicts the *Test power*[7] as a function of the "true" proportion π. Figure 5.1 shows the Power Curve associated with the hypothesis testing of Example 5.3. You can see from Figure 5.1 that as π moves towards the hypothetical value of 0.65, the ability of the testing procedure to reject the test hypothesis decreases (i.e. the Test Power goes down). When π matches the hypothetical value 0.65, the test power becomes identical to the significance level $\alpha = 0.05$.

More importantly, Figure 5.1 also shows that if the true proportion π was below 0.50 or exceeded 0.77, our testing procedure would have rejected the test hypothesis with a high probability that over 0.8. This shows that even if π happens to be different from 0.65, it is very unlikely that it will as small as 0.5 or as large as 0.8. The test procedure of Example 5.3 is weak for detecting

---

[7]The test power as previously defined, represents the probability of rejecting the test hypothesis when it is indeed false.

small discrepancies from the hypothetical value 0.65, but is powerful for detecting large ones. If π is 0.6 or is 0.7 then only 20% of all samples of 100 orders will allow you to reject H_0. The one you have at hand may or may not be one of them. If you have external knowledge about the possible values that π can take, it can be used to decide whether to endorse H_0 or to rejected it.

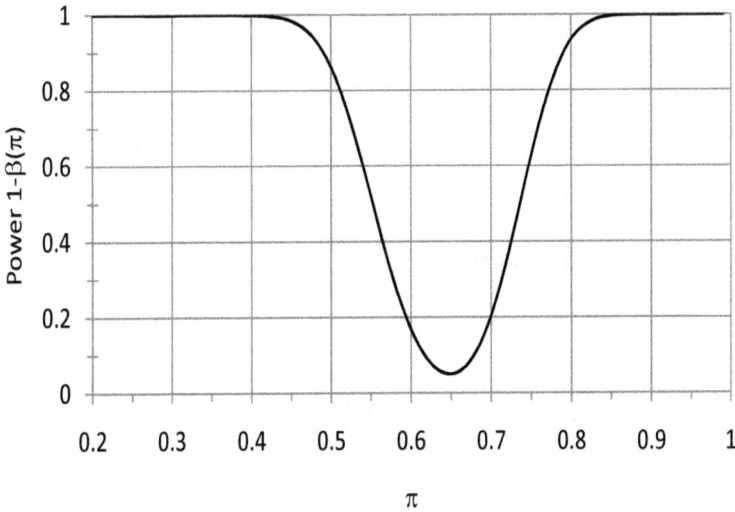

Figure 5.1. Power curves for testing H_0: $\pi = 0.65$ vs H_1: $\pi \neq 0.65$ when the sample size of $n = 100$

How is the Power Curve Created?

The vertical axis in Figure 5.1 represents the test power, which equals $1 - \beta$, where β is the Type II error probability when π (different from π_0) represents the "true" value of the proportion being tested. Therefore, if β as function of π can be specified, it will be all you need to construct the graph in Figure 5.1.

When π is different from π_0 then the variable that follows the standard normal distribution is Z_π defined as $Z_\pi = (p -$

$\pi)/\sqrt{\pi(1-\pi)/n}$. Consequently the test power (i.e. $1-\beta$) is calculated as,

$$\text{Power}(\pi) = 2\big[1 - P(Z_\pi \leq \theta c_\alpha - \Delta_n)\big], \qquad (5.17)$$

where

$$\theta = \frac{\sqrt{\pi_0(1-\pi_0)}}{\sqrt{\pi(1-\pi)}}, \text{ and } \Delta_n = \frac{\pi - \pi_0}{\sqrt{\pi(1-\pi)/n}}.$$

Therefore, the power curve of Figure 5.1 is obtained using equation 5.17.

An Excel Implementation Plan

Using MS Excel, you can easily create 4 columns of data.

▶ The first column for x representing π will vary from 0.2 through 1 (using an increment of 0.05 say)

▶ The third (not the second) column would be used to compute the value of θ follows:

θ =SQRT($\pi_0 * (1 - \pi_0)$)/SQRT($\pi * (1 - \pi)$)

▶ The fourth column would contain Δ_n,

Δ_n =($\pi - \pi_0$)/SQRT($\pi * (1 - \pi)$)

▶ The second column representing the power $1 - \beta(\pi)$ (i.e. the y variable) would be calculated for each π as follows:

=2*(1-NORMSDIST($\theta c_\alpha - \Delta_n$)

▶ Columns 1 and 2 can now be used to create the graph.

In Example 5.1, $c_\alpha = 1.96$, $\pi_0 = 0.65$, and $n = 100$. These are all the elements that are needed to implement the above protocol for creating the power curve of Figure 5.1.

To address question Q_2 of the Power Analysis, which is about knowing how the sample size affects the test power, I needed to create many power curves for different sample sizes as shown in Figure 5.2.

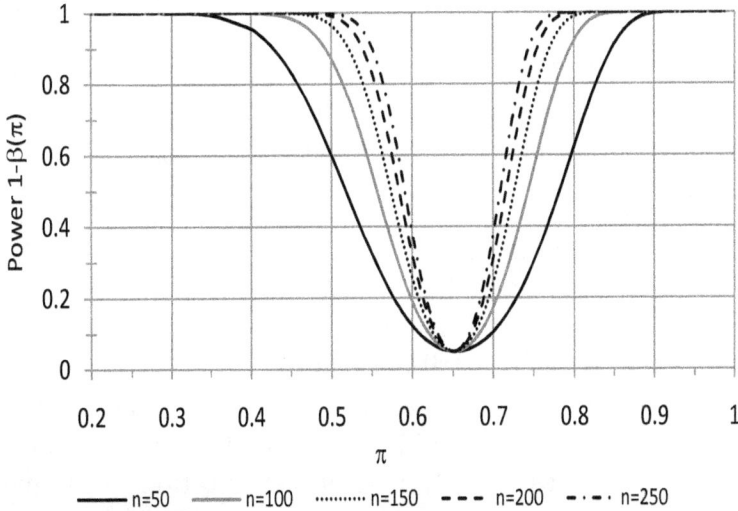

Figure 5.2. Power curves for testing $H_0{:}\pi = 0.65$ vs $H_1{:}\pi \neq 0.65$ for varying sample sizes

All these curves can be created using the same approach described about Figure 5.1 by simply changing the sample size. If follows from Figure 5.2 that increasing the sample size will also increase the test ability to reject the test hypothesis when it is false. For example, if π equals 0.57 then a sample size of 100 is sufficient to reject H_0 with a power of about 0.4; while a sample of size 150 will lead to the rejection of H_0 with a power of 0.60.

If you are going to test the same hypothesis once again at a later time, then you may want to better plan your study by determining the sample size n that you will need in order to reject the test hypothesis if π differs from its hypothetical value π_0 by 5 percentage points for example. The value of n can be determined using the following expression:

$$n = \left(\frac{c_\alpha \sqrt{\pi_0 (1 - \pi_0)} - c_\beta \sqrt{\pi (1 - \pi)}}{\pi - \pi_0} \right)^2, \qquad (5.18)$$

where c_β is defined in such a way that $P(Z_\pi \le c_\beta) = (1 + \beta)/2$. This expression is derived from equation 5.17.

Example 5.3 used the large-sample approach and the standard Normal distribution for implementing the hypothesis testing procedure. In the next example, I will consider a situation where the sample size is small. The law of probability to be used will no longer be the Normal distribution. Instead, it will be the Binomial distribution. The use of a discrete distribution comes with a new set of challenges that will be discussed. The Power analysis particularly requires some special implementation techniques.

Example 5.4 ————————————————————————

Suppose you want to verify the claim of a package delivery service that 96% of packages received are delivered within 3 hours from reception. A sample of 100 packages reveals that 94 of them were delivered within 3 hours of reception. Can you conclude that the claim of 96% of packages delivered within 3 hours is false? Assume for the purpose of this exercise that the risk of rejecting the claim if it is true is the one you want keep under control with a significance level of 5%.

(1) The hypotheses are defined as follows:

$$\begin{cases} H_0: & \pi = 0.96, \\ H_a: & \pi \ne 0.96. \end{cases}$$

(2) The significance level has been set at $\alpha = 5\%$.

(3) The sample of 100 packages is small because $n(1 - \pi_0) = 100 \times (1 - 0.96) = 4$ is smaller than 5. Consequently, the test statistic to use is the following:

$$X = \text{Number of sample packages delivered within 3 hours}$$

(4) You now need to compute the critical values c_1 and c_2 before you can formulate the decision rule. If you are using Excel 2010, this task will be a snap. Both critical values can be obtained as follows:

$$c_1 = \text{BINOM.INV}(100, \pi_0, \alpha/2), \quad and$$
$$c_2 = \text{BINOM.INV}(100, \pi_0, 1 - \alpha/2).$$

However, with Excel 2007, these critical values can be obtained by creating a spreadsheet similar to Table 5.1. The purpose of this table is to determine the values of c_1 and c_2 that will satisfy equation 5.13. To this end, I would try 11 c-values varying from 90 to 100. For each value of c, I will compute the 2 probabilities in columns 2 and 3. The second-column probability $P(X > c)$ is calculated with the following Excel formula =1-BINOMDIST(c,100,π_0,1) , while the third-column probability $P(X \le c)$ is calculated with the Excel formula =BINOMDIST(c,100,π_0,1) .

You want c_1 and c_2 to meet condition 5.13 with $\alpha = 0.05$. Because the law of probability of X is Binomial (i.e. discrete), it is impossible to meet this condition in a strict sense[8]. Therefore your solution for c_1 and c_2 will be a rough approximation determined as follows:

> In this case, c_1 should be the largest value of c that satisfies $P(X \le c) \le 0.025$ and c_2 the smallest value of c that satisfy $P(X > c) \le 0.025$.

---

[8]There are only a limited number of values that X can take, and there are jumps between values. Consequently the probabilities you can compute can also take a limited number of values with the real possibility that no pair of c-values will meet condition 5.13.

The 2 solutions c_1 and c_2 are highlighted in Table 5.1, where $c_1 = 91$ and $c_2 = 99$. The "Type I Error Probability" in this case is actually much smaller than its nominal value of $\alpha = 0.05$. This Type I error probability is $P(X \le c_1) + P(X \ge c_2) = 0.018992 + 0.016870 = 0.03586$.

The decision rule will be (see 5.14),

Decision Rule: Reject H_0 if $X_{obs} \ge 99$ or $X_{obs} \le 91$

It follows from the statement of this problem that $X_{obs} = 94$. Therefore, the test hypothesis cannot be rejected since 94 does not belong to the *Critical Region* defined by $\{X \le 91$ or $X \ge 99\}$.

Table 5.1: Critical Value Calculation Based on Equation 5.13

c	$P(X \ge c)$	$P(X \le c)$
90	0.993156	0.006844
91	0.981008	**0.018992**
92	0.952488	0.047512
93	0.893608	0.106392
94	0.788375	0.211625
95	0.628864	0.371136
96	0.429476	0.570524
97	0.232143	0.767857
98	0.087163	0.912837
99	**0.016870**	0.983130
100	0.000000	1.000000

Since the null hypothesis is not rejected, I will conduct the power analysis to see how powerful my testing procedure is, and whether I should endorse the test hypothesis or not.

The Power Curves Pertaining to Example 5.4

In Example 5.4, I could not reject the test hypothesis H_0, which stipulates that 96% of packages are delivered within 3 hours from reception. But does this really mean that you should endorse the test hypothesis that you have not been able to reject? At this stage of the analysis, it would not be wise to do so. Not rejecting H_0 means that the magnitude of the true proportion π (perhaps too close to the hypothetical value 0.96), combined with the size of the package sample (i.e. $n = 100$ perhaps not sufficiently large) did not make it possible to reject the test hypothesis.

▶ Let me study the values of π that would have caused the rejection of H_0, everything else being held fixed. If π is the true proportion, then the variable X representing the number of packages delivered within 3 hours out of the total of 100 packages, follows the Binomial distribution $\mathcal{B}(100, \pi)$. Consequently, for each value of π I can compute the probability to reject H_0 (i.e. the probability that $X \le 91$ or $X \ge 99$ according to the decision rule)[9]. This probability expressed as a function of π is the test power.

The power curve corresponding Example 5.4 is shown in figure 5.3. This figure indicates that if the "true" proportion π is 0.87 or lower for example, you would have rejected the test hypothesis with a probability that exceeds 0.80. What the power curve shows is that the true value of π is likely to be close to 0.96. Otherwise, the null hypothesis would have been rejected, the test being very powerful for values of π that are very different from 0.96. In practice, this might be all you need to know before actually endorsing the test hypothesis.

---

[9]In Excel, this probability is obtained as =BINOMDIST(91,100,π,1) + =1-BINOMDIST(99,100,π,1).

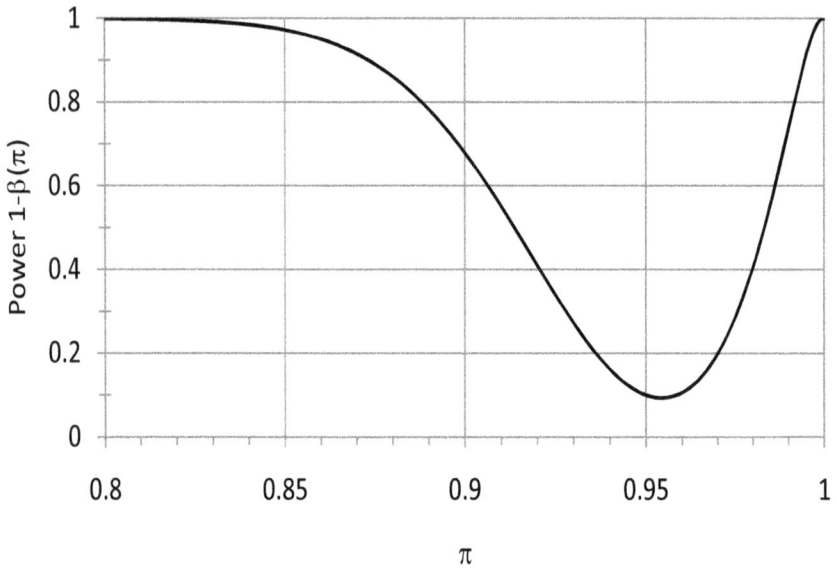

Figure 5.3. Power curve for testing $H_0:\pi = 0.96$ vs $H_1:\pi \neq 0.96$

▶ To investigate how the sample size affects the test, I need to create a graph similar to that of Figure 5.3, for various sample sizes. This analysis allows you to see whether by increasing the sample size, you could have improved your testing procedure in terms of increasing the test power, while maintaining a reasonable significance level. Figure 5.4 shows such a graph. You can see from Figure 5.4 that if the true value of π is away from from the hypothetical value of 0.96, a bigger sample will give you more power. As π gets closer to 0.96, this test gets weaker in the sense a decrease in power. If it is important for you to be able to reject H_0 even if π is as close to 0.96 as 0.93, then only a sample size of 250 or more will give you good power.

To construct the graphs in Figure 5.4, you must determine the critical values c_1 and c_2 associated with each sample size n using the approach shown in the previous example. Then for each value of π, compute $P(X > c_2) + P(X \leq c_1)$ (i.e. the probability to reject H_0) using the Binomial distribution $\mathcal{B}(n, \pi)$. This probability can be computed with Excel as,

$$=1\text{-BINOMDIST}(c_2,n,\pi,1) + \text{BINOMDIST}(c_1,n,\pi,1).$$

Note that the curves of Figure 5.4 are rough approximations because they are based on different significance levels. The significance levels of these curves are 4.9%, 3.6%, 3.4%, 4.7%, 4.0%, and 2.9% for $n = 50, 100, 150, 200,$ and 250 respectively. This details show the additional complexity that characterizes the analysis of small samples. The analyst will need to be mindful when conducting these analyzes.

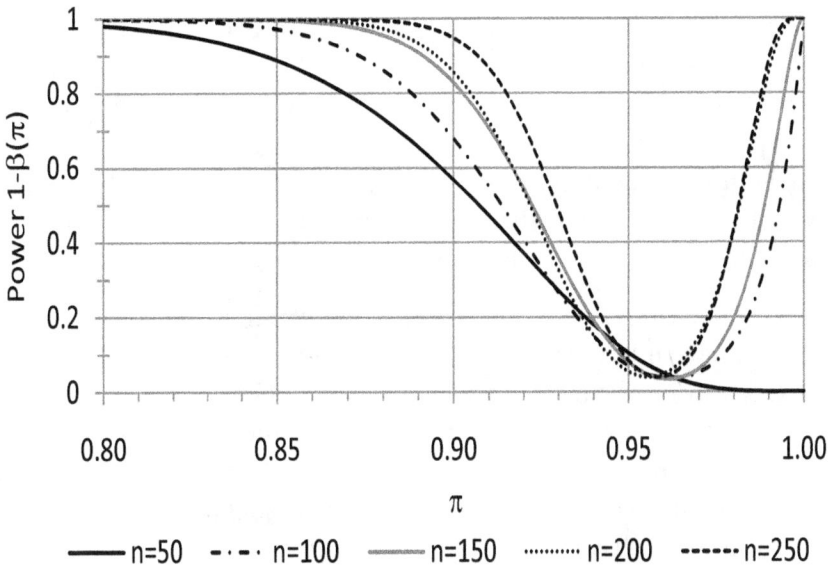

Figure 5.4. Power curves for testing $H_0{:}\pi = 0.96$ vs $H_1{:}\pi \neq 0.96$ for varying sample sizes

5.3.2 The Left-Tailed Test of Hypothesis

(1) The Hypotheses

A test of hypothesis is **left-tailed** when the test and alternative hypotheses have the following form:

$$(Hypotheses) \quad \begin{cases} H_0: & \pi \geq \pi_0, \\ H_a: & \pi < \pi_0. \end{cases} \quad \text{or} \quad \begin{cases} H_0: & \pi = \pi_0, \\ H_a: & \pi < \pi_0. \end{cases} \quad (5.19)$$

You will normally perform the left-tail test of hypothesis when you want to protect yourself against the mistake that consists of concluding that $\pi < \pi_0$ when in reality $\pi \geq \pi_0$. That is, you want to control the risk of rejecting H_0 when it is true. There are indeed situations in practice where such a Type I error may have serious business consequences. Suppose for example that π represents the proportion of your customers who are dissatisfied. You would certainly not want to reject the hypothesis that $\pi \geq 50\%$ when that is indeed the case. Because, then you may continue running your business as usual with a large group of dissatisfied customers who you will eventually lose overtime.

The left-tailed test is primarily determined by the form of the alternative hypothesis H_a shown above. Can you have a left-tailed test with the alternative hypothesis being $\pi \leq \pi_0$ instead of $\pi < \pi_0$? Although the form $\pi < \pi_0$ is the most commonly used, there is nothing wrong specifying it as $\pi \leq \pi_0$ if it happens to be the alternative to the test hypothesis associated with the risk you want to control. But there is no fundamental difference in practice between both formulations. If you endorse the hypothesis that $\pi < \pi_0$, you would automatically have endorsed $\pi \leq \pi_0$ as well. If you endorse $\pi \leq \pi_0$, you would also have endorsed $\pi < \pi_0$, except in the special situation when the endorsement of $\pi \leq \pi_0$ is due to the equality of the true and hypothetical proportions (i.e. $\pi = \pi_0$). Since the test of hypothesis is always based upon a sample, and not the entire population, it is wise to have equality in the test hypothesis so that it can be disproved if there is a large discrepancy between the observed and the hypothetical proportion, rather than having equality in the alternative hypothesis where it will hardly be provable. Consequently I will stick with the standard formulation of the alternative hypothesis shown in 5.19.

Normally for the same alternative hypothesis $H_a:\pi < \pi_0$, you can formulate one of the 2 null hypotheses $\pi \geq \pi_0$ or $\pi = \pi_0$, depending on which one best describes the problem to solve. Here is an example of a practical problem where the null hypothesis must be formulated as $\pi = \pi_0$. Consider a credit card company that usually obtains a 95% nonresponse on its mailing campaign to gain new customers. Suppose that company decides to offer an additional incentive such as 0% interest on new purchases for the first 6 months, and now wants to test if this will result in a decrease in nonresponse rate. It appears that $\pi = \pi_0$ is is a reasonable null hypothesis to consider.

(2) The Significance Level α

If the test hypothesis is H_0: $\pi = \pi_0$ then the significance level α that you set for the test will represent the Type I Error probability. In this case, you will be able to compute that probability because under H_0, π can only take one value, which is π_0, and which will also specify the law of probability of the test statistic.

When the test hypothesis is H_0: $\pi \geq \pi_0$ then the situation is a little problematic because under H_0, the true proportion can take any of the values between 0 and the hypothetical proportion π_0, making it impossible to fully specify any law of probability. Therefore, the Type I error cannot be calculated. By setting $\alpha =$

0.05, you are only setting an upper bound that the Type I error probability should not exceed.

(3) The Test Statistic

▶ If the sample size n is large (i.e. $n\pi_0 \geq 5$ and $n(1-\pi_0) \geq 5$), then the test statistic will be:

$$Z = \frac{p - \pi_0}{\sqrt{\pi_0(1-\pi_0)/n}},$$

All you can say about this test statistic is that if $\pi = \pi_0$, it will follow the standard Normal distribution $\mathcal{N}(0,1)$.

▶ If the sample size n is small (i.e. $n\pi_0 < 5$ or $n(1-\pi_0) < 5$), then the test statistic will be the random variable X defined by,

$$X = Number\ of\ sample\ subjects\ that\ bear\ the$$
$$characteristic\ being\ investigated$$

The law of probability associated with this variable as seen in sub-section 5.3.1 is the Binomial distribution $\mathcal{B}(n,\pi)$.

(4) The Decision Rule

The formulation of hypotheses in step (1) shows intuitively that the test hypothesis H_0 must be rejected when the test statistic of step (3) is "sufficiently" small. The purpose of this step is to determine how small should the test statistic be before you are in a position to reject the test hypothesis. The threshold to be determined here must be consistent with the significance level set in (2).

▶ If the sample size n is large then you find the critical value c_α that meets the following equation:

$$P(Z < c_\alpha) = \alpha, \qquad (5.20)$$

where α is the significance level set in step (2). If for example $\alpha = 0.05$ then you will have to find $c_{0.05}$ such that $P(Z > c_{0.05}) = 0.05$. This is accomplished with Excel 2007 as,

$$c_\alpha \ \text{=NORMSINV}(\alpha)$$

For $\alpha = 0.05$, the implementation of this Excel function with 0.05 replacing α yields $c_{0.05} = -1.645$. The decision rule will then be formulated as follows:

Decision Rule: $\qquad\qquad$ (5.21)
Reject H_0 if $Z < c_\alpha$

To compute c_α I had to assume that $\pi = \pi_0$ so that the test statistic Z can follow the Normal distribution. But will the Type I error still be under control if π is smaller than π_0, which is still a possibility under H_0? The answer to that question is yes, the Type I error probability will be different from 0.05, but will be smaller, which is even better for you.

▶ If the sample size n is small then you find the critical value c_α that satisfies the following condition:

$$P(X < c_\alpha) = \alpha, \qquad (5.22)$$

where X follows the Binomial distribution $\mathcal{B}(n, \pi_0)$. I would calculate the critical value c_α using MS Excel. If you are using Excel 2010, you can have an easy life with the BINOM.INV function, by typing =BINOM.INV(n, π_0, α). But with Excel 2007 or an older version of Excel, the critical value may still be calculated with the Excel function BI-NOMDIST. You will need to do it as discussed earlier. The decision rule will then be formulated as follows:

> **Decision Rule:**
> Reject H_0 if $X < c_\alpha$
>
> (5.23)

(4) The Final Decision

Making the final decision consists of actually computing the test statistic, either Z or X, and applying the appropriate decision rule before deciding whether to reject or not rejected the test hypothesis H_0.

Example 5.5

Surveys are known to reach people with an educational attainment below the high school degree in a proportion that is lower than their actual proportion in the general population. A recent study has revealed that the population with no high school degree represents about 15% of the total population. Suppose that you conducted a survey of 500 people randomly selected from the population, and obtained 12% of respondents with no high school degree. Is it plausible that the "true" proportion of individuals with no high school degree, who would participate in a survey is the same as their proportion in the population? Or is it more reasonable to consider that it below?

(1) Hypotheses

As researcher, it is certainly wise to control the risk of deciding that the lowly educated population will be well represented in the survey, when in reality it will be under-represented. Consequently, the test hypothesis must be H_0: $\pi = 0.15$. Therefore the two Neyman-Pearson hypotheses are formulated as follows:

$$\begin{cases} H_0: & \pi = 0.15, \\ H_a: & \pi < 0.15. \end{cases} \tag{5.24}$$

Therefore the hypothetical proportion is $\pi_0 = 0.15$, and the observed proportion $p_{obs} = 0.12$.

(2) Significance Level

Again, the standard value used is $\alpha = 0.05$. You could experiment with smaller significance levels. Smaller significance levels will bring the test power down, and you may need compensate the loss of power with an increased sample size.

(3) The Test Hypothesis

The sample size of $n = 500$ is large since $n\pi_0 = 500 \times 0.15 = 75 > 5$, and $n(1 - \pi_0) = 500 \times 0.85 = 425 > 5$. Consequently, the test statistic is $Z = (p - \pi_0)/\sqrt{\pi_0(1 - \pi_0)/n} = (p - 0.15)/\sqrt{0.15 \times (1 - 0.15)/500} = (p - 0.15)/0.01597$, p is the abstract proportion of lowly-educated individuals in the sample who participated in the survey.

$$Z = (p - 0.15)/0.01597. \tag{5.25}$$

If H_0 is assumed to be true, then Z will follow the standard normal distribution.

(4) The Decision Rule

Since p is a surrogate for π, you will naturally want to reject H_a when p is "substantially" smaller than 0.15, or Z is "very" small. That is Z will have to be smaller than a critical value c_α that satisfies the relation $P(Z < c_\alpha) = \alpha$. Since $\alpha = 0.05$, this critical value could be calculated using Excel as follows:

$c_{0.05}$ =NORMSINV(0.05)

This Excel expression will lead to the critical $c_{0.05} = -1.645$. Therefore the decision rule will be,

Reject H_0 in favor of H_a if $Z < -1.645$

(5) Final Decision

The observed value Z_{obs} of the test statistic is calculated as $Z_{obs} = (p_{obs} - 0.15)/0.01597 = (0.12 - 0.15)/0.01597 = -1.879$. This observed test statistic is smaller than the critical -1.645. Therefore the test hypothesis H_0 must be rejected. Your survey results tend to confirm that indeed the lowly-educated participate in surveys in a proportion that is lower than their actual representation in the population.

In the previous example, the observed proportion $p_{obs} = 0.12$ was sufficiently low to reject the test hypothesis that the lowly-educated population participates in sample surveys in a proportion that equals their representation in the general population. Three key elements that contributed to this rejection are the chosen significance level, the sample size used in the experiment, and the magnitude of the difference between the true proportion π and the hypothetical one π_0.

If you are planning a test, you may want to know before selecting the sample, what is the minimum difference $\pi - \pi_0$ that will lead to the rejection of the test hypothesis[10], for a given significance level α, and sample size n. This will allow you to adjust the sample size or the significance level. Even if the data has already been collected, you may still want to know how powerful your test is, especially if you have not rejected the test hypothesis, or plan to use the same data for testing other hypotheses. Only the Statistical Power Analysis can help you study the properties of

---

[10]This difference is known in the literature as the *Minimum Detectable Difference (MDD)*

your testing procedure thoroughly.

The Statistical Power Analysis

The probability of rejecting the test hypothesis when π is the true proportion, is calculated using the following equation:

$$
1 - \beta = P(Z < c_\alpha) = \\
P\left(Z_\pi < \frac{(\pi_0 - \pi)\sqrt{n} + c_\alpha\sqrt{\pi_0(1 - \pi_0)}}{\sqrt{\pi(1 - \pi)}} \right), \quad (5.26)
$$

where $Z_\pi = (p-\pi)/\sqrt{\pi(1 - \pi)}$, and Z the test statistic. Equation 5.26 is used to construct the Power Curve, where π is used on the x-horizontal axis, and $1 - \beta$ on the y-vertical axis.

Using Example 5.5 data, and MS Excel, you can compute $1 - \beta$ for each value of π as follows:

$$
\begin{aligned}
&= \text{NORMSDIST}((-1.645*\text{SQRT}(0.15*(1\text{-}0.15)) \\
&\quad - (\pi\text{-}0.15)*\text{SQRT}(500)) \ / \ \text{SQRT}(\pi*(1\text{-}\pi))) \quad (5.27)
\end{aligned}
$$

This will lead to the creation of Table 5.2, which in turn will lead to the Power Curve in Figure 5.5.

Table 5.2 : Power of the Example 5.5 Test as a Function of π

π	$1 - \beta$	π	$1 - \beta$
0.05	1	0.13	0.3384
0.06	1	0.14	0.1472
0.07	1	0.15	0.0500
0.08	0.9998	0.16	0.0135
0.09	0.9958	0.17	0.0029
0.10	0.9615	0.18	0.0005
0.11	0.8368	0.19	0.0001
0.12	0.6013	0.20	0.0000

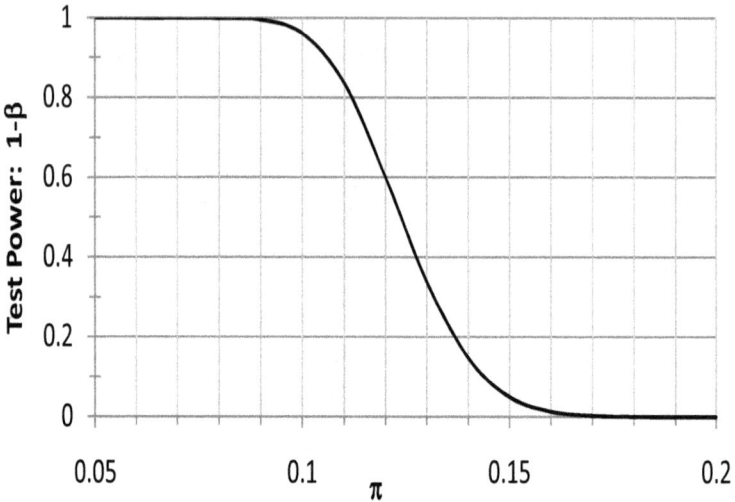

Figure 5.5. Power curve for testing $H_0:\pi = 0.15$ vs $H_a:\pi < 0.15$ for the sample size $n = 500$ and significance level $\alpha = 0.05$

You can see from the Power Curve of Figure 5.5 that the test power increases dramatically as π moves away from its hypothetical value 0.15 towards smaller values. That is, if for example $\pi = 0.11$, then more than 80% of all samples of size 500 will lead to the rejection of the test hypothesis. But if $\pi = 0.14$, only about 15% of the samples will cause the rejection of H_0, which means that you would probably not expect one particular sample of 500 to reject H_0, although that remains a possibility.

Figure 5.6 shows 4 power curves for the sample sizes 500, 200, 100, and 50. These graphs show that for a given value of π the test power increases substantially as the sample size goes up.

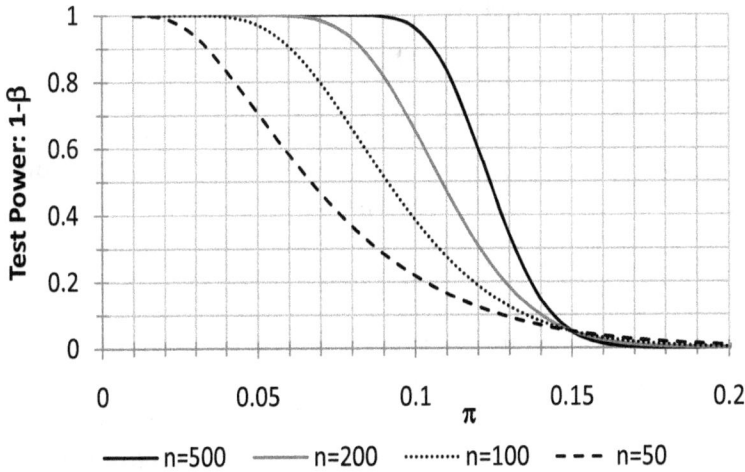

Figure 5.6. Power curves for testing $H_0{:}\pi = 0.15$ vs $H_a{:}\pi < 0.15$ for varying sample sizes, and a significance level $\alpha = 0.05$

Figure 5.7 illustrates the impact the chosen significance level could have on the test power for a given sample size, and value of the true proportion.

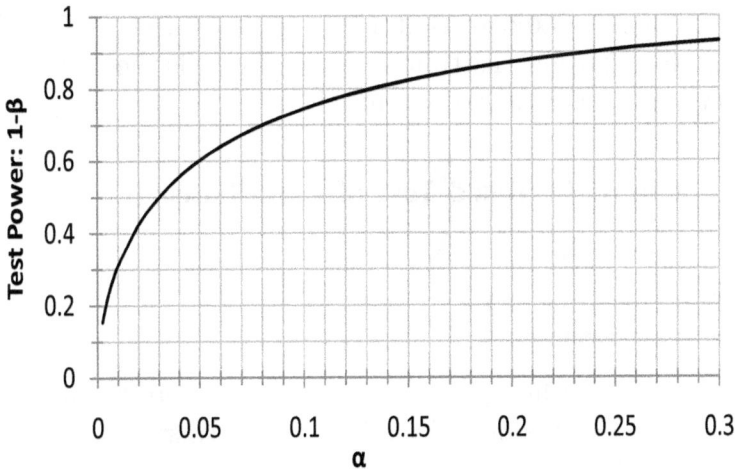

Figure 5.7. Power curve for testing $H_0{:}\pi = 0.15$ vs $H_a{:}\pi < 0.15$ $(n = 500,\ \pi = 0.12)$ for varying significance levels α

If you want to improve the test power without increasing the sample size, one of the few options you have is to increase the significance level. In this case, if you increase α from 5% to 10% your test power will go up from 60% to about 75%. Whether you should improve your chances of proving your case when you have one at the cost of running the risk to wrongfully proving a case when you have none is your call. These debates must take place within the research team before making the final determination.

There are situations in practice where you may want to know for a specified value of the "true" proportion π, the sample size n that will lead to the rejection of the test hypothesis, with a specified power $(1 - \beta)$, at a specified significance level α. This sample size, derived from equation (5.26) is given by:

$$ n = \left(\frac{c_\beta \sqrt{\pi(1 - \pi)} - c_\alpha \sqrt{\pi_0(1 - \pi_0)}}{\pi_0 - \pi} \right)^2, \qquad (5.28) $$

where c_β is calculated so that $P(Z_\pi < c_\beta) = 1 - \beta$, Z_π follows the standard Normal distribution. That is, c_β is the $100(1 - \beta)^{th}$ percentile of the standard Normal distribution.

5.3.3   The Right-Tailed Test of Hypothesis

(1) The Hypotheses

A test of hypothesis is *right-tailed* when the test and alternative hypotheses have one of the following forms:

$$ (Hypotheses) \quad \begin{cases} H_0: & \pi \leq \pi_0, \\ H_a: & \pi > \pi_0. \end{cases} \quad \text{or} \quad \begin{cases} H_0: & \pi = \pi_0, \\ H_a: & \pi > \pi_0. \end{cases} \qquad (5.29) $$

You will perform the right-tailed test of hypothesis when the error you want to avoid the most is one where you conclude that $\pi > \pi_0$ when in reality $\pi \leq \pi_0$. As an example, suppose you are manager at a department store, and you are concerned about the performance of your customer service division that you monitor with the proportion π of satisfied customers. You would certainly not want to reject the hypothesis that $\pi \leq 50\%$ when in reality less than half of your customers are satisfied. By committing that error, you may continue running your business as usual with a large group of dissatisfied customers you will eventually lose them overtime.

(2) The Significance Level α

As previously mentioned the significance level α that you set for the test will represent the Type I Error probability only if the test hypothesis is $H_0: \pi = \pi_0$. Otherwise, it will represent the upper bound of the Type I error probability if the test hypothesis is $H_0: \pi \leq \pi_0$.

(3) The Test Statistic

This step remains essentially the same, whether the test is two-tailed, left-tailed, or right-tailed.

(4) The Decision Rule

The formulation of hypotheses in step (1) shows intuitively that the test hypothesis H_0 must be rejected in favor of the specified alternative H_a when the test statistic of step (3) is "sufficiently" large. The purpose of this step is to determine how large should the test statistic be before you are in a position to reject the test hypothesis. The threshold to be determined here must be chosen in such a way that the Type I error probability will not exceed the nominal significance level of step 2.

▶ If the sample size n is large then you find the critical value c_α that meets the following equation:

$$P(Z > c_\alpha) = \alpha, \qquad (5.30)$$

where α is the significance level set in step (2). If for example $\alpha = 0.05$ then you will have to find $c_{0.05}$ so that $P(Z > c_{0.05}) = 0.05$. This critical value is obtained with MS Excel as follows:

$$c_\alpha \ \ =\text{NORMSINV}(1 - \alpha)$$

For $\alpha = 0.05$, the implementation of this Excel function with 0.05 replacing α yields $c_{0.05} = 1.645$. The decision rule will then be formulated as follows:

Decision Rule:
Reject H_0 if $Z > c_\alpha$ $\qquad (5.31)$

Note that if $H - 0$ then the Type I error probability will not exceed α even if $\pi \leq \pi_0$. This is due to the fact that if $\pi < \pi_0$ then $P(Z > c_\alpha | \pi < \pi_0) \leq P(Z > c_\alpha | \pi = \pi_0)$.

▶ If the sample size n is small then you find the critical value c_α that satisfies the following condition:

$$P(X > c_\alpha) = \alpha, \qquad (5.32)$$

where X follows the Binomial distribution $\mathcal{B}(n, \pi_0)$. I would calculate the critical value c_α using MS Excel. If you are using Excel 2010, you can have an easy life with the BINOM.INV function, by typing =BINOM.INV($n, \pi_0, 1 - \alpha$).

But with Excel 2007 or an older version of Excel, the critical value may still be calculated with the Excel function **BINOMDIST**. You will need to create 2 columns in Excel. The first column will have integer numbers from 0 to n (i.e. $0, 1, 2, \cdots, n$). For each number x in the first column, compute in the second column the probability $P(X > x)$ as follows:

$$=1\text{-BINOMDIST}(x,n,\pi_0,1)$$

The critical value c_α will be the smallest number in column 1 for which the second-column probability exceeds $1 - \alpha$. The decision rule will then be formulated as follows:

> **Decision Rule:**
> Reject H_0 if $X > c_\alpha$

(5.33)

(5) The Final Decision

Making the final decision consists of actually computing the test statistic, either Z or X, and applying the appropriate decision rule before deciding whether to reject or not rejected the test hypothesis H_0.

The Power Analysis

The Power curve depicting the test power $1 - \beta$ as a function of the "true" proportion π can obtained for the right-tailed for a given sample size n, and a hypothetical proportion π_0, using the following relation:

$$1 - \beta = P\left(Z_\pi > \frac{c_\alpha\sqrt{\pi_0(1 - \pi_0)} - (\pi - \pi_0)\sqrt{n}}{\sqrt{\pi(1 - \pi)}}\right),$$

(5.34)

where Z_π follows the standard Normal distribution. To construct the power curve, you need to compute $1 - \beta$ for various values of π. If you want to use Excel, you could compute $1 - \beta$ for a given π as follows:

$$= 1\text{-NORMSDIST}((c_\alpha * \text{SQRT}(\pi_0 * (1 - \pi_0))$$
$$- (\pi - \pi_0) * \text{SQRT}(n)) / \text{SQRT}(\pi * (1 - \pi)))$$

The Sample Size

For a given test power $1 - \beta$ (note that if the desired power is 80%, then $\beta = 20\%$), significance level α, and "true" proportion π, the required sample size is given by:

$$n = \left(\frac{c_\beta \sqrt{\pi(1 - \pi)} + c_\alpha \sqrt{\pi_0(1 - \pi_0)}}{\pi_0 - \pi} \right)^2, \qquad (5.35)$$

where c_β is a positive number that meets the condition $P(Z \le -c_\beta) = \beta$, the law of probability of Z being the standard Normal distribution.

CHAPTER $\boxed{6}$

One-Sample Test of Hypothesis for Means

OBJECTIVE

This chapter discusses all aspects of the statistical test of hypothesis of a single population mean. Large-sample and small-sample testing procedures are discussed, and are all based on the one sample (hence the name one-sample test) selected from the population of interest.

CONTENTS

6.1. Testing A Population Mean

The mean to be tested here is that of a measurement variable such as income, height, or distance. Consider for example a car manufacturer that claims that his X-15 steel-belted radial truck tires would normally be driven 60,000 miles before the tread wears out. The manufacturer's claim can be tested using the principles of statistical inference, by considering the population of X-15 steel-belted radial truck tires, and by testing their lifetime mean mileage μ.

The tire population of interest considered here is the open list of all X-15 steel-belted radial truck tires from this manufacturer. If these tires are manufactured using the same manufacturing processes with minor variations in the final product, then your study of the open population could be effective since any sample of 100 tires will be a good representation of all tires in stock as well as those yet be manufactured. However, if there has been changes in the manufacturing processes at some time in the past then it will be essential to provide a more precise description of the target population of tires.

The tire problem expressed in statistical terms, consists of verifying the conjecture (or hypothesis) that $\mu > 60,000$. As discussed in the previous chapters, the way to test this hypothesis is to select a sample of tires and to observe the number of miles each was driven before it wore out. The population mean μ being tested is unknown[1]. To perform the test, μ will be replaced by a surrogate, which is the mean number of miles the tires in the sample were driven before they wore out. This surrogate will typically be compared to the hypothetical value 60,000,

---

[1]If μ was available, you would not need a statistical test since you will only have to compare it to the hypothetical value 60,000 before accepting or rejecting the conjecture.

and the magnitude of this difference may or may not support your research hypothesis. Let me look more in-depth at how this surrogate for μ will be used to arrive at the testing procedure.

Let n be the number of tires included in the sample, and x_i the number of miles tire i was driven before wearing out. The sample mean denoted by $\bar{x}$ (x bar) will be our best approximation to μ. The hypothetical value for the population mean for the tire problem is $\mu_0 = 60,000$. Even if μ is unknown, I still know that it plays a hidden but important role in the magnitude of $\bar{x}$. Therefore, the difference $\bar{x} - \mu_0$ should provide some information regarding the relationship between μ and μ_0. A sample of tires will yield one number for the difference $\bar{x} - \mu_0$. How large should it be before you reject the hypothesis that $\mu \leq \mu_0$? What should the cut-off point be?

If you base your testing procedure solely on the one sample you select, it will certainly have no validity on a different sample of tires your colleague may select. A serious statistical procedure must borrow strength from what all possible samples of tires have in common. Consider the following ratio:

$$z = \frac{\bar{x} - \mu_0}{s/\sqrt{n}}, \qquad (6.1)$$

where s is the standard deviation of the x variable. Then the law of probability associated with z - albeit the sample size is reasonably large (i.e. n exceeds 30) - is the standard Normal distribution. That is if you select all possible tire samples, and compute the ratio z for all of them, then these z-values will have a known law of probability; and it is the standard Normal distribution. Consequently, for a specific sample at hand, you can claim that the observed difference $\bar{x}_{obs} - \mu_0$ is very large, if the probability for the observed z, defined as $z_{obs} = (\bar{x}_{obs} - \mu_0)/(s_0/\sqrt{n})$ to be exceeded is very small (typically 0.05 or smaller). Here, you have a powerful criteria that is not just based on a single sample, but which instead is based on the law of probability that governs the variation of the mean from all samples that can possibly be selected.

For the sake of testing the research hypothesis, you will still have the option to use the Fisher's Null hypothesis testing approach, or alternatively use the Neyman-Pearson decision theory approach. As I indicated in chapter 5, if your goal is merely to quantify the strength of evidence in your sample in favor of the conjecture and make an educated opinion, I suggest to confine yourself to the Fisher hull hypothesis approach. However, if you want to do a thorough power analysis, in order to determine how effective your testing procedure is for endorsing a wide range of alternative hypotheses, you should definitely opt for the Neyman-Pearson protocol.

6.2. Fisher's Null Hypothesis Approach

In the tire example of the previous section, the research hypothesis believed to be true is $\mu > 60,000$. Therefore the null hypothesis (to be disproved) is H: $\mu \leq 60,000$ (the opposite of the research hypothesis). You will reject H if the difference $\bar{x} - 60,000$ between the sample mean and the hypothetical value is sufficiently large. This is equivalent to having z_{obs} substantially large. Fisher's approach consists of calculating the P-value, which represents the Probability $P(z > z_{obs})$. z_{obs} will be deemed large if the P-value (or the probability of exceeding z_{obs}) is small.

How the P-value must be calculated depends on the following 3 elements:

▶ The sample size n,

▶ The form of the research hypothesis,

▶ If the sample size is small, it will also depend on whether the law of probability governing the data is close to normality or not.

6.2.1 Large Sample Size ($n > 30$)

Research hypothesis: $\mu \neq \mu_0$

The null hypothesis in this case would be:

$$\mathsf{H}: \mu = \mu_0, \tag{6.2}$$

and represents the hypothesis to be rejected. When the sample size is large, the z statistic of equation 6.1 follows the standard Normal distribution under the assumption that the null hypothesis is true[2]. You would reject H if the absolute value of z_{obs} is large. The magnitude of z_{obs} is evaluated by how often it will be exceeded by z. Hence,

$$\text{P-value} = P\big(|z| > |z_{obs}|\big) \tag{6.3}$$

This P-value is obtained using MS Excel as follows:

$$
\begin{aligned}
&\text{Excel 2010:} &&= 2*(1\text{-NORM}\cdot\text{S}\cdot\text{DIST}(\text{ABS}(z_{obs}),\text{TRUE})) \\
&\text{Excel 2007:} &&= 2*(1\text{-NORMSDIST}(\text{ABS}(z_{obs})))
\end{aligned}
\tag{6.4}
$$

---

[2]Because of the Central Limit Theorem, this statement is true regardless of the law of probability associated with the variable X

Research hypothesis: $\mu > \mu_0$

The null hypothesis in this case would be:

$$\textsf{H: } \mu = \mu_0, \text{ or } \textsf{H: } \mu \leq \mu_0. \tag{6.5}$$

Either form of the null hypothesis will lead to the exact same testing procedure. One form may better describe the problem at hand than the other. The P-value is typically calculated under the assumption that $\mu = \mu_0$. With this, if the evidence in the sample happens to be strong (i.e. the P-value is small), then this evidence will remain small for any other value of μ that is smaller than μ_0 (i.e. any μ value consistent with the null hypothesis). I will demonstrate a little later why this is true. For the purpose of calculating the P-value, I will assume that the null hypothesis is in the form $\textsf{H: } \mu = \mu_0$.

Since the sample size is large, the z statistic of equation 6.1 follows the standard Normal distribution under the assumption that the null hypothesis is true. You would reject $\textsf{H}$ in favor of the research hypothesis stated above if the observed sample mean $\overline{x}_{obs}$ (the surrogate for μ) exceeds μ_0 by a large amount. This is equivalent to saying that z_{obs} is large. Hence,

$$\textsf{P-value} = P\left(z > z_{obs}\right) \tag{6.6}$$

This P-value is obtained using MS Excel as follows:

Excel 2010: $= 1\text{-NORM}\cdot\text{S}\cdot\text{DIST}(z_{obs}, \text{TRUE})$

Excel 2007: $= 1\text{-NORMSDIST}(z_{obs})$

(6.7)

Suppose now that the null hypothesis is true, but that $\mu < \mu_0$. The question then becomes how would this affect the P-value of equation 6.6, which is based on the assumption that $\mu = \mu_0$. The answer is that this new P-value will be smaller than that of 6.6. This comes from the fact that if Z is a generic standard Normally distributed variable, then,

$$P\big(Z > z_{obs} \; for \; \mu = \mu_0\big) \geq P\bigg(Z > z_{obs} + \frac{\mu_0 - \mu}{s/\sqrt{n}} \; for \; \mu < \mu_0\bigg).$$

If the left-hand side of the above equation (i.e. the P-value of 6.6) is small then the right-hand side (the P-value when $\mu < \mu_0$) will be small too, and may even be smaller.

> **Research hypothesis**: $\mu < \mu_0$

The null hypothesis in this case would be:

$$\text{H: } \mu = \mu_0, \text{ or H: } \mu \geq \mu_0. \tag{6.8}$$

Once again one form may better describe the problem at hand than the other. For the purpose of calculating the P-value, I will again assume that the null hypothesis is in the form H: $\mu = \mu_0$.

You would reject H in favor of the research hypothesis if the observed sample mean $\bar{x}_{obs}$ is substantially smaller than μ_0. This is equivalent to saying that z_{obs} is negative (and large in absolute value), or that $-z_{obs}$ is a large positive value. The magnitude of $-z_{obs}$ is evaluated by how often it will be exceeded by z. Hence,

$$\boxed{\text{P-value} = P\big(z > -z_{obs}\big)} \tag{6.9}$$

This P-value is obtained using MS Excel as follows:

> Excel 2010: $= 1\text{-NORM·S·DIST}(-z_{obs}, \text{TRUE})$
>
> Excel 2007: $= 1\text{-NORMSDIST}(-z_{obs})$

(6.10)

Example 6.1

A Toyota dealer recommends that new cars be brought in for the first checkup at 3,000 miles. To verify that new Toyota purchasers comply with this recommendation, the company selects 60 recently purchased cars for analysis. This analysis reveals a sample average mileage for initial servicing of 3,250 and a standard deviation of 247 miles. Can you conclude that new Toyota car owners comply with the recommendation?

If μ is the mean mileage at the initial servicing of the population of new Toyota cars sold by the Toyota dealer under study, then you want to see if $\mu \neq 3000$ (since it is supposed to be $\mu = 3000$). Therefore the null hypothesis to be tested is the following:

$$\text{H:}\ \mu = 3,000.$$

▶ Sample size: $n = 60$.

▶ Hypothetical population mean: $\mu_0 = 3,000$.

▶ Sample mean: $\bar{x} = 3,250$ miles

▶ Sample standard deviation: $s = 247$ miles

The sample of 60 cars used in this study is sufficiently large (i.e. it exceeds 30). Therefore, the z statistic 6.1 follows the standard Normal distribution. The Excel expressions 6.4 can be used to compute the P-value. The observed value of z is given by:

$$z_{obs} = \frac{\bar{x}_{obs} - \mu_0}{s/\sqrt{n}} = \frac{3250 - 3000}{247/\sqrt{60}} = 7.84.$$

The P-value is 4.44E − 15 (which for all practical purposes is 0) calculated with Excel 2010 as follows: = 2*(1-NORM.S.DIST(ABS(7.84),TRUE)) . With Excel 2007, this is obtained by, = 2*(1-NORMDIST(ABS(7.84)))

Consequently, the evidence in the sample is strongly in favor of the research hypothesis that new Toyota cars are not brought in for initial servicing at 3,000 miles.

6.2.2 Small Sample Size ($n \leq 30$)

When the sample size is small, the method needed to compute the P-value depends on the general form of the probability law associated with the analytical variable x. If the probability distribution of the variable x is approximately Normal, you should use the t-test, otherwise use the Wilcoxon's signed-rank test. These 2 tests will now be discussed.

1 The t-Test

If the probability distribution of x is believed to be reasonably close to the Normal distribution[3], then the z statistic (6.1) follows the Student distribution with $n - 1$ degrees of freedom. The Student law of probability can then be used to compute the P-value.

Research Hypothesis: $\mu \neq \mu_0$

The P-value in this case is still defined as in equation 6.3 with the exception that this time, the Normal distribution is replaced with the Student T distribution with $n - 1$

---

[3]One simple way to check Normality is to draw a histogram of the data, and to see if it has a bell shape or a shape close to it. There are some Normality tests, which in my opinion should be left to professional statisticians.

degrees of freedom. This P-value would be obtained from Excel as follows:

$$
\begin{aligned}
\text{Excel 2010: } &= 2*(1\text{-T.DIST}(\text{ABS}(z_{obs}), n-1, \text{TRUE})) \\
\text{Excel 2007: } &= 2*\text{TDIST}(\text{ABS}(z_{obs}), n-1, 1) \\
\text{or, } &= \text{TDIST}(\text{ABS}(z_{obs}), n-1, 2)
\end{aligned}
\tag{6.11}
$$

Research Hypothesis: $\mu > \mu_0$

The P-value in this case is still defined as in equation 6.6 except that the Normal distribution is replaced with the Student T distribution with $n-1$ degrees of freedom. This P-value is calculated with Excel as follows:

$$
\begin{aligned}
\text{Excel 2010: } &= 1\text{-T.DIST}(\text{ABS}(z_{obs}), n-1, \text{TRUE}) \\
\text{Excel 2007: } &= \text{TDIST}(\text{ABS}(z_{obs}), n-1, 1)
\end{aligned}
\tag{6.12}
$$

Research Hypothesis: $\mu < \mu_0$

The P-value in this case is still defined as in equation 6.9 although the Normal distribution must be replaced with the Student T distribution with $n-1$ degrees of freedom. This P-value is calculated as follows:

$$
\begin{aligned}
\text{Excel 2010: } &= 1\text{-T.DIST}(\text{ABS}(-z_{obs}), n-1, \text{TRUE}) \\
\text{Excel 2007: } &= \text{TDIST}(\text{ABS}(-z_{obs}), n-1, 1)
\end{aligned}
\tag{6.13}
$$

2 The Wilcoxon's Signed-Rank Test

Suppose now that the law of probability of X is believed to be non-normal (perhaps because of its heavy tails). If you still want to test a hypothesis about the magnitude of μ then you will have to assume that the non-normal distribution of X is at least symmetric. Why? Because the mean of a symmetric distribution is known to be its middle point, just like the median (both quantities are identical for symmetric distributions). For non-symmetric distributions however, the (population) mean becomes an elusive parameter that represents something you do not know. In this situation what you may want to do is test the population median $\tilde{\mu}$ (read mu tilde), by comparing it to a hypothesized population median $\tilde{\mu}_0$.

When the law of probability of the variable X is symmetric and non-Normal, the P-value is calculated using the *Wilcoxon Signed-Rank test* statistic. This test belongs to the class of methods known in the statistical literature as non-parametric[4] methods.

The Method

Let us assume that you have collected n data points $x_1, \cdots, x_n$ to be used for testing a research hypothesis. If μ_0 is the hypothetical mean to which the true population mean μ is being compared then you need to implement the following steps:

(a) Compute the n differences $x_1 - \mu_0, \cdots, x_n - \mu_0$. Some of these differences may be positive, and others.

(b) Compute the absolute values of the differences of step (a).

---

[4]The term "non-parametric" in this context simply means that no predefined law of probability depending on specific parameters is used in the method.

(c) Rank all absolute differences of step (b) in ascending order from 1 to n.

(d) Then sum only the ranks associated with the positive differences, and call that sum S^+. This random variable is the Wilcoxon signed-rank test statistic needed to compute the P-value.

The way to compute the P-value depends as always on the research hypothesis, and is shown in Table 6.1.

Table 6.1 : P-values for the Wilcoxon's Signed-Rank Test

Research Hypothesis	P-Value
$\mu \neq \mu_0$	$P(S^+ \geq s_{obs}^+) + P(S^+ \leq n(n+1)/2 - s_{obs}^+)$
$\mu > \mu_0$	$P(S^+ \geq s_{obs}^+)$
$\mu < \mu_0$	$P(S^+ \geq n(n+1)/2 - s_{obs}^+)$

As usual, the smaller the P-value the stronger the evidence in the sample in favor of the research hypothesis. To understand why this procedure works, consider a research hypothesis of the form $\mu > \mu_0$. The corresponding P-value will be small if the observed S^+ denoted s_{obs}^+ is large. And s_{obs}^+ can have a large value only if the number of observations that exceeds μ_0 (i.e. $x - \mu_0 > 0$) substantially outnumber those that are smaller than μ_0. This can happen frequently if $\mu > \mu_0$.

You should also note that $n(n+1)/2$ is the maximum value that s_{obs}^+ can take. It represents the sum $1 + 2 + \cdots + n$. Consequently, the quantity $n(n+1)/2 - s_{obs}^+$ is very large only when s_{obs}^+ is very small. And a small s_{obs}^+ is an indication that the number of observations that exceed μ_0 is outnumbered by the number of observations smaller than μ_0. This will be the case when $\mu < \mu_0$, justifying the formulation of the P-value for the research hypothesis $\mu < \mu_0$.

After calculating s_{obs}^+, the P-values are obtained using one of the Tables A.1 through A.5 of Appendix A.

Example 6.2

The average rate of electrical consumption of households in a given city was 25.45 Kilowatt-hours per day in the month of January. You want to know whether that consumption went down in the month of March of the same year. A sample of 8 households revealed the following average rates of electrical consumption in kilowatt-hours in the month of March:

Household	1	2	3	4	5	6	7	8
KWH	17.00	27.12	22.56	30.44	20.24	32.92	13.76	18.84

The research question here is to determine whether this data provides a strong enough evidence in favor of a decrease in electricity consumption for the month March.

Solution:

Let μ be the mean electricity consumption in March, and $\mu_0 = 25.45$ the hypothetical mean. The research hypothesis in this problem is $\mu < 25.45$ while the null hypothesis associated with this problem is defined as H: $\mu = 25.45$. The implementation of the Wilcoxon signed-rank test is described in Table 6.2.

Table 6.2 : Calculation of s_{obs}^+

x	$x - 25.45$	$|x - 25.45|$	Rank	Signed Rank
17.00	-8.45	8.45	7	-7
27.12	1.67	1.67	1	[1]
22.56	-2.89	2.89	2	-2
30.44	4.99	4.99	3	[3]
20.24	-5.21	5.21	4	-4
32.92	7.47	7.47	6	[6]
13.76	-11.69	11.69	8	-8
18.84	-6.61	6.61	5	-5
s_{obs}^+				10

It follows from Table 6.1 that for the type of research hypothesis used in this example, the P-value is calculated as,

$$P\left(S^+ \geq n(n+1)/2 - s^+_{obs}\right),$$

where $n = 8$ (the sample size), and $s^+_{obs} = 10$ (as calculated in Table 6.2). Consequently, the P-value is calculated as $P(S^+ \geq 26)$ (note that $26 = 8 \times (8+1)/2 - 10$). This probability is obtained from Table A.1 using the column labeled as $n = 8$ and the c value of 26. It follows that P-value $= 0.16$, which is very large, indicating that the evidence in favor of the research hypothesis ($\mu < 25.45$) is very weak. Therefore, you cannot reject the null hypothesis despite the fact that $\bar{x}_{obs} = 22.86$. The small sample size of 8 used here may have played an important role in the non-rejection of the null hypothesis.

The Wilcoxon Signed-Rank tables of Appendix A can only provide P-values for the sample sizes of 3 through 26. If your study uses a sample size of 27 or higher, then instead of using the test statistic S^+, I suggest to use the z statistic defined as follows:

$$z = \frac{S^+ - n(n+1)/4}{\sqrt{n(n+1)(2n+1)/24}}. \tag{6.14}$$

The law of probability associated with the Z statistic of equation 6.14 is approximately the standard Normal distribution. The P-values based of this Z statistic can be obtained from MS Excel as shown in section 6.2.1.

6.3. Neyman-Pearson Decision Approach

In this section, I will present the Neyman-Pearson five-step procedure for testing the population mean μ of a measurement variable x. As a second objective, I will also show how the important and often neglected power analysis of the tests can be conducted. The particular procedure to be used depends upon the size of the sample at hand. The situations to be discusses are those when the sample size n is large (i.e. n exceeds 30), and when thee sample size is small with n being below 30.

The main difference between the large-sample procedures and the small-sample procedures lies in the form taken by the test statistic used to formulate the decision rule. As far as hypothesis testing is concerned, life is always easier when the sample size exceeds 30. In this situation, the law of probability associated with the test statistic is the standard Normal distribution. For small samples, the situation gets more complex with a possible use of the Student distribution, and non-parametric methods.

6.3.1 Large Sample Size ($n > 30$)

Five-Step Hypothesis Testing Procedure

Step 1. Definition of Hypotheses

The test and alternative hypotheses H_0 and H_a will be defined here. Your pair of hypotheses will be one of the following three:

Two-tailed test: $H_0 : \mu = \mu_0$ and $H_a : \mu \neq \mu_0$,
Right-tailed test: $H_0 : \mu \leq \mu_0$, and $H_a : \mu > \mu_0$,
Left-tailed test: $H_0 : \mu \geq \mu_0$, and $H_a : \mu < \mu_0$.

Note that the test hypothesis H_0 for the right-tailed as well as for the left-tailed test could also be $\mu = \mu_0$ as previously discussed.

Step 2. The Significance Level

This significance level will be determined by the researcher, as it represents the threshold that the probability of Type I error must not exceed. A value commonly used by practitioners is $\alpha = 0.05$.

Step 3. The Test Statistic

The *Test Statistic* is a summary statistical measure used to evaluate the discrepancy between the data and the test hypothesis. In this case, it will be the same z statistic of equation 6.1.

Step 4. Decision Rule

The decision rule specifies the threshold, which defines the range of values of z (i.e. the Critical Region) that will cause the rejection of the test hypothesis H_0. This ranges of values is determined in such a way that the Type I error probability does not exceed the significance α of step 2. Let c_α be determined so that $P(z > c_\alpha) = \alpha$. The decision rule is formulated as follows:

▶ If H_a: $\mu > \mu_0$ then **Reject** H_0 if $z_{obs} > c_\alpha$
▶ If H_a: $\mu < \mu_0$ then **Reject** H_0 if $z_{obs} < -c_\alpha$
▶ If H_a: $\mu \neq \mu_0$ then **Reject** H_0 if $|z_{obs}| > c_{\alpha/2}$.

The critical value c_α is calculated with MS Excel as follows:

▶ Excel 2010: c_α =NORM.S.INV(1-α)
▶ Excel 2007: c_α =NORMSINV(1-α)

Step 5. Final Decision

This step consists of actually computing the observed test statistic z_{obs} using the data gathered, and applying the decision rule of step 4 before deciding whether or not the null hypothesis must be rejected.

Example 6.3 _____

Let us consider the Toyota dealer problem previously described in Example 6.1. The problem is to determine whether owners of new Toyota cars comply with the manufacturer's recommendation that requires the first checkup to take place at 3,000 miles. To answer this question, you may choose to consider $\mu \neq 3,000$ as the research hypothesis. As a car dealer, you have a number of ways for defining your test hypothesis, each depending on your business objective.

▶ Suppose you want to avoid spending time and money unnecessarily reminding your customers to have their first car checkup at 3,000 miles when they are already inclined to comply without a reminder. Then you must protect yourself against reaching the conclusion that new car owners do not bring their cars for service at 3,000 when in reality they do. Then your test hypothesis should be H_0:$\mu = 3,000$.

▶ On the other hand, you may find it more problematic to conclude that car owners do their first checkup at 3,000 or earlier when in reality they have been exceeding that limit regularly. In this case your test hypothesis should be H_0:$\mu \geq 3,000$ (i.e. if H_0 is true then you want to eventually reject it, only with a small probability).

For this example, I will assume that you want to avoid having to deal with a problem that does not exist.

$$H_0: \ \mu = 3,000 \quad \textit{(The Test Hypothesis)},$$
$$H_a: \ \mu \neq 3,000 \quad \textit{(The Research Hypothesis)},$$

Therefore, it will be a two-tailed test that will be implemented.
(2) The Significance Level
I assume here that you want the likelihood for mistakenly rejecting H_0 when it is true, not to exceed $\alpha = 5\%$.
(3) The Test Hypothesis

The sample size in this example is $n = 60$, and is sufficiently large for the Central Limit Theorem to apply. Therefore the test statistic is z of equation 6.1

(4) Decision Rule

It follows from step (4) of the five-step procedure of section 6.3.1 that for a two-tailed test (i.e. alternative hypothesis is in the form $\mu \neq \mu_0$), you will need the critical value $c_{\alpha/2}$ with $\alpha = 0.05$. Using Excel 2007, you would compute $c_{\alpha/2} = c_{0.05/2} = 1.96$ =NORMSINV(1-0.05/2) . With Excel 2010, you would obtain the same result by typing =NORMSINV(1-0.05/2) . Here is the decision rule:

Reject H_0 if the absolute value of z_{obs} exceeds 1.96.

(5) Final Decision

It follows from example 6.1 that $z_{obs} = 7.84$, which exceeds the critical value 1.96. Therefore, the test hypothesis H_0 must be rejected. Consequently, the data collected provide sufficient evidence that new car owners do not follow the recommendation to have their first checkup at 3,000 miles.

In Example 6.3, you have rejected the test hypothesis, and concluded that car owners do not have their first checkup at 3,000 miles as recommended. There are nevertheless quite a bit of issues that can be debated here before you as manager can use this analysis as basis for making management decisions. The data indicate that the average mileage of new cars at their first checkup is 3,250, which is only 250 miles over the recommended limit. Should you consider this a serious enough problem that warrants action? Maybe not. Even when the data has already been collected, I would recommend conducting a Power Analysis to determine how the true value of μ affects the propensity of the test to reject H_0 for the sample size used.

Power Analysis of the z-Test

In previous chapters, I defined the Test Power as being the probability of rightfully rejecting the test hypothesis H_0 when it is indeed false. When H_0 is false however, which amounts to saying that H_a is true, then the population mean μ may take many different values, each of which leading to a different value of the test power. Figure 6.1 depicts the power of the test used in Example 6.3 as a function of the "true" mean mileage of cars coming to the dealership for the first checkup.

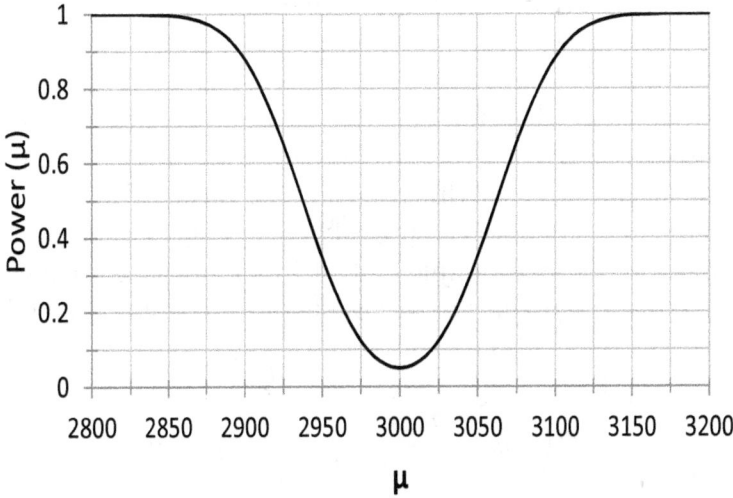

Figure 6.1. Power curve for testing $H_0{:}\mu = 3{,}000$ vs $H_a{:}\mu \neq 3{,}000$ ($n = 60$, $\alpha = 0.05$)

The startling fact about Figure 6.1 is that even if the "true" mean mileage departs from the hypothetical value $\mu_0 = 3{,}000$ by a mere 100 miles, the test used in Example 6.3 will reject the test hypothesis with a probability that exceeds 0.90. That is, more than 90% of all possible samples of 60 new car owners will lead to the rejection of the test hypothesis. You will certainly not want your test to be so powerful if you do not see drivers exceeding the recommended 3,000-mile limit by only 100 miles as a serious concern. If you are to redo this study in light of what you know now, I would recommend choosing a sample size that will reject the test hypothesis with a 90% power only when μ exceeds the hypothetical mean by 250 miles or more for example[5]. To determine the appropriate sample size, you need to know first how the Power curve is constructed.

How to Construct the Power Curve?

I assume here that your hypothesis is being tested at a significance level α. Let c_α be defined so that $P(z > c_\alpha) = \alpha$ (z is a random variable whose law of probability is the Standard Normal distribution). I showed previously how this number is easily calculated with Excel. Moreover, n is the sample size, s the standard deviation, and μ_0 the hypothetical mean. For the sake of simplicity, let us denote the probability $P(z \leq a)$ and $\Phi(a)$. Now the power of the one-sample test of hypothesis for means when the true mean is μ and the sample large is calculated as follows:

$$\text{Power}(\mu) = \begin{cases} 1 - \Phi\left(c_\alpha - \dfrac{\mu - \mu_0}{s/\sqrt{n}}\right) & \text{if } H_a : \mu > \mu_0, \\[2mm] \Phi\left(-c_\alpha + \dfrac{\mu_0 - \mu}{s/\sqrt{n}}\right) & \text{if } H_a : \mu < \mu_0, \\[2mm] 1 - \Phi\left(c_{\alpha/2} + \dfrac{\mu_0 - \mu}{s/\sqrt{n}}\right) + \Phi\left(-c_{\alpha/2} + \dfrac{\mu_0 - \mu}{s/\sqrt{n}}\right) \\ \qquad \text{if } H_a : \mu \neq \mu_0. \end{cases}$$

In the above expression, $c_{\alpha/2}$ is implicitly defined as $P(z > c_{\alpha/2}) = \alpha/2$.

To construct the power curve of Figure 6.1, I used the third

---

[5]This assumes there is an expert opinion suggesting that driving a new car 250 miles or more over the 3,000 mile limit without the first service may put the engine at risk.

of the 3 power functions shown above since the test is two-tailed (i.e. $H_a : \mu \neq \mu_0$). For values of μ varying from 2,800 to 3,200 by an increment of 50, I calculated the corresponding Power function with Excel as follows:

▶ For Excel 2007

$$=1\text{-NORMSDIST}(c_{\alpha/2}+(\mu_0 - \mu)/(s/\text{SQRT}(n)))+$$
$$\text{NORMSDIST}(-c_{\alpha/2}+(\mu_0 - \mu)/(s/\text{SQRT}(n)))$$

▶ For Excel 2010

$$=1\text{-NORM.S.DIST}(c_{\alpha/2}+(\mu_0 - \mu)/(s/\text{SQRT}(n)),\text{TRUE})+$$
$$\text{NORM.S.DIST}(-c_{\alpha/2}+(\mu_0 - \mu)/(s/\text{SQRT}(n)),\text{TRUE})$$

The Minimum Detectable Difference

If you have a specific difference $\mu - \mu_0$ between the true mean and the hypothetical one, that you would like to detect with a given power, you can calculate the corresponding sample size n that will make that detection happen. This is achieved by using the following equation:

$$n = \begin{cases} \left(s\dfrac{c_\alpha + c_\beta}{\mu_0 - \mu} \right)^2, & \text{for a one-tailed test,} \\ \left(s\dfrac{c_{\alpha/2} + c_\beta}{\mu_0 - \mu} \right)^2 & \text{for a two-tailed test.} \end{cases} \qquad (6.15)$$

where s is the standard deviation (which must be known at least approximately), c_β is implicitly defined as $P(z > c_\beta) = 1 - \text{Power}$[6]. In equation 6.15, the one-tailed test could be either the left-tailed or the right-tailed test.

---

[6]Note that 1-Power represents the Type II error, and is often denoted in the statistical literature by the Greek character β; hence the notation c_β

Example 6.4

Suppose now that in Example 6.3, you want to be able to reject the test hypothesis H_0 at the 5% significance level, and with a power of 0.90 when the difference in absolute value between μ and μ_0 is 250 miles or more. You want to know what sample size n will achieve this goal.

I assume that the standard deviation $s = 247$ is the same, as well as the significance level $\alpha = 0.05$. Since the desired power is 0.90, $P(z > c_\beta) = 1 - 0.90 = 0.10$ or $P(z \le c_\beta) = 0.90$. Using Excel, I can calculate $c_\beta = 1.282$ =NORMSINV(0.9) (Excel 2007) or =NORM.S.INV(0.9) (Excel 2010). You saw in Example 6.3 that $c_{\alpha/2} = 1.96$. Since the test of Example 6.3 is two-tailed, I will use the second equation of 6.15 to compute the required sample size as follows:

$$ n = \left(247 \times \frac{1.96 + 1.282}{250} \right)^2 = 10. $$

That is, a small sample of 10 new cars is sufficient to detect with a 90% power, a departure from the recommended mileage $\mu_0 = 3{,}000$ that exceeds 250 miles. This is due to the standard deviation of 247 miles, which indicates only a small variation in the data. The sample size of 10 calculated here is valid only under the assumption that the law of probability of the z statistic will remain approximately Normal even with such a small sample size.

The purpose of this example was to illustrate how the power analysis can be used to design the statistical test that fits our practical needs. Although the Fisher's P-value approach provides a quick way for calculating the strength of evidence in the sample, it clearly does not provide anything comparable to the power analysis, and is not part of a formal system of statistical inference.

In this subsection, I have discussed the Neyman-Pearson method for testing a population mean when the sample size if sufficiently large to ensure the applicability of the Central Limit Theorem, and the validity of the Normal distribution as the law of probability associated with the test statistic. I now like to turn my attention to the case where the sample size is too small to meet the conditions of the Central Limit Theorem. The problem now is to find a test statistic, and the associated law of probability.

6.3.2 Small Sample Size ($n < 30$) with Approximately Normal Distribution

As seen in section 6.2, when the sample size is small (i.e. smaller than 30) the method for testing the population mean depends on whether the law probability associated with the variable under investigation is at least approximately Normal, or is downright non-Normal. Note that in either case, the law of probability is assume to be symmetric[7].

Approximately Normal Distribution

The five-step hypothesis testing procedure that is applicable here is similar to the that of sub-section 6.3.1, except for the steps 3 and 4 that are related to the test statistic, and the decision rule. These are the only 2 steps I will describe here.

---

[7]In case of asymmetry, you should test the median not the mean. Asymmetry can take too many different forms that affect the location of the mean, making it impossible to develop a small-sample test

Step 3. Test Statistic

The test statistic is defined by,

$$t = \frac{\overline{x} - \mu_0}{s/\sqrt{n}}. \tag{6.16}$$

This t statistic has a form that is identical to that of the z statistic of equation 6.1. Because the sample size in this case is small, the law of probability associated with the t statistic is the Student distribution with $n - 1$ degrees of freedom, and not the standard Normal distribution.

Let α be the significance level of the test, and $t_\alpha(n - 1)$ the critical value implicitly defined as $P[t > t_\alpha(n - 1)] = \alpha$, where t follows the Student distribution with $n - 1$ degrees of freedom.

Step 4. Decision Rule

▶ If H_a: $\mu > \mu_0$ then **Reject** H_0 if $t_{obs} > t_\alpha(n - 1)$
▶ If H_a: $\mu < \mu_0$ then **Reject** H_0 if $t_{obs} < -t_\alpha(n - 1)$
▶ If H_a: $\mu \neq \mu_0$ then **Reject** H_0 if $|t_{obs}| > t_{\alpha/2}(n - 1)$.

For a given significance level α and sample size n, the critical value $t_\alpha(n - 1)$ can be be obtained with MS Excel as follows:

$$\begin{array}{l} \text{Excel 2010: } t_\alpha(n - 1) \text{ =T.INV(1-}\alpha\text{,}n - 1) \\ \text{Excel 2007: } t_\alpha(n - 1) \text{ =TINV(2} * \alpha\text{,}n - 1) \end{array} \tag{6.17}$$

Example 6.5

A car manufacturer is advertising a new car model for its average gasoline mileage performance of 30 mpg[8]. To verify this claim, 10 drivers using each a car of the same model, conduct a performance test over the same distance. The test reveals the following miles per gallon values for the 10 cars: 24.58, 31.47, 28.42, 27.85, 30.45, 26.33, 33.42, 29.82, 31.22, 26.74. The research question is whether the sample data support the manufacturer's claim.

Solution:

The parameter that you are interested in is,

$$\mu = \text{``true''} \text{ average miles per gallon of the test car.}$$

Step 1: The hypotheses

One way to approach this problem, is to consider that as a car manufacturer, you do not want your test to suggest that $\mu < 30$ when in reality $\mu \geq 30$. This would mean being deceptive to your customers with the risk of losing their business overtime. Therefore, this mistake will be your Type I error. Hence, the 2 hypotheses:

$$H_0: \mu \geq 30, \text{ versus } H_1: \mu < 30.$$

Step 2: The significance level

I will assume that this test will be conducted at the 5% significance level. That is, $\alpha = 5\%$.

Step 3: The test statistic

The sample size representing the number of cars used for testing is $n = 10$. This number is too small for the Central Limit

---

[8]mpg is an abbreviation of Miles Per Gallon, which corresponds approximately to 0.4251 Kilometers Per Liter

Theorem to be used. Instead of using the z statistic, I will use the t statistic (I am clearly assuming here without verification that the number of miles per gallon follows approximately the Normal distribution).

Step 4: The decision rule

Since this is a left-tailed test, the critical value is $-t_\alpha(n-1)$, which is in fact $-t_{0.05}(9)$. Using one of Excel expressions (6.17), you will obtain $-t_{0.05}(9) = -1.833 = \texttt{-TINV(2*0.05,10-1)}$ (using Excel 2007). Therefore, the decision rule is,

$$\text{Reject } H_0 \text{ if } t_{obs} < -1.833$$

Step 5: Final decision

In order to compute the observed value t_{obs} of the t statistic, you need to compute the sample mean $\overline{x}$, and the sample standard deviation s from the 10 mileage per gallon values. These values are $\overline{x} = 29.03$, and $s = 2.724$. Therefore,

$$t_{obs} = \frac{\overline{x} - \mu_0}{s/\sqrt{n}} = \frac{29.03 - 30}{2.724/\sqrt{10}} = -1.126.$$

It appears that $t_{obs} = -1.126$, a value that exceeds the critical value of -1.833. According to the decision rule of step 4, you cannot reject the test hypothesis H_0.

The Power Analysis of the t-Test

In Example 6.5 the test hypothesis H_0 that the mean mileage per gallon is 30 or over was not rejected due to insufficient evidence in the sample of 10 cars. Ending the hypothesis testing activity at this stage will make it incomplete. I am then going to study the power of this test for various values of the true mean μ. This entails evaluating the probability to reject H_0 when the alternative hypothesis H_a is true.

When the alternative hypothesis H_a is true, the law of probability associated with the t-statistic of equation 6.16, is the noncentral t distribution[9] with $n - 1$ degrees of freedom, and a noncentrality parameter δ defined as follows:

$$\delta = \sqrt{n}\,\frac{\mu - \mu_0}{s}, \tag{6.18}$$

where s is the standard deviation. Consequently, for a given sample size n, and a given significance level, the test power will be a function of the standardized mean difference $d = (\mu - \mu_0)/s$.

Let $F_{\nu,\delta}(x)$ be the cumulative probability distribution of the noncentral t distribution with ν degrees of freedom, and noncentrality parameter δ. One can prove that

$$\mathsf{Power}(d) = \begin{cases} 1 - F_{\nu,|\delta|}(t_{\alpha,\nu}), & \textit{for one-tailed tests,} \\ 1 - F_{\nu,|\delta|}(t_{\alpha,\nu}) + F_{\nu,|\delta|}(-t_{\alpha,\nu}), & \textit{for} \\ & \quad \textit{2-tailed tests,} \end{cases} \tag{6.19}$$

where $\delta = \sqrt{n}\,d$, and $\nu = n - 1$. This equation comes from the fact that a noncentral t distribution with ν degrees of freedom, and δ as noncentrality parameter satisfies the following equation:

$$F_{\nu,\delta}(-t_{\alpha,\nu}) = 1 - F_{\nu,-\delta}(t_{\alpha,\nu}) \tag{6.20}$$

In order to create the Power curve, you need to compute $\mathsf{Power}(d)$ for various values of d. This requires being able to compute $F_{\nu,\delta}(t_{\alpha,\nu})$ for various values of δ. Unfortunately, this cannot be done in MS Excel. The noncentral Student's distribution is not implemented in Excel 2007 nor in Excel 2010. Therefore, it is necessary to use a statistical package to accomplish this task. I have used the R package for this purpose. R calculates

---

[9]This probability distribution is briefly discussed in section 2.3.8 of chapter 2.

$t_{\alpha,\nu}$ as `t.alpha <- qt(1-`α`,`ν`)`, while $F_{\nu,|\delta|}(t_{\alpha,\nu})$ is obtained as `f <- pt(t.alpha,`ν`,`δ`)`.

The R script in Program 6.1 produces the Power values as a function of d as shown in Table 6.3. The standardized mean difference d varies from 0 to 3.5, while corresponding power values vary from 0.05 (the significance level) to 1. This table provides all the values needed to create the Power curve of Figure 6.2.

```
01   alpha <- 0.05
02   n <- 10
03   t.alpha <- qt(1-alpha,n-1)
04   d <- seq(0,3.5,0.25)
05   power.d <- 1-pt(t.alpha,n-1,sqrt(n)*d)
06   d.power<-cbind(d,power.d)
07   d.power
08   write.csv(d.power, file =
"C :/Users/OWNER/Desktop/power.csv", row.names
= FALSE)
```

Program 6.1. R Script for Power Calculation

Note that in Example 6.5, $d = (\mu - \mu_0)/s = (\mu - 30)/2.724$. That is, $\mu = 2.724d + 30$. Figure 6.2 shows that the test used will reject the test hypothesis H_0 with a power of 0.90 or more only when d exceeds 1 in absolute value, which corresponds to $d < -1$ since the test is left-tailed. But $d < -1$ means that $\mu < 27.28$ (obtained as $30 - 2.724$). Therefore, the non-rejection of the test hypothesis is a strong indication that the true value of μ is likely to be larger than 27.28.

Table 6.3: Power of Example 6.5's Test as a function of the absolute value of the standardized mean difference d

	$|d|$	power.d
[1,]	0.00	0.0500000
[2,]	0.25	0.1806979
[3,]	0.50	0.4272898
[4,]	0.75	0.7065914
[5,]	1.00	0.8975170
[6,]	1.25	0.9766499
[7,]	1.50	0.9966153
[8,]	1.75	0.9996921
[9,]	2.00	0.9999825
[10,]	2.25	0.9999994
[11,]	2.50	1.0000000
[12,]	2.75	1.0000000
[13,]	3.00	1.0000000
[14,]	3.25	1.0000000
[15,]	3.50	1.0000000

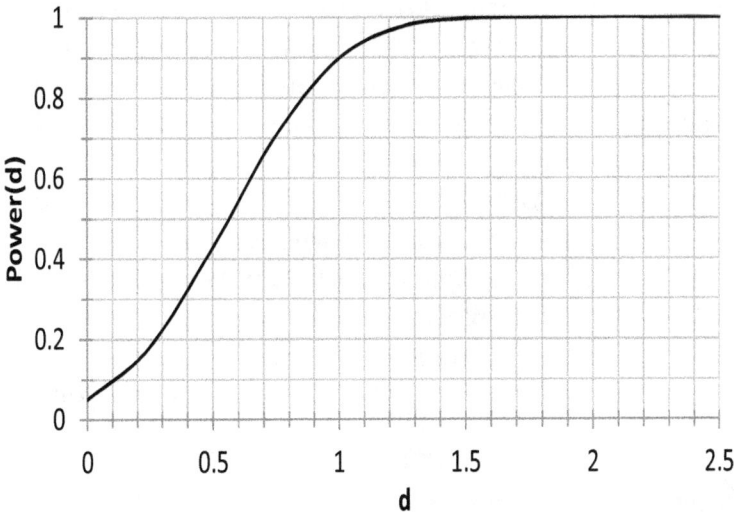

Figure 6.2. Power curve of Example 6.5

Sample Size Calculation for Study Planning

As previously mentioned, there are times when you want to know what sample size you should use when testing a hypothesis. You can solve this problem only if you take the time to think about what you want to accomplish with your analysis. The sample size can be determined if you know all of the following 3 elements:

▶ The significance level α of your test.

▶ The minimum difference between μ and μ_0 that you want to be able to detect with high probability.

▶ The probability (or power) P with which you want to reject the test hypothesis when the minimum detectable difference is reached or exceeded.

There is no close mathematical expression for computing the sample size n once these 3 elements are at your disposal. What you need to do is use the family of power curves associated with the significance level α you have chosen. Figure 6.3 shows an example of family of power curves associated with a significance level of 5%. Although all curves of Figure 6.3 are based on $\alpha = 5\%$, each curve is associated with a specific sample size n (note that $n = df + 1$, where df is the number of degrees of freedom). The intersection of the desired power P, and the minimum detectable difference d, will identify a single curve on Figure 6.3. This curve will be associated with the sample size n you need in your study.

Other families of power curves can be found in Appendix B for other significance levels and test types.

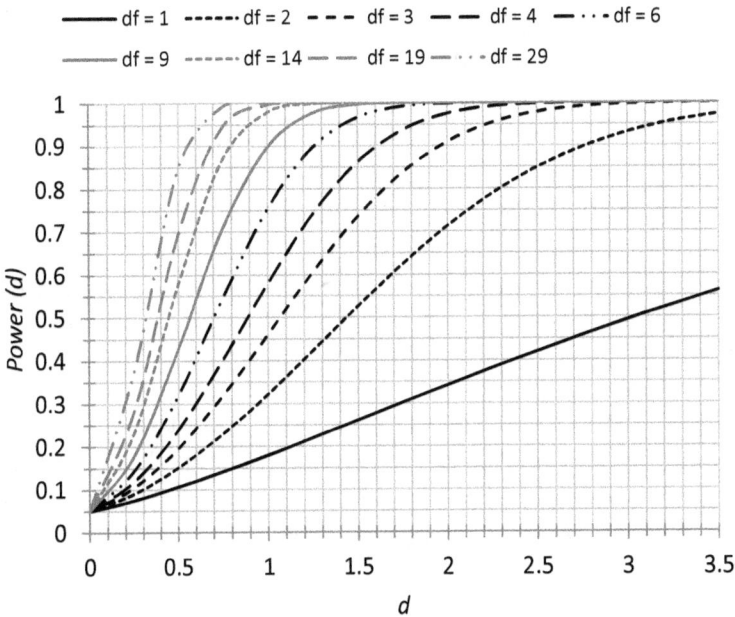

Figure 6.3. *t*-test Power curves

Example 6.6

Suppose that in Example 6.5, you want to be able to reject the test hypothesis with a power of 0.80 if the true mean μ smaller than 30 by more than 2.0 mpg, using a 5% significance level. I assume that the standard deviation is still $s = 2.724$ (perhaps obtained from a previous study).

To calculate the sample size n required to achieve the goal stated above, you need to proceed as follows:

▶ Compute $d =$ Minimum Difference$/s = 2.0/2.724 = 0.73$.

▶ Using Figure 6.3, locate the power curve that is closet to the point (0.73, 0.80). In this case, it will be the curve associated with the number of degrees of freedom $\mathsf{df} = 14$, which corresponds to a sample size $n = 15$ since $df = n-1$.

6.3.3   Small Sample Size ($n < 30$) with Symmetric Non-normal Distribution

In this section, I will address the problem of testing a population mean μ when the number of observations is small, and the law of probability associated with the variable under investigation is symmetric but non-normal. Due to lack of normality nor approximate normality, the t-test of section 6.3.2 cannot be used. The recommended solution is to use the Wilcoxon's Signed-Ranked Test of the family of nonparametric tests.

The five-step hypothesis testing procedure to be used is similar to that of sub-section 6.3.1, except for steps 3 and 4 that are related to the test statistic, and the decision rule. Again, these are the only 2 steps I will describe here.

Step 3. Test Statistic

> The test statistic needed here is the S^+ statistic. Suppose $x_1, \cdots, x_n$ are the n observations being analyzed, μ_0 the hypothetical mean, and r_i the rank of the absolute value of the difference $x_i - \mu_0$ (i.e. $r_i = \mathsf{Rank}(|x_i - \mu_0|)$. The S^+ statistic as defined in section 6.2.2 represents the sum of the r_i's for all observations for which the difference $x_i - \mu_0$ is positive.

To formulate the decision rule of step 4, you need to know the law of probability associated with the S^+ statistic. The cumulative distribution function of this statistic is tabulated in a series of tables in Appendix A, which will be used in step 4.

Let c_α be the number implicitly defined by $P(S^+ \geq c_\alpha) = \alpha$. This means that $c_{\alpha/2}$ will satisfy the equation $P(S^+ \geq c_{\alpha/2}) = \alpha/2$. If the α represents the significance level, then the decision rule of the Wilcoxon's signed-rank test is formulated as follows:

$\boxed{\textbf{Step 4. Decision Rule}}$

- ▶ If H_a: $\mu > \mu_0$ then Reject H_0 if $s^+_{obs} \geq c_\alpha$
- ▶ If H_a: $\mu < \mu_0$ then Reject H_0 if $s^+_{obs} \leq n(n+1)/2 - c_\alpha$
- ▶ If H_a: $\mu \neq \mu_0$ then Reject H_0 if $s^+_{obs} \geq c_{\alpha/2}$ or $S^+ \leq n(n+1)/2 - c_{\alpha/2}$.

Example 6.7 ———————————————————

Consider the data of Example 6.5, and suppose that the law of probability miles per gallon data is symmetric and substantially different from the normal distribution. (1) Therefore, you decide to test the hypothesis H_0: $\mu \geq 30$ versus H_1: $\mu < 30$ using the Wilcoxon signed-rank test.
(2) The significance level of the test is assumed to be $\alpha = 0.05$
(3) The test statistic as indicated earlier is S^+.
(4) It follows from step (1) that this is a left-tailed test. Consequently, the test hypothesis is rejected if the observed value s^+_{obs} of the S^+ statistic is smaller than $n(n+1)/2 - c_\alpha = 10(10+1)/2 - 44 = 11$ (note that $c_\alpha = 44$, and this value is obtained from the "n=10" column of Table A.2, and represents the value that corresponds to $P = 0.05$).
(5) Finally, you need to compute s^+_{obs} in order to apply the decision rule of step 4. This calculation is done in Table 6.4. It follows from Table 6.4 that $s^+_{obs} = 17$, and this value exceeds the critical value 11 of step 4. Consequently you cannot reject the test hypothesis H_0, which confirms the conclusion previously reached with the t-test.

The first column of Table 6.4 is the series of raw observations. The second column contains the deviations from the hypothetical mean μ_0, while column (3) shows their absolute values. Column (4) contains the ranks associated with column-three values. Finally column (5) is similar to column (4) with the only exception that the ranks associated with negative deviations are replaced with zeros. The sum of all column-five ranks yields the observed value s_{obs}^+ of the S^+ statistic.

Table 6.4: Calculation of s_{obs}^+ based on the 10 observations of Example 6.5

x (1)	$x - \mu_0$ (2)	$\lvert x - \mu_0 \rvert$ (3)	r_i (4)	r_i^+ (5)
24.58	-5.42	5.42	10	0
31.47	1.47	1.47	4	4
28.42	-1.58	1.58	5	0
27.85	-2.15	2.15	6	0
30.45	0.45	0.45	2	2
26.33	-3.67	3.67	9	0
33.42	3.42	3.42	8	8
29.82	-0.18	0.18	1	0
31.22	1.22	1.22	3	3
26.74	-3.26	3.26	7	0
s_{obs}^+				17

You must have noticed by now that I am a passionate proponent of power analysis, since it always provides further insight into the testing procedure. However, conducting the power analysis with the signed-rank is a very complicated problem that does not have an easy answer. This is due to the difficulty to determine the law of probability associated with the S^+ statistic when the test hypothesis H_0 is false. For that reason, I will not address this problem here. If doing a power in this situation is essential for what you are doing, I advise to seek help from a professional statistician.

CHAPTER 7

Two-Sample Test of Hypothesis for Proportions

OBJECTIVE

This chapter discusses the statistical test of hypothesis involving two population proportions. Various large-sample and small-sample testing procedures are presented for dependent and for independent samples.

CONTENTS

7.1. Comparing Two Proportions: Introduction

You will encounter the two-sample test of hypothesis for proportions in situations where two proportions associated with two populations, or with the same population on two occasions (or two points in time) must be compared. In the latter case, you may still consider the problem as being that involving two populations, the population in time 2 being different from the same population in time 1. If you are comparing two means of a measurement variable, then you should read chapter 8, which is devoted to that topic. Proportions represent the relative number of individuals that bear a particular trait that you want to investigate (e.g. the relative number of car owners in the US).

When studying two distinct populations, it is common practice to select 2 samples (independently) from each population, and to use a testing procedure that are valid for independent samples. It will be a *"two-sample test of hypothesis for proportions based on independent samples."* The study of a single population on two occasions often requires the use of a single sample to observe change from time 1 to time 2. The test procedures recommended in this case must be valid for dependent samples. It will be a *"two-sample test of hypothesis for proportions based on dependent samples."*

7.1.1 Two-sample Test of Hypothesis for Proportions based on Independent Samples

Consider the following statement "The proportion of unlisted telephone numbers in the US has risen nearly 27 percent in four years, from 22 percent of households in 1984 to 28 percent in 1988." There are two distinct populations in this problem, which are the population of 1984 telephone numbers and that of 1988 numbers. One way to test this claim would be to have two random samples of telephone numbers selected independently from the 1984 and the 1988 populations. Such samples may be organized as shown in Table 7.1, where samples of 500 and 600 telephone numbers selected from 1984 and 1988 lists are distributed according as they are listed or unlisted.

Table 7.1 : 1984 and 1988 Distributions of Listed and Unlisted Numbers in the US

Samples	Unlisted Numbers	Listed Numbers	Sample Size
1984 Sample	100	400	500
1988 Sample	170	430	600
Total	270	830	1, 100

The 2 proportions that you will primarily be interested in from Table 7.1 are $p_{1984} = 100/270 = 37.04\%$ and $p_{1988} = 170/270 = 62.96\%$ representing the proportions of unlisted telephone numbers in 1984 and 1988 respectively.

In general, the data used for testing hypotheses in problems involving two independent samples will be organized as shown in Table 7.2, where N_1 and N_2 represent the number of individuals in samples 1 and 2 respectively. Of these subjects, n_1 and n_2 of samples 1 and 2 respectively present the trait you are interested in, for a total of $n = n_1 + n_2$ subjects bearing the trait for the pooled sample. Moreover, the two proportions of interest associated with samples 1 and sample 2 will be denoted by p_1 and p_2 respectively, where $p_1 = n_1/n$ and $p_2 = n_2/n$. Table 7.3 is the representation of Table 7.2 in the form of column percentages.

Table 7.2 : Distribution of Populations 1 & 2's Samples by Trait Presence

Samples	Individuals With Trait	Individuals Without Trait	Sample Size
Sample 1	n_1	$N_1 - n_1$	N_1
Sample 2	n_2	$N_2 - n_2$	N_2
Total	n	$N - n$	N

Table 7.3 : Percentages of Sample Units Conditionally on the Trait Presence and Absence

Samples	Individuals With Trait	Individuals Without Trait	Sample Size
Sample 1	p_1	g_1	P_1
Sample 2	p_2	g_2	P_2
Total	1	1	1

The two populations considered will be labeled as 1 and 2. Therefore, the two population parameters under comparison are defined as follows:

$\pi_1 = $*Proportion of Population 1 Subjects who Possess the Characteristic of Interest,*

$\pi_2 = $*Proportion of Population 2 Subjects who Possess the Characteristic of Interest.*

The numbers 1 and 2 to designate the two populations represent a generic notation that you may want to change to adapt to your specific context. If for example you are comparing the 1984 and the 1988 populations of telephone numbers as mentioned earlier, you may want to use π_{1984} and π_{1988}, which are more descriptive.

In the beginning of this section, I gave the example of proportion of unlisted telephone numbers that increased by 6 percentage points in 4 years. The test hypothesis suitable for this problem would be $H_0: \pi_{1988} = \pi_{1984}$, and the research hypothesis would be $\pi_{1988} > \pi_{1984}$, which you may want to reject with a high power if π_{1988} exceeds π_{1984} by 6% or more. The more general formulation of the null hypothesis for 2 arbitrary populations 1 and 2 would be $H_0: \pi_1 = \pi_2$. The statistical procedure for testing this hypothesis will be based on the following z statistic:

$$z = \frac{p_1 - p_2}{\sqrt{\dfrac{p_1 q_1}{n_1} + \dfrac{p_2 q_2}{n_2}}}, \tag{7.1}$$

where $q_1 = 1 - p_1$ and $q_2 = 1 - p_2$.

7.1.2 Two-sample Test of Hypothesis for Proportions based on Dependent Samples

Table 7.4 shows the distributions of 100 US residents by their attitude concerning the use (or abuse) of animals in medical research. The attitudes of these individuals were observed before and after the viewing of a TV series highly critical of the use of animals in research. The purpose of the experiment is to see how the TV series could influence public opinion in the US.

Table 7.4 : Distribution of 100 Subjects by their Opinion on the use of Animals in Research Before and After the Test

		After Test		
		Against	In favor of	Row Total
Before Test	Against	10	⑬	23
	In favor of	㊋ 41	36	77
	Column Total	51	49	100

In other words, you want to see if the TV series has affected the percentage of the US population that is against the use of animals in research. Note that this question can be answered simply by looking at the 2 numbers circled in Table 7.4. For the percentage of interest to remain the same, these 2 numbers must be identical. Therefore, the **McNemar Chi-Square Test** (c.f. McNemar, 1947) to be used for dependent samples, aims to test the equality of these 2 numbers. A good reference of McNemar's test is Fleiss et al. (2003, Chap. 13). In the example of Table 7.4, it appears that the TV series has dramatically increased the number of subjects against the use of animals in research. The reason why these circled cells are used as opposed to marginals was dictated by the availability of the law of probability for the test statistic.

More generally, the observations associated with dependent samples will be organized as shown in Table 7.5, where subjects may change group from one point in time to the next. For example 100 children in need for mental health services at various degrees could be classified into 2 categories ("Extensive Support", "No to Limited Support") at the time they enter a government assistance program (i.e. time 1). A follow-up evaluation in time 2 will be carried out to determine the program's success rate. To test the 2 circled numbers b and c, you will compute the associated proportions $p_b = b/(b+c)$, and $p_c = c/(b+c)$, upon which the **McNemar Chi-Square Test** statistic is based. It is when the difference between these 2 proportions exceeds a certain threshold that you will be in the position to reject the hypothesis of equality.

Note that the hypothesis testing procedures for independent as well as for dependent samples will have two versions, which are the large-sample version and the small-sample version. These issues will be discussed in greater details in subsequent sections. In section 7.2, I will discuss about the testing of hypotheses using the Fisher's P-value approach, while section 7.3 will be devoted to the test of hypothesis using the Neyman-Pearson approach.

Table 7.5 : Distribution of Populations 1 & 2's Samples by Group and Time Point

		Occasion 2		
		Group 1	Group 2	Row Total
Occasion 1	Group 1	a	$\boxed{b}$	$a+b$
	Group 2	$\boxed{c}$	d	$c+d$
	Column Total	$a+c$	$b+d$	n

7.2. Fisher's Null Hypothesis Approach

The Fisher's P-value approach to hypothesis testing in this case, will be based on a null hypothesis in the form,

$$H_0: \pi_1 = \pi_2, \qquad (7.2)$$

which will also represent the test hypothesis when using the Neyman-Pearson decision approach. As mentioned in previous chapters, the test hypothesis H_0 may take other forms such as $\pi_1 \leq \pi_2$ or $\pi_1 \geq \pi_2$. In either case, the testing procedure will still be based on the same test statistic associated with the hypothesis of equation 7.2.

What criterion should you use for deciding whether or not hypothesis 7.2 must be rejected? I now want to discuss in broad terms the logic behind the tests presented in this chapter. More formal procedures are described in subsequent sections. Since the population proportions π_1 and π_2 are attributes of their respective populations, the truthfulness of hypothesis 7.2 cannot be ascertained simply by looking at the samples described in Tables 7.2 and 7.3. However, you can observe the sample proportion difference $p_1 - p_2$. If this difference is "large" to the point of exceeding the magnitude you would expect the error margin alone to create under the conditions of the hypothesis, then you could safely conclude that this hypothesis must be rejected. For reasons extensively discussed in previous chapters, the empirical difference $p_1 - p_2$ will not be practical since its law of probability depends upon the unknown quantities π_1 and π_2. The way to resolve this problem depends on the size of both samples, and whether these samples are dependent or not.

7.2.1 Fisher's Approach for Large Independent Samples: The z-Test

For large independent samples[1], the test statistic is defined as,

$$z = \frac{p_1 - p_2}{\sqrt{pq(1/n_1 + 1/n_2)}}, \tag{7.3}$$

where p is given by,

$$p = \frac{n_1}{n_1 + n_2}p_1 + \frac{n_2}{n_1 + n_2}p_2, \tag{7.4}$$

and $q = 1 - p$. When hypothesis 7.2 is true, and both samples are large, then z is known to follow the Standard Normal distribution. With the law of probability of z known, qualifying the magnitude of its observed value z_{obs} for deciding about the rejection of the test hypothesis becomes possible. The denominator of z in equation 7.3 represents the standard error of the difference $p_1 - p_2$ under the assumption that hypothesis 7.2 is true. Note that if you assume that $\pi_1 = \pi_2$, then both proportions have a common value π that is best estimated by combining both

---

[1]Both samples are large and independent if (1) $n_1 p_1 > 5$, (2) $n_1(1-p_1) > 5$, (3) $n_2 p_2 > 5$, and (4) $n_2(1 - p_2) > 5$ and the samples are selected independently from one another.

samples to obtain the pooled proportion estimate p of equation 7.4.

The P-value for testing the null hypothesis 7.2 depends on the research hypothesis, and is described in Table 7.6.

Table 7.6 : P-Value Calculation in Two-Sample Hypothesis Testing for Proportions

Research Hypothesis	P-value	Excel 2010 Formula[a]
$H_a: \pi_1 > \pi_2$	$P(z > z_{obs})$	$=$1-NORM.S.DIST(z_{obs},TRUE)
$H_a: \pi_1 < \pi_2$	$P(z \leq z_{obs})$	$=$NORM.S.DIST(z_{obs},TRUE)
$H_a: \pi_1 \neq \pi_2$	$2P(z > |z_{obs}|)$	$=$2*(1-NORM.S.DIST(z_{obs},TRUE))

[a]For equivalent instructions with Excel 2007, R, and OpenOffice Calc see Appendix C

Example 7.1 _____

A US nationwide sample of Republicans and Democrats was asked as part of a comprehensive survey, whether they favored lowering environmental standards. The results are summarized in Table 7.7.

Table 7.7 : Distribution of voters by political affiliation & opinion

Affiliation	In Favor	Sample Size
Republicans	200	1,000
Democrats	168	800
Total	368	1,800

Can you conclude that there is a larger proportion of Democrats in favor of lowering the standards? Determine the p-value.

Solution:

Note that in the question *"Can you conclude that there is a larger proportion of Democrats in favor?"* the word proportion refers to the population proportion π_d of democrats and the population proportion π_r of republicans. Therefore, the research hypothesis is formulated as $\pi_d > \pi_r$, which leads to the following null hypothesis:

$$H_0: \pi_d = \pi_r.$$

The observed proportions based of democratic and republican samples described in Table 7.7 are given by $p_{r(obs)} = 200/1000 = 20\%$ and $p_{d(obs)} = 168/800 = 21\%$. The corresponding sample sizes are given by $n_r = 1,000$ for the number of republicans, and $n_d = 800$ for the number of democrats.

To compute the P-value, I will first compute the observed value of z based on equation 7.3. If I assume that the null hypothesis is true, then the proportion of voters in favor of lowering environmental standards is the same for democrats as well as for republicans and the whole population (democrats and republican combined). Consequently using the whole population to compute the common proportion π will yield a more precise estimation. The common proportion p of voters in favor is given by (see 7.4),

$$p = (n_r p_r + n_d p_d)/(n_r + n_d),$$
$$= (1,000 \times 0.2 + 800 \times 0.21)/(1,000 + 800) = 0.2044.$$

The observed value of z can now be calculated as follows:

$$z_{obs} = (p_{d(obs)} - p_{r(obs)})/\sqrt{pq(1/n_d + 1/n_r)},$$
$$= (0.21 - 0.20)/\sqrt{0.2044 \times 0.7956(1/800 + 1/1,000)},$$
$$= 0.5228.$$

Note that $0.7956 = 1 - 0.2044$. The research hypothesis in this problem is $\pi_d > \pi_r$, therefore it is a right-tailed test and it follows from Table 7.6 that the P-value can be obtained from

Excel 2010 as 0.300 =1-NORM.S.DIST(0.5228,TRUE). This P-value is very large, indicating weak evidence in the sample in favor of the research hypothesis.

7.2.2 Fisher's Approach for Small Independent Samples: The Fisher-Irwin Exact Test

For small samples (i.e. the conditions $n_1 p_1 > 5$, $n_1(1 - p_1) > 5$, $n_2 p_2 > 5$, and $n_2(1 - p_2) > 5$ are not all met), the z statistic cannot be used primarily because the law of probability associated to it will no longer be the Standard Normal distribution. Instead, it is the exact law of probability of the difference $p_1 - p_2$ that is used. The resulting procedure is known as the **Fisher exact test**, or the **Fisher-Irwin test**. I found it very convenient to describe the logic of this test with a simple example before generalizing it.

Table 7.8 contains data about 17 seeds of 2 types that were planted, and the number of those that germinated within five weeks recorded. The research problem is to see whether type 2 seeds tend to germinate more than type 1 seeds.

Table 7.8: Distribution of seeds by type and germination status *(Row Percentages are in Parentheses)*

Seed Type	Germinated	Failed to Germinate	Total
1	3 *(0.30)*	7 *(0.70)*	10 *(1.0)*
2	5 *(0.71)*	2 *(0.29)*	7 *(1.0)*
Total	8 *(0.47)*	9 *(0.53)*	17 *(1.0)*

Table 7.8 features 2 populations, which are the population of type 1 seeds, and the population of type 2 seeds. The observations indicate that while 71% (i.e. $p_{2(obs)} = 0.71$) of type 2

seeds germinated, only 30% (i.e. $p_{1obs} = 0.30$) of type 1 seeds did. Therefore, the observations are consistent with the research hypothesis that $\pi_2 > \pi_1$ where π_1 and π_2 represent respectively the proportions of type 1 and type 2 seeds that germinated. *If the observations are inconsistent with the research hypothesis, then the null hypothesis cannot be rejected in favor of this research hypothesis.*

Once the data is deemed consistent with the research hypothesis, the question to be asked becomes whether the observed difference $p_{2obs} - p_{1obs} = 0.71 - 0.30 = 0.41$ is sufficiently large for you to reject the null hypothesis H_0: $\pi_1 = \pi_2$ in favor of the research hypothesis $\pi_1 < \pi_2$. To answer this question, I need to evaluate the probability to have the difference $p_2 - p_1$ larger than or equal its observed value 0.41 based on the current experiment, under the fixed condition that the marginal totals[2] remain unchanged. Now, it is essential to realize that if you keep marginal totals fixed, the only other ways you can obtain $p_2 - p_1$ equal or larger than 0.41 is in the following 2 scenarios described in Table 7.9:

Table 7.9: Distribution of seeds by type and germination status

Seed Type	G[a]	No G[b]	Total		Seed Type	G	No G	Total
1	2	8	10		1	1	9	10
2	6	1	7		2	7	0	7
Total	8	9	17		Total	8	9	17

[a]G: Seed Germinated
[b]G: Seed Failed to Germinate

To define the P-value, let **D** be a random variable with the

---

[2]Table 7.8's marginal totals are the 2 column totals 8, and 9; and the the 2 row totals 10 and 7.

same law of probability as $p_2 - p_1$. Therefore, the P-value is defined as follows:

$$\text{P-value} = P\big(\mathsf{D} \geq p_{2(obs)} - p_{1(obs)} \,|\, (10,7); (8,9)\big), \qquad (7.5)$$

which represents the probability for D to exceed its currently observed value given the fixed row and column marginal totals $(10,7)$ and $(8,9)$. This amounts to calculating the probability of observing any of the 3 tables shown in Table 7.8 or Table 7.9 if tables are generated randomly with the constraint that the row and column marginals are $(10,7)$ and $(8,9)$ respectively. It follows from Table 7.9 that when the marginal totals are fixed, the knowledge of the first table cell (i.e. circled number) determines the configuration of the whole table. Consequently, for the Table 7.8 example, the P-value is given by:

$$\text{P-value} = P(1) + P(2) + P(3), \qquad (7.6)$$

where $P(2)$ for example represents the probability for 2 seeds of type 1 to germinate. I will now show how these probabilities can be calculated.

Consider the following experiment:

▶ You have at your disposal a list of m subjects, n of which are known to belong to group 1. Necessarily $m - n$ of them will belong to group 2.

▶ The random experiment consists of selecting a random sample of m_1 subjects out of the initial m subjects, and to categorize them into population 1. The remaining $m_2 = m - m_1$ subjects will be categorized into population 2. With this information, Table 7.10 can be completed.

▶ The problem now is to compute the probability that you obtain the m subjects distributed as shown in contingency table 7.10

Table 7.10: Distribution of m subjects by Group and Population

Population	Group 1	Group 2	Total[a]
1	n_1	$m_1 - n_1$	m_1
2	n_2	$m_2 - n_2$	m_2
Total	n	$m - n$	m

[a]Note that $n = n_1 + n_2$, and $m = m_1 + m_2$

Note that n_1 in this experiment is a random variable that follows the hypergeometric distribution $\mathcal{H}(m_1, n, m)$, where m is the total number of units, n is the number of units of interest (or number of successes), and m_1 the sample size of the Hypergeometric experiment. The probability of observing n_1 is defined as follows:

$$P(n_1) = \frac{\binom{n}{n_1}\binom{m-n}{m_1 - n_1}}{\binom{m}{m_1}}, \qquad (7.7)$$

where $\binom{n}{n_1}$ is the number of combinations of n_1 items out of a total of n items. Note that equation 7.7 is the Probability Mass Function of the the Hypergeometric distribution.

To compute the P-value, you need to proceed as follows:

(a) Use Table 7.11 to determine the specific probability (or probabilities) to compute based on your research hypothesis.

(b) For each probability to compute, identify all values of n_1 that will meet the condition defining the event whose probability must be computed. Then compute and sum all the $P(n_1)$ as shown in equation 7.7. For example, if you must compute $P(D \geq D_{obs})$ then there is a set of values for n_1 for which $D \geq D_{obs}$ that you must identify.

In practice, you should not have to evaluate equation 7.7 manually, since this probability is easily obtained with MS Excel, OpenOffice Calc, or R.

▶ Excel 2007: $P(n_1)$ =HYPGEOMDIST(n_1,m_1,n,m)

▶ Excel 2010: $P(n_1)$ =HYPGEOM.DIST$(n_1,m_1,n,m,\text{FALSE})$

▶ OpenOffice Calc: $P(n_1)$ =HYPGEOMDIST$(n_1;m_1;n;m)$

▶ R: $P(n_1)$ = <-dhyper$(n_1,n,m-n,m_1)$

If you choose to use R for this task, it is essential to use the correct parameters. Note that only Excel 2010, and the R package would give you the cumulative distribution function (see Appendix C for the corresponding functions).

Table 7.11 : P-Values in Two-Sample Hypothesis Testing for Proportions & Small Samples

Alternative Hypothesis	P-value
$H_a: \pi_1 > \pi_2$	$P(D \geq D_{obs})$
$H_a: \pi_1 < \pi_2$	$P(D \leq D_{obs})$
$H_a: \pi_1 \neq \pi_2$	$P(D \geq D_{obs}) + P(D \leq -D_{obs})$

Example 7.2 ⎯⎯⎯⎯⎯⎯⎯⎯⎯⎯⎯⎯⎯⎯⎯⎯⎯⎯⎯⎯⎯⎯

Let us consider the seed germination data of Table 7.8. The two populations under investigation are those type 1 and type 2 seeds. The corresponding proportions of interest are defined as follows:

π_1 = Proportion of type 1 seeds that have germinated,

π_2 = Proportion of type 2 seeds that have germinated.

The research question is to determine whether the sample provides sufficient evidence to conclude that the two populations of seeds are different with respect to their proportions of seeds likely to germinate. That is, you like to conclude whether $\pi_1 \neq \pi_2$ or not.

Solution:

Let $D = p_2 - p_1$ where p_1 and p_2 are the pre-experimental (i.e. unobserved) sample proportions of Type 1 and Type 2 seeds that will germinate, and $p_{1(obs)}$ and $p_{2(obs)}$ the corresponding post-experimental (or observed) proportions. It follows from Table 7.8 that $D_{obs} = p_{2(obs)} - p_{1(obs)} = 0.41$. Therefore the P-value associated with the research hypothesis (i.e. $\pi_1 \neq \pi_2$) is given by (see Table 7.711):

$$\text{P-value} = P(D \geq 0.41) + P(D \leq -0.41).$$

A close look at Table 7.8 suggests the following 2 things:

▶ $D \geq 0.41$ only if $(n_1, n_2) \in \{(3, 5), (2, 6), (1, 7)\}$
▶ $D \leq -0.41$ only if $(n_1, n_2) \in \{(5, 3), (6, 2), (7, 2), (8, 0)\}$

Consequently, if H is a random variable that follows the Hypergeometric distribution $\mathcal{H}(10, 8, 17)$, then the P-value is calculated as follows:

$$\text{P-value} = P(\mathsf{H} \leq 3) + P(\mathsf{H} \geq 5),$$
$$= P(\mathsf{H} \leq 3) + 1 - P(\mathsf{H} \leq 4).$$

Using Excel 2010, this P-value can be evaluated as follows:

$$\text{P-value} = \text{HYPGEOM.DIST}(3,10,8,17,\text{TRUE})$$

$$+1-\text{HYPGEOM.DIST}(4,10,8,17,\text{TRUE}),$$

$$= 0.117 + 1 - 0.419 = 0.698.$$

This P-value is very high, suggesting weak evidence in the sample in favor of the research hypothesis that the two seed types are different with respect to their propensity for germination. Therefore, the null hypothesis of equality cannot be rejected.

7.2.3 P-Value Approach for Large Dependent Samples: The McNemar Chi-Square Test

The tests of hypothesis discussed in sections 7.2.1, and 7.2.2, require an independent selection of a random sample from each of the 2 populations under investigation. An important problem often encountered in practice involves observing the exact same population on two different occasions using a unique sample to monitor change. If you observe the same subjects at two different points in time, you will not really be observing the exact same subjects. Instead, you may be observing transformed subjects in time 2. It is in that sense that I will talk about 2 dependent samples.

Table 7.4 provides a good example of a scenario involving 2 dependent samples. Using Table 7.5, I will now describe the McNemar Chi-square test, and will illustrate it afterwards using Table 7.4 data. Let π_b be the population proportion of subjects that changed from group 1 to group 2 among those who experienced change. Likewise, π_c will be the population proportion of subjects that moved from group 2 to group 1 among those who experienced a change. The null hypothesis is defined as follows:

$$H_0: \pi_b = \pi_c. \tag{7.8}$$

The test statistic to be used is defined as follows:

$$\chi^2 = \frac{(b-c)^2}{b+c}. \tag{7.9}$$

The notation χ^2 used to designate the McNemar's test statistic is purely symbolic, and simply reflects the fact that the law of probability associated with this statistic is the **Chi-Square** distribution with one degree of freedom, when the sample is large (i.e. a, b, c, and d are all greater than or equal 5). You may well use a different symbol for χ^2. Table 7.12 shows the form of the P-value for different research hypotheses.

Table 7.12 : P-Values for the McNemar Test of Hypothesis based on Large Dependent Samples

Research Hypothesis	P-value
$\pi_b > \pi_c$	$^a P(\chi^2 \geq \chi^2_{obs})/2$
$\pi_b < \pi_c$	$^b P(\chi^2 \geq \chi^2_{obs})/2$
$\pi_b \neq \pi_c$	$P(\chi^2 \geq \chi^2_{obs})$

[a]This represents the strength of evidence in the sample only if $b > c$. The two-tailed P-value must be divided by 2, because it includes the 2 situations where $b > c$ and $b < c$.

[b]This represents the strength of evidence in the sample only if $b < c$.

Example 7.3

Consider the animal research example of Table 7.4. You want to know whether the anti-animal-research TV series has been effective in turning public opinion against the use of animals in medical research. For the TV program to be considered successful, you would expect the number of proponents of animal research turned opponents to exceed the number of opponents turned proponents, by a substantial margin. The research hypothesis is $\pi_b < \pi_c$, and the null hypothesis is given by $H_0: \pi_b = \pi_c$.

The observed value of the McNemar's chi-square test statistic is given by:

$$\chi^2_{obs} = \frac{(b-c)^2}{b+c} = \frac{(41-13)^2}{41+13} = \frac{28^2}{54} = 14.52.$$

The P-value associated with this statistic is $P(\chi^2 \geq 14.52)/2 = 0.00014/2 = 0.00007$, and is obtained from Excel 2007 as follows:

P-value $= 0.0000693$ =CHIDIST(14.52,1)/2

This P-value is very small and indicates a very strong evidence in the sample in favor of the research hypothesis $\pi_b < \pi_c$. Consequently, the null hypothesis must be rejected and the TV series critical of the use of animals in research has been a success.

7.2.4 P-Value Approach for Small Dependent Samples: The McNemar Exact Binomial Test

The McNemar's chi-square test of section 7.2.3 relies upon the assumption that the sample size is large, which means that cell counts b and c in particular are each greater than 5. If one of these cells is small then the Binomial distribution $\mathcal{B}(a + b, \pi_b)$ will provide a better approximation to the law of McNemar's test statistic χ^2, than the Chi-square distribution with 1 degree of freedom.

To see how this Exact McNemar's binomial test works, consider Table 7.4 where cell counts are $a = 10$, $b = 13$, $c = 41$, and $d = 36$. The P-value is defined as follows:

P-value $= P(p_c - p_b \geq p_{c(obs)} - p_{b(obs)}) = P(p_c - p_b \geq 0.5185)$.

Note that $p_c - p_b$ will equal or exceed 0.5185 only if c equals or exceeds 41 while the sum $b + c$ remains constant at 54. However, if the null hypothesis is true (i.e. $\pi_b = \pi_c = 0.5$ then c will follow the binomial distribution $\mathcal{B}(54, 0.5)$. Consequently, the P-value is obtained by calculating $P(c \geq 41)$, which is equivalent to $1 - P(c \geq 40)$. Using Excel 2007, this probability is obtained as follows:

P-value $=$1-BINOMDIST(40,54,0.5,TRUE) ,

$= 0.0000876.$

This P-value is reasonably close to the one calculated in section 7.2.3. However, the P-value calculated in this section is more accurate since it is based on the exact probability distribution of the cell counts.

The McNemar's test has a limitation that you should be aware of. This test is solely based on the 2 numbers 13 and 43 of Table 7.4 and ignores the other 2 numbers 10 and 36 entirely. The TV series being tested is deemed effective if the number of new opponents exceeds the number of proponents. To understand the nature of the limitation of the McNemar's test consider the data displayed in Table 7.13. A quick look at this table is sufficient to see that this TV series has not really led that many watchers to change their minds. The question then becomes to see whether there is even a need to conduct McNemar's test, which may well lead to the conclusion that the Tv series is effective, when it is pretty obvious that it is not.

Regarding the use of the McNemar test, my recommendation is for you the practitioner to decide upfront about the minimum number of subjects whose views must change before you decide whether or not a formal statistical test must be conducted. This minimum number of opinion switchers may represent for example 50% of the program participants or more. This number percentage will be left at your discretion. I advise against consider a preliminary statistical test of any kind before deciding whether the McNemar's test should be implemented or not. The overall testing procedure may not methodological sound.

Table 7.13: Distribution of 5,154 Subjects by their Opinion on Animal Research Before and After the Test

		After Test		
		Against	In favor of	Row Total
Before Test	Against	1,500	13	1,513
	In favor of	41	3,600	3,641
	Column Total	1,541	3,613	5,154

7.3. Neyman-Pearson Approach

In this section, I will discuss the Neyman-Pearson approach to the two-sample test of hypothesis for proportions. I will confine myself to the situation where the samples are reasonably large so that I can focus on exposing the basic power analysis techniques. Power analysis techniques are typically based on the assumption that the sample size will be reasonably large. Moreover, as previously indicated the main advantage of the Neyman-Pearson approach relies upon the power analysis. If you are dealing with a small sample, then I would advise to use Fisher's P-value approach of the previous section.

Throughout this section, I will consider the null hypothesis to be defined by equation 7.2 for independent samples, and by equation 7.8 for dependent samples.

7.3.1   The z-Test for Large Independent Samples

In this section, I consider the situation where you want to compare two population proportions π_1 and π_2 associated with two populations labeled as Population 1 and Population 2 as described in section 7.1.1. The testing procedure uses 2 independent samples described in Tables 7.2 and 7.3. The 5 steps of the Neyman-Pearson hypothesis testing procedure for proportions based on independent samples are the following:

1 *The Hypotheses.*

The null hypothesis H_0 and the alternative hypothesis H_a will take one of the following forms, depending on the research problem being investigated.

Two-tailed test: $H_0 : \pi_1 = \pi_2$ versus $H_a : \pi_1 \neq \pi_2$
Right-tailed test: $H_0 : \pi_1 \leq \pi_2$ versus $H_a : \pi_1 > \pi_2$
Left-tailed test: $H_0 : \pi_1 \geq \pi_2$ versus $H_a : \pi_1 < \pi_2$

As usual, it is the error you want to protect yourself against that will determine the form of the null and alternative hypotheses.

2 *The Significance Level* α.

Set the ceiling α that you do not want the type I error probability to exceed, typical values being 1%, 5%, or 10%.

3 *The Test Statistic.*

The test statistic is the z-statistic of equation 7.3

4 *The Decision Rule.*

Depending on the nature of the research hypothesis, the decision rule would formulated differently as follows:

Alternative Hypothesis	Decision Rule
Two-tailed test	Reject H_0 if $z_{obs} \leq -z_{\alpha/2}$ or $z \geq z_{\alpha/2}$
Left-tailed test	Reject H_0 if $z_{obs} \leq -z_{\alpha}$
Right-tailed test	Reject H_0 if $z_{obs} \geq z_{\alpha}$

where z_{α} is the $(1-\alpha)^{th}$ percentile of the Standard Normal distribution, also known as the *Critical Value* of the test. The critical value is calculated with Excel 2007 as follows: z_{α} =NORMSINV(1-α) or as z_{α} =NORM.S.INV(1-α) with Excel 2010.

5 *The Final Decision.*

The final decision regarding the null hypothesis is made after computing the observed test statistic z_{obs} and comparing it to the critical value.

Example 7.4 _____

Consider once again the problem described in Example 7.1 where a sample of members of the US Democratic party and a sample of members from the US Republican party are asked whether they would support lowering environmental standards. The data obtained is summarized in Table 7.7, and the problem is to compare π_d and π_r the relative number of proponents of this reform among democrats and republicans respectively. The objectives you set for this study will dictate the form of the test and the research hypotheses. I will assume that you want to prove that π_d exceeds π_r. The testing procedure will then be carried out as shown below.

Solution

- ▶ **The Hypotheses.** The null and alternative hypotheses are respectively defined as, H_0: $\pi_d = \pi_r$ versus H_a: $\pi_d > \pi_c$

- ▶ **The Significance Level.** The significance level $\alpha = 0.05$ will normally be given in the problem statement, or will be set by the researcher.

- ▶ **The Test Statistic.** The test statistic is z defined by equation 7.3 where p_1 is replaced with p_d (the sample proportion of democrats), and p_2 with p_r (the sample proportion of republicans).

- ▶ **The Decision Rule.** To formulate the decision rule, you need to first compute the critical value associated with the right-tailed test. **Step 4** of the five-step hypothesis testing procedure stipulates that the critical value in this case is $z_{0.05}$, which may be obtained from Excel 2007 as $z_{0.05} = 1.645$ =NORMSINV(1-0.05) . Consequently the decision rule is formulated as follows:

    Reject H_0 if z_{obs} exceeds 1.645

- ▶ **The Final Decision.** It follows from example 7.1 that $z_{obs} = 0.5228$ which is below the critical value of 1.645 by a substantial margin. It follows from the decision rule that the test hypothesis H_0 cannot be rejected. Consequently, there is insufficient evidence in the sample to support the claim that democrats are more in favor of lowering environmental standards than republicans. You can conclude that *based on current data, democrats are no more likely to support lowering environmental standards than republicans.* Could they bee less likely? It is possible. You and I simply do not know.

The non-rejection of the null hypothesis in example 7.4 raises some legitimate questions. Is it because the number of study participants (democrats & republicans) was too small that the observed difference was not statistically significant? Is it because one of the two groups does not support low environmental standards than the other? Or perhaps it is a combination of both factors. Note that with a larger number of study participants, any difference between parties however trivial it might be, will eventually be found statistically significant and cause the rejection of the null hypothesis. But the reality of life often leads us to only care about differences of a certain magnitude. The power analysis to be discussed next, sheds some light into some of these issues.

Power Analysis of the z-Test

Let $d = \pi_1 - \pi_2$ be the "true" difference between the two population proportions being compared. I like to show how to compute the power of the z tests, which is the probability of rejecting the null hypothesis when it is false. The two problems I like to address are the calculation of the power of a testing procedure that has already been implemented, and the sample size determination for the planning future testing procedures.

The Power of a Post-Experiment Test

If $\pi_1 = \pi_2$ then both populations share a common proportion π that is estimated from the samples by $\bar{p}$ calculated as $\bar{p} =$

$(n_1 p_1 + n_2 p_2)/(n_1 + n_2)$. It follows that $1 - \bar{p}$, denoted by $\bar{q}$ is given by $\bar{q} = (n_1 q_1 + n_2 q_2)/(n_1 + n_2)$. The standard error s of the sample difference $p_1 - p_2$ is given by:

$$s = \sqrt{\frac{p_1 q_1}{n_1} + \frac{p_2 q_2}{n_2}}. \tag{7.10}$$

Let z be an arbitrary percentile (to be specified) of the standard Normal distribution, and c_d a function of z defined as follows:

$$c_d(z) = \frac{z\sqrt{\bar{p}\,\bar{q}(1/n_1 + 1/n_2)} - d}{s}. \tag{7.11}$$

The power of the z-test is defined as follows:

Alternative Hypothesis	Power of the Test
$\pi_1 > \pi_2$	Power $= 1 - P\big(Z \leq c_d(z_\alpha)\big)$
$\pi_1 < \pi_2$	Power $= P\big(Z \leq c_d(-z_\alpha)\big)$
$\pi_1 \neq \pi_2$	Power $= 1 - P\big(Z \leq c_d(z_{\alpha/2})\big) +$ $P\big(Z \leq c_d(-z_{\alpha/2})\big)$

where Z is a variable that follows the standard Normal distribution. These equations allows you to graph the test power as a function of the difference d between the "true" population proportions π_1 and π_2.

Example 7.5

To illustrate the calculation of the power associated with a statistical test, I like to show you how to compute the power of the right-tailed test discussed in Example 7.4. Populations 1 and 2 are represented by democrats (d) and republicans (r) respectively, and the observations used in the test are from Table 7.7. It follows from Table 7.7 that $p_d = 0.21$, and $p_r = 0.20$, which leads to $\bar{p} = (168 + 200)/(800 + 1,000) = 0.2044$, and $\bar{q} = 1 - \bar{p} =$

0.7956. Moreover, the standard error s of the difference $p_d - p_r$ is given by:

$$s = \sqrt{\frac{p_d q_d}{n_d} + \frac{p_r q_r}{n_r}} = \sqrt{\frac{0.21 \times 0.79}{800} + \frac{0.2 \times 0.8}{1000}} = 0.0192.$$

Since the test being discussed is right-tailed, its power is calculated as $1 - P(Z \leq c(z_\alpha))$ where z_α is $(1 - \alpha)^{th}$ percentile of the standard Normal distribution. Since the test in Example 7.4 was conducted at the significance level of $\alpha = 0.05$, $z_\alpha = z_{0.05} = 1.645$, and using equation 7.11 one obtains $c(z_{0.05})$ as follows:

$$c_d(1.645) = \frac{1.645 \times \sqrt{0.2044 \times 0.7956(1/800 + 1/1000)} - d}{0.0192},$$

$$= (0.03147 - d)/0.0192.$$

Therefore, the power curve associated with the right-tailed of this example has the following equation:

$$\mathsf{Power}(d) = 1 - P(Z \leq c_d(z_{0.05})), \qquad (7.12)$$

where $c_d(z_{0.05}) = (0.03147 - d)/0.0192$. Equation 7.12 is then used to create Table 7.14 for different values of d. Table 7.14 in turn, is used to create the power curve of Figure 7.1. For each value of d in this table, I obtained $\mathsf{Power}(d)$ using Excel 2010 as follows =1-NORM.S.DIST(d,TRUE) .

It follows from Figure 7.1 that if indeed democrats are more supportive of the lowering of environmental standards than republicans in proportions, then a difference of 6 percentages points for example would cause the rejection of the null hypothesis with a probability that exceeds 0.9. This suggests that even if π_1 and π_2 are different, the fact that equality was not rejected in Example 7.4 suggests that this difference must be substantially smaller than 0.06.

However, a difference of 3 percentage points or less will cause that rejection with a probability that is below 0.5. In this situation, you will need considerable luck for your test to reveal such a small difference. In the end, you need to decide about the magnitude of the difference between the 2 groups you really care about. If you are a politician, then any difference however small that can win an election is welcome. On the other hand, if you are a researcher who studies phenomena, only "substantial" differences may matter to you, and this test may just be what you want.

Table 7.14 : Test Power for Various Values of d

d	Power(d)	c_d
0	0.0506	1.6391
0.01	0.1317	1.1182
0.02	0.2751	0.5974
0.03	0.4695	0.0766
0.04	0.6716	-0.4443
0.05	0.8328	-0.9651
0.06	0.9314	-1.4859
0.07	0.9776	-2.0068
0.08	0.9943	-2.5276
0.09	0.9988	-3.0484
0.1	0.9998	-3.5693

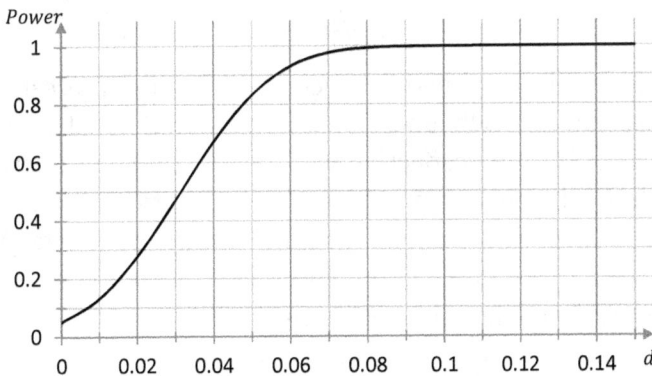

Figure 7.1. Power Curve of the Right-tailed z-test of Example 7.4

Power Analysis for Sample Size Determination

In Example 7.5, I calculated the power of the right-tailed test used in Example 7.4. In this case, the samples were already selected, and I simply illustrated the properties of the test with this powerful tool that is the power curve. Now, I want to consider the situation where you are planning a new test, and knows what magnitude the proportion difference d must have to be considered meaningful. Your problem is to determine the sizes n_1 and n_2 of the two samples you must select from the two populations.

Although there is no problem having 2 samples of different sizes (i.e. $n_1 \neq n_2$), you could well select two samples of the same size n if you are the one making that determination. This common sample size is easier to determine. If you want to select 2 samples of different sizes, then there will be numerous possible solutions that will all yield the specified power, and significance level. But a clear guidance for deciding about the best one is not available. Therefore, I am assuming that you want to select 2 samples of the same size n. This size is given by the following equation:

$$n = \frac{\left(z_\alpha \sqrt{(p_1 + p_2)(q_1 + q_2)/2} + z_\beta \sqrt{p_1 q_1 + p_2 q_2}\right)^2}{d^2}, \quad (7.13)$$

where d is the detectable difference that you are interested in, z_α and z_β are the $(1-\alpha)^{th}$ and $(1-\beta)^{th}$ percentiles of the standard Normal distribution respectively, β being the type II error (generally derived from the power using the relation $\beta = 1 - \textbf{Power}$. Note that once d is known, you only need to know p_1 since $p_2 = p_1 - d$. To recapitulate, you need the following 3 things before computing the required sample size:

▶ d, the desired detectable difference between π_1 and π_2,

▶ α the significance level at which the test will conducted,

▶ The Power that you want your test to have. This allows you to get the type II error β and z_β.

▶ The last thing you need is p_1. A priori the knowledge of p_1 seems to be a major constraint. In many applications it is not. In fact, p_1 can be seen as a baseline value against which any improvement (or progress) will be evaluated. For example, p_1 may represents the current car seat belt wear rate by drivers, and d the minimum increase in wear rate necessary to consider a new seat belt regulation successful.

To illustrate the use of equation 7.13, I created Tables 7.15 and 7.16 that show the required sample size n for difference values of d, p_1, α, and Power. More detailed tables can be found in Appendix D. It follows from these tables that the required sample size goes down with the desired power. Moreover, detecting small differences such as 0.01 with powerful tests (e.g. power=0.90) will require a large sample size. If the baseline value is 0.3 then you will need samples of size 400 in order to detect a difference of 0.09 with a power of 0.90. With a power of 0.80 (see Table 7.16) the required sample size goes down to 289.

Table 7.15 : Sample size as a function of d and p_1 for $\alpha = 0.05$ and **Power** $= 0.90$

	p_1						
d	0.05	0.1	0.3	0.5	0.7	0.9	0.95
0.01	7,359	14,724	35,619	42,813	36,305	16,094	8,900
0.02	1,642	3,505	8,815	10,699	9,157	4,190	2,413
0.03	641	1,478	3,876	4,752	4,104	1,935	1,155
0.04	310	786	2,156	2,670	2,327	1,129	695
0.05	N/A	473	1,364	1,707	1,501	747	473
0.06	N/A	308	936	1,183	1,050	536	348
0.07	N/A	211	679	868	777	406	270
0.08	N/A	149	513	663	599	320	217
0.09	N/A	108	400	522	476	260	180
0.1	N/A	79	319	422	388	216	152

Table 7.16: Sample size as a function of d and p_1 for $\alpha = 0.05$ and **Power** $= 0.80$

d	p_1						
	0.05	0.1	0.3	0.5	0.7	0.9	0.95
0.01	5,313	10,630	25,715	30,909	26,210	11,619	6,426
0.02	1,186	2,531	6,364	7,724	6,611	3,025	1,742
0.03	463	1,068	2,799	3,431	2,963	1,397	834
0.04	224	568	1,557	1,928	1,681	815	502
0.05	N/A	342	985	1,232	1,084	540	342
0.06	N/A	223	676	855	758	387	252
0.07	N/A	152	490	627	561	294	195
0.08	N/A	108	371	479	433	232	157
0.09	N/A	78	289	378	344	188	130
0.1	N/A	58	231	305	280	157	110

7.3.2   The McNemar's Test for Large Dependent Samples

In this section, I consider the situation where you want to compare population proportions using 2 dependent samples. The 5 steps of the Neyman-Pearson hypothesis testing procedure are the following:

▶ Step 1: The Hypotheses

The null hypothesis H_0 and the alternative hypothesis H_a will take one of the following forms:

$$\text{Two-tailed test:} \quad H_0 : \pi_b = \pi_c \text{ versus } H_a : \pi_b \neq \pi_c$$
$$\text{Right-tailed test:} \quad H_0 : \pi_b \leq \pi_c \text{ versus } H_a : \pi_b > \pi_c$$
$$\text{Left-tailed test:} \quad H_0 : \pi_b \geq \pi_c \text{ versus } H_a : \pi_b < \pi_c$$

▶ Step 2: The Significance Level α

Set the ceiling α that you do not want the type I error probability to exceed.

▶ Step 3: The Test Statistic The McNemar's test statistic was previously defined in equation 7.9, and is given by

$$\chi^2 = (b - c)^2/(b + c).$$

As previously mentioned, the law of probability associated with statistic is the chi-square distribution with 1 degree of freedom.

▶ Step 4: The Decision Rule

Two-tailed test: Reject H_0 if $\chi^2_{obs} > c_\alpha$ where c_α is the critical value and implicitly defined by $P(\chi^2 > c_\alpha) = \alpha$.

Right-tailed test: Reject H_0 if the following 2 conditions are satisfied:

$$\begin{cases} (a) & n_b > n_c, \\ (b) & \chi^2_{obs} > c_{\alpha/2}, \end{cases} \tag{7.14}$$

where the critical value $c_{\alpha/2}$ is implicitly defined by $P(\chi^2 > c_{\alpha/2}) = \alpha/2$.

Left-tailed test: Reject H_0 if the following 2 conditions are satisfied:

$$\begin{cases} (a) & n_b < n_c, \\ (b) & \chi^2_{obs} > c_{\alpha/2}. \end{cases} \tag{7.15}$$

The critical value of the McNemar test can be obtained from Excel 2007 as follows: c_α =CHIINV(α,1) . To compute $c_{\alpha/2}$, one simply needs to replace α in the Excel formula with $\alpha/2$.

▶ Step 5: Final Decision

In this step, you will actually compute the observed value χ^2_{obs} of the test statistic before applying the decision rule of step 4.

Example 7.6 —————————————————————————

Consider once again Table 7.4 data on the use of animals in research. The problem is to test the extent to which a TV series that is highly critical of this practice, affects the opinion of 100 human subjects. Note that your expectation of what the test might accomplish heavily depends on your own opinion on this issue.

If you are the promoter of the TV series, you will naturally want the statistical test to demonstrate that the TV series is effective. The error you do want to avoid is having to conclude that the TV series is "Ineffective" when in reality it is quite effective, and this will form the basis for defining the null hypothesis. The null and alternative hypotheses in this case would be H_0: $\pi_b \geq \pi_c$, and H_a: $\pi_b < \pi_c$. Note that the TV series is seen as ineffective if the proportion π_b of opponents turned proponents exceeds the number of proponents turned opponents after viewing the series.

On the other hand, if you are a radical opponent of the use of animals in research, you will certainly not want the statistical test to reveal that the TV series is effective, when in fact it is not. The statistical test in this case may incite you to rest on your two laurels thinking that the TV program has turned public opinion around. To limit the likelihood of this error occuring, you will consider it to be your type I error by defining your hypotheses as follows: H_0: $\pi_b \leq \pi_c$, and H_a: $\pi_b > \pi_c$.

If you are neutral by not taking side in this debate, perhaps you may simply want to see if the TV series is having any effect at all in one way or another. Your ideal test will be two-sided.

For this example, I will assume that you are neutral, and primarily interested of knowing whether the TV series has any impact at all on public opinion. Moreover, you want to avoid concluding that the series has an impact on public opinion when it has none. That is you want this to occur no more that 5% of the time.

Solution

▶ **The Hypotheses.** The null and alternative hypotheses are respectively defined as, $H_0: \pi_b = \pi_c$ versus $H_a: \pi_b \neq \pi_c$

▶ **The Significance Level.** The significance level is $\alpha = 0.05$.

▶ **The Test Statistic.** The statistics of the McNemar test is always χ^2 of equation 7.9

▶ **The Decision Rule.** To formulate the decision rule, you need to first compute the critical value associated with the two-tailed test. Step 4 of the five-step hypothesis testing procedure stipulates that the critical value in this case is $\chi^2_{0.05}$, which may be obtained from Excel 2007 as $c_{0.05} = 3.84$ =CHIINV(0.05,1) . Consequently the decision rule is formulated as follows:

$$\text{Reject } H_0 \text{ if } \chi^2_{obs} \text{ exceeds } 3.84$$

▶ **The Final Decision.** It follows from example 7.3 that $\chi^2_{obs} = 14.52$ which exceeds the critical value of 3.84 substantially. It follows from the decision rule that the test hypothesis must be rejected. Consequently, the TV series is considered efficient.

In Example 7.6, I subjectively chose to study the two-tailed test. You could well attempt to replicate this example by playing the role of the TV series producer, and conduct the one-tailed test. This should not pose any particular difficulty except that one of the equations 7.14 or 7.15 will have to be used in the formulation of the decision rule.

It follows from Table 7.4 that the number of proponents of the use of animals in research who became opponents following the broadcasting of the TV series (41) exceeds the number of opponents turned proponents (13) by a ratio of 3 to 1. But this ratio[3] (that I denote by δ) is based on the sample of 100 individuals who participated in the study, and is expected to be different from the "true" but unknown ratio based on the whole population. An important question that may be of interest to the research is how likely is this testing procedure to reject the test hypothesis for a given value of δ. If the procedure can reject with high probability, the test hypothesis H_0 for small values of δ that are meaningless to you, then rejection will not carry as much value to you than if only large and meaningful values of δ are rejected with high probability. The Probability to reject H_0 as a function of δ is what is referred to as the power curve of the McNemar test. I now show you how this curve would be obtained.

Power Analysis of the McNemar Chi-Square Test

You as researcher may have 2 motivations for conducting the power analysis for the McNemar chi-square test. Either you want to compute the power of the test you have already conducted, or you want to determine the sample size required to achieve a predetermined power prior to the data collection phase. The techniques presented in this section were previously discussed by Connett et al. (1987) and in Zar (1999, pp.171-172). You may also want to look at Daniel (1990) and Fleiss et al. (2003) for an in-depth and more technical discussion on the power of the McNemar's test.

---

[3]Note that for the two-tailed test the number that quantifies departure from the test hypothesis is the maximum of the 2 ratios b/c and c/b. In the previous example, that number is $c/b = 3.15$.

The Power of a Post-Experiment Test

I am assuming here that you have already implemented the Mc-Nemar's chi-square test, and that you want to evaluate the probability for your test to reject the null hypothesis if the difference between b and c reaches or exceeds a predetermined threshold called "Minimum Detectable Difference" (MDD), and denoted by δ. The MDD is generally expressed in the form of a ratio of the 2 numbers b and c. As an example, consider $\delta = 3$. If your test is right-tailed then $\delta = 3$ represents the situation where $b/c = 3$ (i.e. b is 3 times higher than c). For the left-tailed test the MDD represents $c/b = 3$. For the two-tailed test however, $\delta = 3$ represents one of the 2 situations $b/c = 3$ or $c/b = 3$.

Let z_α be the $(1 - \alpha)^{th}$ percentile of the standard Normal distribution, and c_α a related quantity defined as follows:

$$c_\alpha = \begin{cases} z_\alpha & \text{if the test is two-tailed,} \\ z_{\alpha/2} & \text{if the test is one-tailed.} \end{cases} \tag{7.16}$$

Consider the following quantity z_δ defined as follows:

$$z_\delta = \frac{(\delta - 1)\sqrt{np} - c_\alpha\sqrt{\delta + 1}}{\sqrt{(\delta + 1) - p(\delta - 1)^2}}, \tag{7.17}$$

where n, p, and d are defined as,

$$\begin{cases} n &= \textit{sample size,} \\ p &= \min(b/n, c/n), \\ d &= \textit{minimum detectable difference.} \end{cases}$$

The power of the McNemar's test is given by,

$$\text{Power} = P(Z \leq z_\delta), \tag{7.18}$$

where Z is a random variable that follows the standard Normal distribution.

Example 7.7 _____

I now want to obtain the power curve associated with the Mc-Nemar test of example 7.6. The test used in that example is two-tailed based on a significance level $\alpha = 0.05$. Its power curve is depicted in Figure 7.2, where the McNemar's power (on the vertical axis) is displayed as a function of the MDD δ.

Note that $\delta = 1$ refers to the test hypothesis situation where the proportion of new proponents equals the number of new opponents (i.e. the TV series is not working). In this case, the Power is 0.05, which is equivalent to the significance level (i.e. the test has almost no power). The McNemar test becomes more powerful in detecting the TV series success as δ goes up. The real question for you at this stage is to know what you care about the most. Figure 7.2 indicates that the test will detect an MDD δ of 1.75 or smaller with a low power of 0.5 or less. Therefore, by rejecting H_0, there is a good chance that the true value of δ would exceed 1.75 or even 2.0.

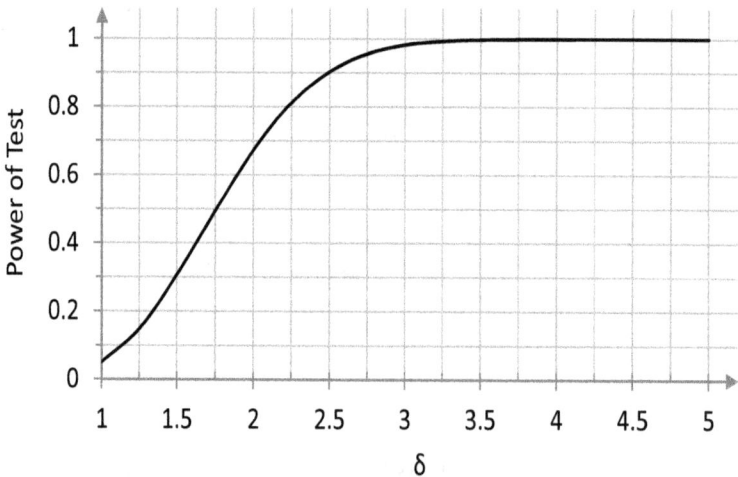

Figure 7.2. Power curve of the McNemar Test of Example 7.6

If the power curve indicates that the test may detect a small MDD of 1.2 with high probability, then you will know that a mere rejection of the null hypothesis does not indicate that the TV series has any meaningful impact.

Power Analysis for Sample Size Determination

If you want to determine at the planning stage, the sample size n that will give you the desired power for the McNemar test, you may accomplish this by using the following sample size expression:

$$n = \frac{\left(c_\alpha \sqrt{\delta + 1} + z_\beta \sqrt{(\delta + 1) - p(\delta - 1)^2}\right)^2}{p(\delta - 1)^2}, \tag{7.19}$$

where c_α is given by equation 7.16, δ is the MDD, $p = \min(b, c)/n$, and z_β is the βth percentile of the standard Normal distribution, with $\beta = 1 - Desired\ Power$.

CHAPTER 8

Two-Sample Test of Hypothesis for Means

OBJECTIVE
This chapter presents various statistical methods for testing the magnitude of the difference between two population means of a measurement variable.

CONTENTS

8.1. Comparing Two Population Means

In this chapter, I will present methods for testing the magnitude of the difference between two population means. The means I am concerned about are those of measurement variables such as income, height, or distance. In chapter 7, I discussed several methods for testing the magnitude of the difference between two population proportions. These methods will not translate to measurement variables because the laws of probability that describe the distribution of sample proportions are generally different from those that describe the distribution of sample means. This is the case even though proportions represent special means of 0-1 dichotomous variables. Therefore, I will discuss new testing methods in this chapter not previously discussed in earlier chapters.

In chapter 6, I discussed the one-sample test of hypothesis for means where one population mean μ was compared to a predetermined hypothetical value μ_0. I then used the example of a car manufacturer's claim that his truck tires would be driven $\mu_0 = 60,000$ miles before the tread wears out. The population of truck tires from that manufacturer was the only population under consideration, and the lifetime mean mileage μ of these tires, the unique parameter of interest.

The basic problem I will focus on in this chapter is a more general one. As an example, suppose that you want to determine whether Duracell batteries have a longer lifetime (in number of hours) than Eveready batteries. The variable under investigation is the battery lifetime expressed in number of hours, which is a measurement variable. The two populations of interest are the population of Duracell batteries, whose (unknown) mean lifetime is denoted by μ_D (D for Duracell) and the population of Eveready batteries whose mean lifetime is μ_E (E for Eveready).

In addition to the mean lifetime, the Duracell population has a standard deviation σ_D and the Eveready population, a standard deviation σ_E. I will need both standard deviations or their approximated values when defining the test statistics. Did I need to use standard deviations in chapter 7 to develop two-sample tests of hypothesis for proportions? The answer is yes. However this may not have been perceptible, primarily because the standard deviation of a proportion solely depends on the proportion itself. For measurement variables however, the standard deviation is a function of the mean.

Regarding the battery lifetime example of the previous paragraph, you may want to verify the claim that Duracell batteries last longer than Eveready batteries or $\mu_D > \mu_E$. You will never be able to test all Duracell and Eveready batteries available in the world at a given point in time. Therefore, you should forget about ever obtaining the exact values of the 2 means, which would have allowed you to make a direct comparison between them and reach a definitive conclusion without having to worry about statistical science. This problem will be tackled by selecting a sample of n_D Duracell batteries, and another sample of n_E Eveready batteries. From the samples selected, you will obtain the sample means $\overline{x}_D$ (from the Duracell sample), and $\overline{x}_E$ from the Eveready sample. These sample means will serve as surrogates to the true population means μ_D and μ_E that are unknown to us.

As previously mentioned, a straight comparison between the true means μ_D and μ_E (if available) is sufficient for making the final determination regarding the lifetimes of the 2 battery types. However, such a direct comparison is impossible when using the surrogates $\overline{x}_D$ and $\overline{x}_E$. In fact, each of the surrogates is subject to a sampling error, and so is their difference $D(\overline{x}_D, \overline{x}_E) = \overline{x}_D - \overline{x}_E$. Consequently, the law of probability associated with the difference is far more valuable than one observed value tied to 2 specific samples. But, it was discovered that the ratio of $D(\overline{x}_\mathrm{D}, \overline{x}_\mathrm{E})$ to its standard error, also called the standardized difference, has a known and well documented law of probability. *An observed standardized difference will be considered large, only if the probability for it to be exceeded is small.* This ratio will later in this chapter be called the test statistic. Depending on the formulation of the research hypothesis, it will be supported or not by the data according as the test statistic is large or small in a probabilistic sense.

8.2. Types of Tests

There are many different data in practice involving two populations that will call for different hypothesis testing procedures. First of all, a clear distinction must be made between *Paired data* and *Unpaired data* when conducting your study. If you want to compare the lifetime of Duracell batteries to the lifetime of Eveready batteries, then you will select two samples *independently* from each population. Because these are independent and unrelated samples, the information obtained is known as unpaired data. On the other hand, paired data is very common in many experimental situations where two measurements are obtained from the exact same group of subjects to demonstrate the effectiveness of an intervention program. For example, in order to test the effectiveness of an SAT[1] preparation program, a group of 30 students are asked to take an SAT test before and after going through the program. The 2 series of SAT scores are paired to match the same students and represent two dependent samples, which will require a special testing technique.

---

[1]The SAT (previously known as the Scholastic Assessment Test) is a globally recognized college and university admission test in the US. Applicants to US colleges or universities are often asked to provide their SAT test scores

For Unpaired data, you will need to treat the case of large samples (i.e. each of the 2 samples has 30 subjects or more) differently from the case of small samples where one of the 2 samples has fewer than 30 subjects. In the large-sample case, the law of probability associated with the test statistic is the standard Normal distribution regardless of the probability distribution of the raw data. In the case of small samples however, the test statistic will follow the Student's t distribution only if the Normal distribution provides a reasonable approximation to the law of probability associated with the raw data. If the samples are small, and the raw data is not even approximately Normal, but is reasonably symmetric at the minimum, then you will need to use the Wilcoxon Rank-Sum test also known in the literature as the Mann-Whitney test.

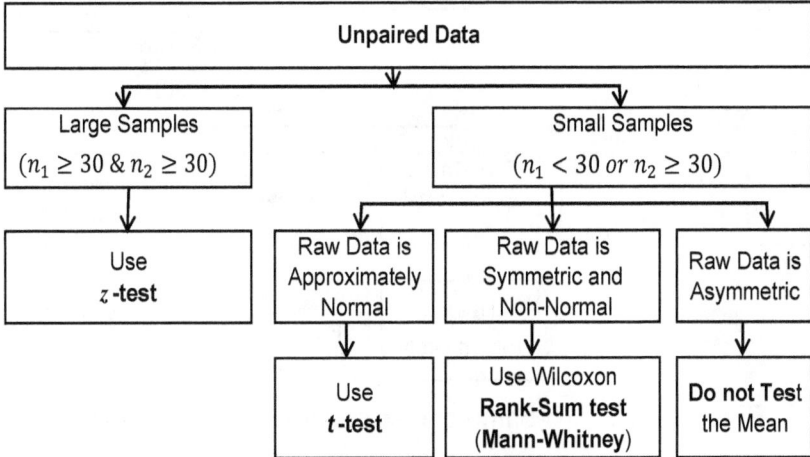

Figure 8.1. Two-Sample Test of Means for Unpaired Data

Figure 8.1 summarizes the different situations you may encounter in practice when analyzing unpaired data, and indicates which statistical technique to use in each of them.

In the next few sections, I will discuss successively the (ordinary) z-test, the t-test, and the Mann-Whitney test. When the samples are small and the underlying distribution of data known to be asymmetric, then I would recommend to test the median instead of the mean[2].

For paired data, you will be dealing with a single sample of subjects that were scored twice on two occasions. This will still require a separate treatment for large and small samples. As shown in Figure 8.2, the paired versions of the unpaired tests will be implemented. The paired z-test, the paired t-test, and the paired Wilcoxon signed-rank test will be discussed in subsequent sections of this chapter.

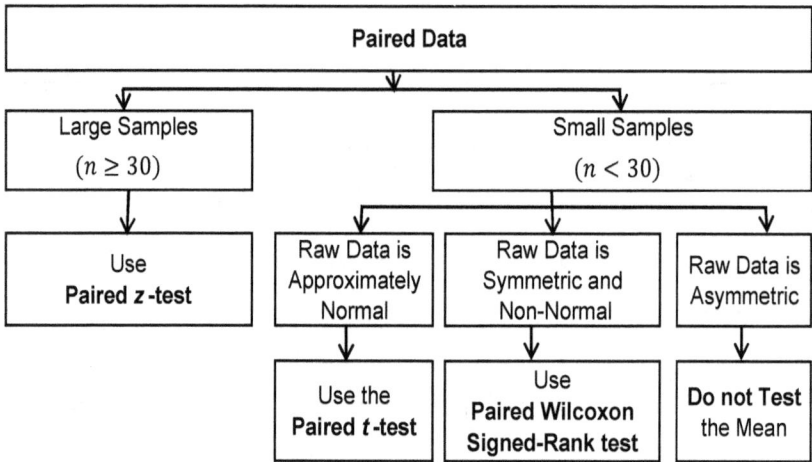

Figure 8.2. Two-Sample Test of Means for Paired Data

In sections 8.3, 8.4, 8.5, and 8.6, I will discuss the different tests presented in Figures 8.1 and 8.2 using both the P-value approach

---

[2]Note that for small samples, the sample mean will not have a symmetric distribution when the raw data do not, and will therefore be a poor surrogate to the population mean

of Fisher, and the decision-theory approach of Neyman-Pearson. Whether you decide to simply calculate the P-value associated with a particular null hypothesis following the Fisherian approach, or whether you decide you opt for the more comprehensive system of statistical inference of Neyman and Pearson, you will need the test statistic. Because of its pivotal role, I will always start each of the subsequent sections with a description of the test statistic, the associated law of probability, and its validity conditions.

Some of the test statistics discussed have unfriendly mathematical formulas. This should not deter you from using them, because many software solutions free or not are available for computing all of these statistics. I will show you with all the details how these calculations can be done.

I assume that you want to compare two means, which are μ_1 from population 1, and μ_2 from population 2 (both means may well refer to the same populations at two points in time). The null hypothesis you are dealing with has the following form:

$$H_0\colon \mu_1 - \mu_2 = \Delta_0, \qquad (8.1)$$

where Δ_0 is a hypothetical value used as reference for testing the magnitude of the difference in population means (in many applications, the value of Δ_0 will be 0). In other words, the null hypothesis assigns a hypothetical value to the magnitude of the difference in means. If the sample means $\overline{x}_1$ and $\overline{x}_2$ are used as surrogates to the unknown population means μ_1 and μ_2, then the difference $(\overline{x}_1 - \overline{x}_2) - \Delta_0$ is a statistic to look at since it gives us a sense of the consistency between the observations and the null hypothesis. In reality the raw difference $(\overline{x}_1 - \overline{x}_2) - \Delta_0$ will not be used, since its law of probability is not known. The standard deviation of that difference will be calculated, and the test statistic will be the ratio of the raw difference in means to its standard deviation. This ratio is in fact, the observed difference measured in number of standard deviations.

8.3. Large Unpaired Samples (z-test)

This section is divided into three sub-sections. The first sub-section describes the test statistic, and discuss the conditions of its validity. The second sub-section describes the test itself using both the Fisher, and the Neyman-Pearson approaches, while the third sub-section is devoted to the power analysis.

8.3.1 The z-Statistic

Let s_1^2 and s_2^2 be the sample variances, and n_1 and n_2 the sizes associated with the 2 samples selected for the study. Then the test statistic for large unpaired samples is defined as follows:

$$z = \frac{(\overline{x}_1 - \overline{x}_2) - \Delta_0}{\sqrt{s_1^2/n_1 + s_2^2/n_2}}. \tag{8.2}$$

The law of probability associated with the z statistic is the standard Normal distribution, if the validity conditions listed below are met.

Validity Conditions

(1) The data being analyzed is of interval or ratio type.

(2) Each sample is independently selected randomly from the population it represents

(3) Each of the two sample sizes n_1 and n_2 exceeds 30.

8.3.2 Description of the z-test

▶ The Fisher's P-value Approach

Let $\delta = \mu_1 - \mu_2$. The expression necessary to compute the P-value depends on the research hypothesis H_a associated with your study, as shown in Table 8.1.

Table 8.1: P-Value for testing 2 Population Means with z-test for Large Unpaired Samples - H_a *represents the research hypothesis -*

(H_a)	P-value	Excel 2010 Formula[a]
$\delta > \Delta_0$	$P(z > z_{obs})$	=1-NORM.S.DIST(z_{obs},TRUE)
$\delta < \Delta_0$	$P(z > -z_{obs})$	=1-NORM.S.DIST($-z_{obs}$,TRUE)
$\delta \neq \Delta_0$	$2P(z > |z_{obs}|)$	=2*(1-NORM.S.DIST(z_{obs},TRUE))

[a]For equivalent instructions with Excel 2007, R, and OpenOffice Calc see Appendix C

▶ The Neyman-Pearson Approach

I indicated in the previous chapters that the Neyman-Pearson approach to hypothesis testing is implemented by first identifying the error you want to protect yourself against. This error will help formulate the test hypothesis H_0, the opposite of which becoming the alternative hypothesis H_a. Although the test hypothesis can take one of the 3 forms H_0: $\mu_1 - \mu_2 = \Delta_0$, H_0: $\mu_1 - \mu_2 \geq \Delta_0$, or H_0: $\mu_1 - \mu_2 \leq \Delta_0$, the Neyman-Pearson testing procedure is always constructed using the test hypothesis H_0: $\mu_1 - \mu_2 = \Delta_0$. This is the reason why I will use the latter form. If one of the other 2 forms describes the error to protect yourself against more accurately, then the significance level α should be interpreted not as the type I error probability; instead it should be interpreted as the ceiling that cannot be exceeded by the type I error probability.

The test and alternative hypotheses you will defined based on your problem will take one of the three forms described in Table 8.2.

Table 8.2 : The Possible Hypotheses

Null Hypothesis	Alternative Hypothesis	Test Type
H_0: $\mu_1 - \mu_2 = \Delta_0$	H_a: $\mu_1 - \mu_2 \neq \Delta_0$	*(Two-tailed)*
	H_a: $\mu_1 - \mu_2 > \Delta_0$	*(Right-tailed)*
	H_a: $\mu_1 - \mu_2 < \Delta_0$	*(Left-tailed)*

The Neyman-Pearson hypothesis testing procedure is implemented as follows:

(1) The Hypotheses	H_0: $\mu_1 - \mu_2 = \Delta_0$ vs H_a: *(specify)*
(2) Significance Level	α
(3) Test Statistic	*z-statistic of equation 8.2*
(4) Decision Rule	
Alternative Hypothesis	Decision
H_a: $\mu_1 - \mu_2 \neq \Delta_0$	*Reject H_0 if $z \geq z_{\alpha/2}$ or $z \leq -z_{\alpha/2}$*
H_a: $\mu_1 - \mu_2 < \Delta_0$	*Reject H_0 if $z < -z_{\alpha}$*
H_a: $\mu_1 - \mu_2 > \Delta_0$	*Reject H_0 if $z > z_{\alpha}$*
(5) Final Decision	*Compute z_{obs} and Apply the Decision Rule*

In practice, the way you implement the z-test for large independent samples depends on the nature of your input data, and the tool you have at your disposal. There are situations in practice where your input data is in the form of summary statistics such as means, and standard deviations. Oftentimes your input data is in the form of raw data such a the hourly wage rate data of plumbers and electricians of Table 8.4. The next two examples will illustrate these two scenarios.

Example 8.1 _____

Table 8.3 contains descriptive statistics on the populations of Plumbers and Electricians based on a sample of 33 plumbers and a sample of 31 electricians. If you are going to implement a two-sample z-test for independent samples, the minimum amount of information you will need to accomplish this task is shown in Table 8.3. The two samples are independent, and the both sample sizes exceed 30. Therefore the z-test for independent samples is applicable.

Table 8.3 : Sample Statistics on Plumbers and Electricians

	Population	
Statistic Description	*Plumbers*	*Electricians*
Sample Size	33	31
Sample Mean	30.160	29.699
Standard Deviation	1.638	1.271

Although there are several ways to proceed from here depending on the software you are using, my preferred choice for such a small problem is Excel or the free OpenOffice Calc. Because I can visualize everything I do.

If you want to use Fisher's P-value approach, and that you are interested in a straight comparison between plumbers' and electricians' wages (i.e. $\Delta_0 = 0$), then your null hypothesis will be that the plumbers' hourly wage rate is identical to that of electricians. That is, H_0: $\mu_P = \mu_E$. If you suspect that plumbers might be charging higher hourly wage rates, then your research hypothesis would be H_a: $\mu_P > E$ (or $\delta = \mu_P - \mu_E > 0$). It follows from Table 8.1 that the corresponding P-value is given by the probability $P(z > z_{obs})$, where z_{obs} is obtained using equation

8.2 and Table 8.3 as follows:

$$z_{obs} = (30.160 - 29.699)/\sqrt{1.638^2/33 + 1.271^2/31} = 1.262$$

If Excel 2010 is what you want to use to compute the P-value, then Table 8.1 provides the necessary formula. That is,

P-value = 0.1035 =1-NORM.S.DIST(1.262,TRUE) .

This P-value appears high, suggesting a weak evidence in the sample to support your research hypothesis. Do not hesitate to look at the tables in Appendix C for equivalent formulas in OpenOffice Calc, Excel 2007, or an R function. For example Table C.1 shows that the Excel 2007's equivalent (NORMSDIST($\cdot$)) of the Excel 2010 function NORM.S.DIST. Table C.2 shows the OpenOffice Calc's equivalent (NORMSDIST($\cdot$)), and the R's equivalent (pnorm($\cdot$)).

Example 8.1 shows a simple way of implementing Fisher's P-value approach for testing two population means with large independent samples, when the input data is in the form of summary statistics. However, the data to be analyzed will often be in the form of a long list of raw data as in Table 8.4. I propose three options for dealing with this problem:

(a) From the raw data of Table 8.4, you may produce summary sample statistics as in Table 8.3, and use the same approach of Example 8.1 for computing the P-value. This approach will even allow you to implement the Neyman-Pearson, provided you can compute the critical values[3] z_α or $z_{\alpha/2}$)depending on whether you conduct the one-tailed test or the two-tailed test) needed in step 4. I would use either Excel of OpenOffice Calc to implement this approach.

---

[3]I showed in Chapter 7 how this can be done with Excel

(b) The second and more elegant solution can be obtained with an Excel Add-In named "Analysis ToolPak," which comes with MS Office for Windows (not for Mac). However, you need to first install the Analysis ToolPak following the instructions I provide in section E.1 of Appendix E.

The biggest advantage of using the Excel's Analysis Tool-Pak lies in its ability to assist with the implementation of both the Fisher and the Neyman-Pearson approaches to hypothesis testing. Here, Excel will automatically compute the P-value, and the critical values for the one- and two-tailed tests. It will be up to you to interpret Excel output properly. I will give an example below.

(c) The third approach I would recommend is the use of the R built-in function `t.test(x,...)`, which requires that you have R installed.

Example 8.2 _____

Let us consider the hourly wage rate data of plumbers and electricians previously analyzed in example 8.1. What is new in this example is that you must analyze the raw data given in Table 8.4. I will use the Neyman-Pearson approach to hypothesis testing here for illustration purposes.

Let us assume that you would like to protect yourself against the error of concluding that plumbers earn less than electricians, when it is the opposite that is true. Consequently, your test hypothesis is H_0: $\mu_P \leq \mu_E$. Therefore, you will be conducting a right-tailed test where the alternative hypothesis is H_a: $\mu_P > \mu_E$.

To use the two-sample z-test implemented in Excel, you are required to supply the "true" variances of the two populations under investigation. You do not have them. No panic, for large independent samples, the z-test, and the t-tests will produce the same results. Therefore you may use the two-sample t-test assuming equal[4] variances also implemented in Excel. If you are interested, please check section E.2 of appendix E to see how I used Excel 2007 and Table 8.4 data to implement the two-sample t-test with equal variances to obtain the output shown in Figure 8.3.

	D	E	F
t-Test: Two-Sample Assuming Equal Variances			
		Plumbers	Electricians
Mean		30.160303	29.699032
Variance		2.68421553	1.615589
Observations		33	31
Pooled Variance		2.16713819	
Hypothesized Mean Difference		0	
df		62	
t Stat		1.25273861	
P(T<=t) one-tail		0.10750266	
t Critical one-tail		1.66980416	
P(T<=t) two-tail		0.21500532	
t Critical two-tail		1.99897152	

Figure 8.3. Excel 2007 Output of the Two-Sample t-Test Assuming Equal Variances

Using Figure 8.3 results, here is how I resolve my hypothesis testing problem:

▶ *The Hypotheses.* H_0: $\mu_P \leq \mu_E$ versus H_0: $\mu_P > \mu_E$.

▶ *The Significance Level.* I assume that you want to test your hypotheses at the $\alpha = 5\%$ significance level.

▶ *The Test Statistic.* The test statistic in this case will naturally be the z-statistic.

▶ *The Decision Rule.* This is a right-tailed test, and the Neyman-Pearson approach suggests the following decision rule: Reject H_0 if $z_{obs} > 1.667$. Note that the critical value

---

[4]See section 8.4 for a detailed discussion on this assumption. I later recommend to use the unequal variances option whenever uncertain about the variance homogeneity assumption.

$z_{0.05} = 1.667$ comes from Figure 8.3 and is found in the row labeled as "t Critical one-tail." If you conduct a two-tailed test then you would use the critical value for two-tailed test 1.999 in the row labeled as "t Critical two-tail." (Note: for left-tailed tests, you need to add a negative sign to the critical value that Excel provides, which is always that of the right-tailed test). *Note that when you formulate your hypotheses with plumbers' population mean μ_P mentioned first, then you need to ensure that the labels Plumbers and Electricians appear in the output of Figure 8.3 in that order. That is the input column data in Excel will also be listed in that order.*

▶ *The Final Decision.* In Figure 8.3, the value of t Stat of 1.253 should be used as the observed z statistic. That is $z_{obs} = 1.253$. It is smaller than 1.67, and according to the decision rule, the test hypothesis cannot be rejected. I will discuss the other elements of Figure 8.3 in the sections devoted to t-tests.

Table 8.4 : Hourly Wage Rates of a Sample of 33 Plumbers and a Sample of 33 Electricians

Obs	Plumb	Elec	Obs	Plumb	Elec	Obs	Plumb	Elec
1	$29.80	$28.76	12	$29.35	$29.07	23	$31.15	$31.28
2	$30.32	$29.40	13	$29.42	$28.79	24	$25.85	$32.60
3	$30.57	$29.94	14	$29.78	$29.54	25	$30.55	$31.36
4	$30.04	$28.93	15	$29.60	$29.60	26	$28.95	$30.00
5	$30.09	$29.78	16	$30.60	$30.19	27	$33.75	$30.25
6	$30.02	$28.66	17	$30.79	$28.65	28	$30.79	$33.54
7	$29.60	$29.13	18	$29.14	$29.95	29	$29.14	$27.75
8	$29.63	$29.42	19	$29.91	$28.75	30	$29.91	$31.86
9	$30.17	$29.29	20	$28.74	$29.21	31	$30.82	$29.21
10	$30.81	$29.75	21	$27.36	$28.75	32	$29.95	
11	$30.09	$28.05	22	$33.60	$29.21	33	$35.00	

Before I move on, I like to address an issue of curiosity. If the *t*-test can replace the *z*-test for large samples, then what was the point developing the *z*-test, which by the way also requires a large sample? The reason is that the *z*-test is based on the Normal distribution, which is easier to handle than the Student's *t* distribution that underlies the *t*-test. This advantage was very important a few decades ago when the availability of computers was very limited. Nowadays, computers handle both distributions with ease, and you may well decide to always use the *t*-test in place of the *z*-test.

Using the R Package with Example 8.2 Data

Figure 8.4 shows the output that I obtained after using the R package to implement the two-sample right-tailed *z*-test with Table 8.4 data. *If interested, you may see all the details about the use of R to obtain this output in section F.1 of appendix F.*

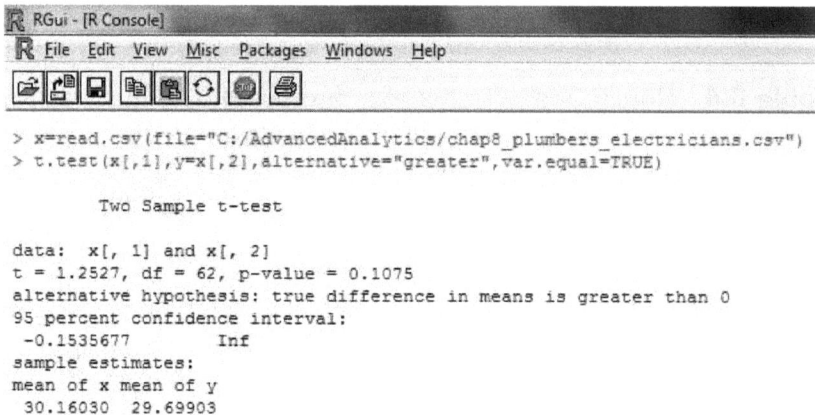

```
R RGui - [R Console]
R File  Edit  View  Misc  Packages  Windows  Help

> x=read.csv(file="C:/AdvancedAnalytics/chap8_plumbers_electricians.csv")
> t.test(x[,1],y=x[,2],alternative="greater",var.equal=TRUE)

        Two Sample t-test

data:  x[, 1] and x[, 2]
t = 1.2527, df = 62, p-value = 0.1075
alternative hypothesis: true difference in means is greater than 0
95 percent confidence interval:
 -0.1535677       Inf
sample estimates:
mean of x mean of y
 30.16030  29.69903
```

Figure 8.4. R Output for the Analysis of Example 8.2 Data

The file "chap8_plumbers_electricians.csv" is a two-column **csv** file (i.e. a special text file where columns are delimited by a special character, which is the comma in my case). Such a file is easily read into R as shown in Figure 8.4, where it is assigned to x (a two-column matrix). R's **t.test** function generated the output, where $t = 1.2527$ is the observed value of the z statistic (identical to t in this case), and the p-value is 0.1075. These results match those of Figure 8.3.

Using Calc and Excel to Obtain P-value

If your analytical goal is limited to obtaining the P-value associated with the z-test of Example 8.2 with Table 8.4 data, then you do not need to use the Analysis ToolPak as I did in Example 8.2. You could use the **ttest** function implemented in Calc and in Excel. The only difference between the two implementation is that in Calc, the function arguments are separated with semicolons (the syntax is TTEST(Array1; Array2; Tails; Type)), while they are separated with commas in Excel (the syntax is TTEST(Array1, Array2, Tails, Type)). The arguments are defined as follows:

▶ Array1 is the dependent array or range of data for the first record.

▶ Array2 is the dependent array or range of data for the second record.

▶ Tails = 1 calculates the one-tailed test, Mode = 2 the two-tailed test.

▶ Type is the kind of t-test to perform. Type 1 means paired. Type 2 means two samples, equal variance (homoscedastic). Type 3 means two samples, unequal variance (heteroscedastic).

To obtain the one-tailed test P-value of Example 8.2 (i.e. P-value=0.1075) with Calc, I used the following Calc formula:

=TTEST(A2:A34;B2:B32;1;2)

The same number is obtained with Excel as follows:

$$=\text{TTEST}(A2:A34,B2:B32,1,2)$$

8.3.3   The Power Analysis of the z-test

In Example 8.2, I was unable to reject the test hypothesis in favor of the alternative hypothesis that plumbers earned a higher hourly wage than electricians. Typically there is one main reason for the non-rejection of the null hypothesis. It is that the current sample sizes are two small to detect the actual difference between the two population means. Here are two questions that I believe you as analyst should ask:

(i) What is the minimum difference in wage rate (or Minimum Detectable Difference) that the current sample sizes will allow your testing procedure to detect? Is that minimum too low or too high in the sense of being meaningful business wise?

(ii) What are the minimum sample sizes that will allow your statistical procedure to detect the smallest difference in wage rates that you care about at a specified power?

The power analysis is what you need to conduct in order to answer these two questions. You will answer the first question by constructing the Power Curve, which depicts the probability of rejecting the test hypothesis when the alternative is true with the difference in means having a specific value d. This probability represents the test Power. That is, each value of d yield a different power for the test. As for the second question, you need to be able to link the sample and the test power as will be shown shortly.

The Power Curve

Table 8.5 provides an expression of the z-test power for each possible form of the alternative hypothesis. The power in this table is expressed in the form of the probability of rejecting the test hypothesis when it is false. Φ represents the cumulative distribution function of the standard Normal distribution, z_α the $100(1 - \alpha)^{th}$ percentile of the standard Normal distribution, d a variable referring to the current value of the population mean difference (i.e. $\Delta = \mu_1 - \mu_2$), and s the standard deviation of the difference of sample means.

I will now compute the power of the right-tailed test that I implemented in Example 8.5. Once again I consider populations 1 and 2 to be those of plumbers and electricians respectively. Since this test consists of a straight comparison between the 2 population means, $\Delta_0 = 0$. It follows from the numbers in Figure 8.3 that s is obtained as follows:

$$s = \sqrt{s_\text{P}^2/n_\text{P} + s_\text{E}^2/n_\text{E}} = \sqrt{2.6842/33 + 1.6156/31} = 0.3653.$$

Since the significance level is $\alpha = 0.05$, then the 95th percentile of the Standard Normal distribution is $z_{0.05} = 1.645$. Since $\Delta_0 = 0$, the power curve associated with the right-tailed test of Example 8.2 has the following equation:

$$\text{Power}(\Delta) = 1 - \Phi(1.645 - \Delta/0.3653). \qquad (8.3)$$

This power function can be graphed using any tool that can compute the cumulative standard Normal distribution for different values of d. One option is to use Excel as I did here to obtain the graph in Figure 8.5. I created 2 adjacent columns of data. The first column labeled as d with numbers that vary from 0 to 2.5 by step of 0.05, and a second column labeled as "Power". The numbers in that second column are obtained with the formula `=1-NORM.S.DIST(1.645-J8/0.3653,TRUE)` (note that cell J8 simply represent a value for d)

Table 8.5 : Power of the two-sample z-test for large unpaired samples

Alternative Hypothesis	*Power of the z-test*
$H_a\colon \mu_1 - \mu_2 > \Delta_0$	$1 - \Phi\left[z_\alpha - (\Delta - \Delta_0)/s\right]$
$H_a\colon \mu_1 - \mu_2 < \Delta_0$	$\Phi\left[-z_\alpha - (\Delta - \Delta_0)/s\right]$
$H_a\colon \mu_1 - \mu_2 \neq \Delta_0$	$1 - \Phi\left(z_{\alpha/2} - \dfrac{\Delta - \Delta_0}{s}\right) + \Phi\left(-z_{\alpha/2} - \dfrac{\Delta - \Delta_0}{s}\right)$

where $s = \sqrt{s_1^2/n_1 + s_2^2/n_2}$ is the standard error of the difference of means $\overline{x}_1 - \overline{x}_2$

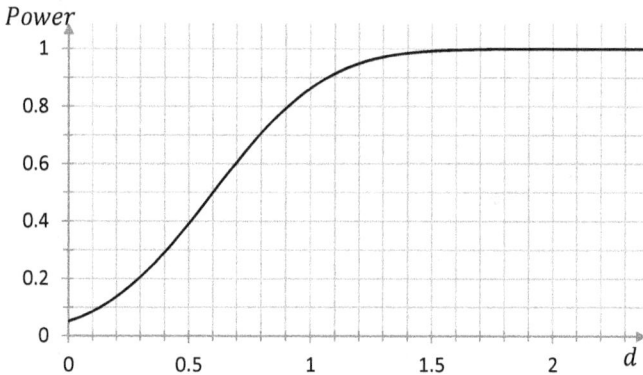

Figure 8.5. The Power Curve of the Example 8.2 Test

It follows from Figure 8.5 that a difference of population means of only $0.90 will lead to the rejection of the test hypothesis with a probability of about 0.8. Should you be satisfied with the power of this test ? The answer to this question depends on whether you see a difference in mean wages of $0.90 as being meaningful or not. If you consider a difference of $0.50 to be meaningful, and want it to be detectable then Figure 8.5 indicates that the current procedure will be able to detect it with a low 0.4 probability. It means you will need to increase the sample sizes before obtaining a procedure that can detect a difference of \$0.50 with a higher power of your choice. Developing a useful hypothesis testing procedure requires an in-depth understanding of the problem at hand and a clear vision of the objectives to be achieved.

Suppose that you want to be able to reject the test hypothesis with a probability of 0.85 if plumbers' wage rate exceeds that of electricians by \$0.50 or more. Here is a situation where you use the test power for the purpose of calculating the sample size required to meet specific goals.

The Required Sample Size

Once you know about (i) the significance level α at which you want to conduct your test, (ii) the your Minimum Detectable Difference d, and (iii) the Power P with which you want to detect it, then the sizes n_1 and n_2 of the two samples can be determined. Both sizes need not be the same. However, I suggest to let them be identical since equation 8.4 gives a straightforward expression for the common sample size in that situation. Unless one of the two populations has higher data collection costs than the other, I do not see a justification for not selecting two samples of the same size. The common sample size that I denote by n is calculated as follows:

$$n = \frac{(s_1^2 + s_2^2)(c_\alpha + z_{(1-\text{P})})^2}{(\Delta - \Delta_0)^2}, \tag{8.4}$$

where P is the desired power for the test, $c_\alpha = z_\alpha$ for one-tailed tests, and $c_\alpha = z_{\alpha/2}$ for two-tailed tests.

Let me consider again the problem of comparing wage rates between plumbers and electricians discussed in Example 8.2. I want to compute n so that I can reject the null hypothesis with a probability of P=0.85 (the test power), if plumbers' mean wage exceeds that of electricians by $0.50. The test is being conducted at the significance level $\alpha = 0.05$. I know from Figure 8.3 that $s_P^2 = 2.6842$ and $s_E^2 = 1.6156$ are the two sample variances[5]. The 2 percentiles of the Normal distribution needed here are $z_\alpha = z_{0.05} = 1.645$, and $z_{1-P} = z_{1-0.85} = z_{0.15} = 1.0364$. The desired sample size is then determined as follows:

$$n = \frac{(2.6842 + 1.6156)(1.645 + 1.0364)^2}{0.5^2} = 124.$$

The above sample size was rounded up to the nearest integer. As you could see, detecting a small difference in wage rate of $0.50 between plumbers and electricians will require a substantial increase in sample size. That is 124 plumbers as well as 124 electricians will need to be sampled in order to make it happen.

8.4. Small Unpaired Samples with Approximate Normal Distribution (t-test)

When the size of one of the two samples being analyzed falls below 30, you cannot use the z-test of the previous section because the Normal distribution is no longer a good approximation of the law of probability of the mean. Most statistics textbooks in this case, will separately treat the situation when the population variances are identical, and the situation when they are different. When you assume homogeneity of variance, the test statistic becomes slightly easier to compute than when you assume different variances. Such simplifications were important in

---

[5]These numbers may have to come from a pilot or a past study, if rough approximations of the population variances cannot be found elsewhere.

the pre-computer era when these methods were developed. Given that researchers almost never know for sure whether the population variances are identical or not, I would recommend in this post-computer era to always use the t-test with the assumption that the variances are different (if you have to choose). I will say more on this later in this section.

I will still present the t-test for equal variances (also known as the Pooled t-test) as well as the t-test for unequal variances, since such a separation is still widespread in the literature. The "Unequal-variance t-Test" has several names in the literature. It is sometimes referred to as the Welch test due to the work of Welch (1938, 1947). Others refer to this test as the Unpooled-variance t-test, the Welch-Satterwaite test, after Satterwaite (1946), the Smith-Satterwaite test, or Smith-Welch-Satterwaite test to acknowledge the work of Smith (1936)

8.4.1   The t-Statistic

▶ *If the two populations have the same variances*
$$(i.e.\ \sigma_1^2 = \sigma_2^2),$$

Under this assumption, the two populations under investigation will have a common variance, that I will denote by σ^2. After selecting a sample of data from each of the two populations, an efficient way to compute the common variance is to pool the data from both samples. The test statistic in this case is defined as follows:

$$t = \frac{(\overline{x}_1 - \overline{x}_2) - \Delta_0}{s_p\sqrt{1/n_1 + 1/n_2}}, \tag{8.5}$$

where s_p represents the square root of s_p^2, which is the estimated value of the common variance σ^2, also known as the

Pooled Variance. This pooled variance is defined as follows:

$$s_p^2 = \frac{(n_1 - 1)s_1^2 + (n_2 - 1)s_2^2}{n_1 + n_2 - 2} \qquad (8.6)$$

The law of probability of the t statistic of equation 8.5 is the Student's distribution (or t-distribution) with $n_1 + n_2 - 2$ degrees of freedom

▶ *If the two populations have the different variances*
$$(i.e.\ \sigma_1^2 \neq \sigma_2^2),$$

After selecting a sample of data from each of the two populations, each sample will be used to estimate the variance of the population it represents. The test statistic in this case is defined as follows:

$$t^* = \frac{(\overline{x}_1 - \overline{x}_2) - \Delta_0}{\sqrt{s_1^2/n_1 + s_2^2/n_2}}, \qquad (8.7)$$

where s_1^2, and s_2^2 are the sample variances obtained using the samples selected respectively from populations 1 and 2. The test statistics of equations 8.2 and 8.7 have the same form, but are governed by different laws of probability. It is why I used different notations. *The law of probability of the t* statistic of equation 8.7 is the Student's distribution (or t-distribution) with a number ν of degrees of freedom given by:*

$$\nu = \frac{(s_1^2/n_1 + s_2^2/n_2)^2}{\dfrac{(s_1^2/n_1)^2}{n_1 - 1} + \dfrac{(s_2^2/n_2)^2}{n_2 - 1}} \qquad (8.8)$$

Validity Conditions

(1) The data being analyzed is of interval or ratio type.

(2) Each sample is selected randomly from the population it represents.

(3) The distribution of data in each population is approximately Normal.

8.4.2 Description of the t-Statistic

As usual, I will successively present Fisher's P-value approach followed by the Neyman-Pearson approach based on decision theory.

Fisher's P-value Approach

Let $\delta = \mu_1 - \mu_2$ be the difference between the means of populations 1 and 2 respectively. Table 8.6 defines the P-value for each alternative hypothesis type H_a, and provides Excel 2010 formulas for computing it.

Table 8.6 : P-Value for testing 2 Population Means with t-test for Small Unpaired Samples - H_a *represents the research hypothesis*

(H_a)	P-value	Excel 2010 Formula[a]
$\delta > \Delta_0$	$P(t > t_{obs})$	=1-T.DIST(t_{obs},DF[b],TRUE)
$\delta < \Delta_0$	$P(t > -t_{obs})$	=1-T.DIST($-t_{obs}$,DF[b],TRUE)
$\delta \neq \Delta_0$	$2P(t > \lvert t_{obs} \rvert)$	=2*(1-T.DIST(t_{obs},DF[b],TRUE))

[a]For equivalent instructions with Excel 2007, R, and OpenOffice Calc see Appendix C

[b]DF=Number of degrees of freedom = $n_1 + n_2 - 2$

The Neyman-Pearson Approach

Let t_{α, n_1+n_2-2} be the $100(1-\alpha)^{th}$ percentile of the Student's t distribution with $n_1 + n_2 - 2$ degrees of freedom. The 5 steps of the Neyman-Pearson hypothesis testing procedure are described as follows:

(1) The Hypotheses	H_0: $\mu_1 - \mu_2 = \Delta_0$ vs H_a: *(specify)*
(2) Significance Level	α *(to be specified)*
(3) Test Statistic	*t-* or *t*-stat. (equation 8.5 or 8.7)*
(4) Decision Rule	
Alternative Hypothesis	Decision
H_a: $\mu_1 - \mu_2 \neq \Delta_0$	$\triangleright$ *Reject H_0 if $t \geq t_{\alpha/2, n_1+n_2-1}$ or*
	$t \leq -t_{\alpha/2, n_1+n_2-1}$
H_a: $\mu_1 - \mu_2 < \Delta_0$	$\triangleright$ *Reject H_0 if $t \leq -t_{\alpha, n_1+n_2-2}$*
H_a: $\mu_1 - \mu_2 > \Delta_0$	$\triangleright$ *Reject H_0 if $t \geq t_{\alpha, n_1+n_2-2}$*
(5) Final Decision	*Compute t_{obs} or t^*_{obs} and apply the decision rule*

In practice, you will generally not have to compute the t statistic using equations 8.5 or 8.7. The software you are using will do the calculations for you. You will however need to interpret the results properly. Let me consider the following example:

Example 8.3 ───────────────────────────────

A group of 9 men and another group of 7 women took a biology exam are obtained the following scores:

Group	Scores								
Men	72	69	98	66	85	76	79	80	77
Women	81	67	90	78	81	80	76		

You like to know whether the women mean score is higher than that of men. The test will be done at the 1% significance level. The first thing I want to do is to use Excel as discussed in section E.2 of Appendix E (specify 0.01 in the Alpha text box) in order to the perform successively the equal-variance and unequal variance t-tests. The results of these two analyzes are depicted in Figures 8.6 and 8.7.

Let μ_M and μ_W be the mean scores of men and women populations respectively. If you have a claim that women score higher than men (i.e. $\mu_M < \mu_F$), and want to test it, then this can be achieved by calculating the P-value, assuming equality of population means represents the null hypothesis.

Under the assumption of variance homogeneity, the P-value is 0.4091 (see Figure 8.6). If you assume unequal variances, this P-value becomes 0.4053 (see Figure 8.7). In both situations the P-value is very large, which suggests weak evidence in the samples in favor of the claim that women perform better than men.

The Neyman-Pearson approach under the assumption of variance homogeneity will yield the following procedure:

(1) The hypotheses are H_0: $\mu_M \geq \mu_W$ versus H_a: $\mu_M < \mu_W$. *The reasoning behind the definition of these hypotheses is as follows: if you expect women to perform better than men, then the error you will want to protect yourself against is concluding that men perform better than women when in reality, it is the opposite that is true. That error becomes by definition your Type I error, and determines what your null hypothesis will be.*

(2) The significance level of the test has been set at $\alpha = 1\%$.

(3) The test statistic is t of equation 8.5.

(4) The decision rule is formulated as follows *(see the Neyman-Pearson five-step procedure above)*:

Reject H_0 if $t_{obs} < -2.6245$. *(The critical value of -2.6245 comes from Figure 8.6 "t Critical one-tail." A negative sign is added because it is a left-tailed test. Note that Excel's Data Analysis ToolPak never requests the type of hypothesis test you are performing. Therefore, you must expect to occasionally have to adjust some of the results)*

(5) It follows from Figure 8.6 that $t_{obs} = -0.234$ (labeled as "t Stat" by Excel). Because $-0.234 > -2.6245$ you cannot reject the test hypothesis. Therefore the samples that were analyzed do not provide sufficient evidence in favor of the alternative hypothesis that women have a higher mean score than men.

Figure 8.7 indicates that even if the men and women population variances are assumed unequal, it is still not going to be possible to reject the test hypothesis. You may want to know that in Figure 8.6, the quantity labeled as "Pooled Variance" is the pooled variance of equation 8.6, while the "Hypothesized Mean Difference" is the quantity Δ_0 that appears in the expression of the t-test statistic.

F	G	H	I	J	K
t-Test: Two-Sample Assuming Equal Variances				Men	Women
				72	81
	Men	*Women*		69	67
Mean	78	79		98	90
Variance	90	47.333		66	78
Observations	9	7		85	81
Pooled Variance	71.714			76	80
Hypothesized Mean Difference	0			79	76
df	14			80	
t Stat	-0.234			77	
P(T<=t) one-tail	0.4091				
t Critical one-tail	2.6245				
P(T<=t) two-tail	0.8181				
t Critical two-tail	2.9768				

Figure 8.6. Excel Output of the Pooled t-Test Analysis of Example 8.3 Data

	A	B	C
1			
2	t-Test: Two-Sample Assuming Unequal Variances		
3			
4		*Men*	*Women*
5	Mean	78	79
6	Variance	90	47.333
7	Observations	9	7
8	Hypothesized Mean Difference	0	
9	df	14	
10	t Stat	-0.2443	
11	P(T<=t) one-tail	0.4053	
12	t Critical one-tail	2.6245	
13	P(T<=t) two-tail	0.8106	
14	t Critical two-tail	2.9768	
15			

Figure 8.7. Excel Output of the Unequal-Variance t-Test Analysis of Example 8.3 Data

Although I have used Excel to analyze Example 8.3 data, you may well use R or OpenOffice Calc to obtain the same results as shown in section 8.3. As previously indicated R's t.test function will provide the observed t-statistic and the p-value. If you need the critical value you will have to calculate it using the qt() function (see "R" column of Table C.4 in Appendix C) along with the p-value definition of Table 8.6.

How to deal with the variance homogeneity issue ?

Each time the t-test must be used, there is always this recurrent question as to whether you must assume equal or unequal variances. My recommendation is to *always assume unequal variances* with the following two rare exceptions:

(a) You are doing your analysis manually or with a hand-held calculator. Assuming equal variances may slightly simplify the calculations, and you may want to take advantage of it.

This convenience of calculation is negligible when you are using a computer.

(b) If the data you are analyzing were simulated in a controlled experiment where the experimenter could set the magnitude of the variances (e.g. a Monte-Carlo simulation). Then you will know for sure whether the variances are identical or different.

A statistical test performs well when its nominal significance level α (set by you) is sufficiently close to its true one[6], and when at a given significance level, the test shows good power (i.e. a good capability of rejecting the test hypothesis when it is the alternative that is true). Coombs et al. (1996) have demonstrated with a simulation study that the true and nominal significance levels of the unequal-variance t-test are always very close, whether the variances are equal or not, while the equal-variance or Student's t-test may perform poorly when the variances are unequal. Moreover, Moser et al. (1989), Moser and Stephen (1992), and Coombs et al. (1996) have showed that the power of the unequal variance t-test is similar to that of Student's t-test even when the variances are equal.

If the unequal-variance t-test is superior to its equal-variance counterpart, then why are both versions still being presented these days in most statistics textbooks? Student's t-test was the first of the two tests to be discovered (in Student, 1908a, 1908b), and it is possible that the statistical community does not want to stop paying tribute to William Sealy Gosset (the real name of Student) for his contribution to the development of statistical science. Otherwise, there is no other apparent reason why a single t-test could not be presented. It would not even be labeled as

---

[6]The true significance level represents the actual relative number of times the null hypothesis is rejected when it is true. This could be obtained by simulation

unequal-variance since it will work in all situations.

8.4.3 The Power Analysis of the *t*-Test

In this section, I will show you how to compute the power of a two-sample *t*-test that you have already implemented, and how the sizes of the two samples may be estimated so as to obtain a hypothesis testing procedure that can detect a given mean difference with a specified power.

The Power Curve

In section 6.3.2, I described the construction of the power curve of the one-sample *t*-test, and presented a series of curves in appendix B that could be used for calculating the power. Each of these power curves depicts the power of the *t*-test as a function of the difference d between the true mean μ and its hypothetical value μ_0 normalized by the standard deviation s. The same curves can still be used for calculating the power of the two-sample *t*-tests discussed in this section.

▶ For Student's (or equal-variance) *t*-test, the power can be computed using the power curves of appendix B with d calculated as follows:

$$d = \frac{\Delta - \Delta_0}{s_p} \sqrt{\frac{n_1 n_2}{(n_1 + n_2)(n_1 + n_2 - 1)}}. \qquad (8.9)$$

▶ For the unequal-variance *t*-test, the power can be computed using the power curves of appendix B with d calculated as follows:

$$d = \frac{\Delta - \Delta_0}{\sqrt{(\nu + 1)(s_1^2/n_1 + s_2^2/n_2)}}, \qquad (8.10)$$

where ν is the number of degrees of freedom of the unequal-variance *t*-test statistic (see equation 8.8).

The Sample Size Determination

As previously indicated on a few occasions, when determining the sizes of the two samples at the planning stage, you could conveniently choose two samples of the same size n. I am assuming here that you have a minimum difference $\Delta - \Delta_0$ that you want to detect at a specified power P, and that your test will be conducted at a significance level α. The size n common to both samples could be calculated according to the following procedure:

▶ **Student's (or equal-variance) t-test**

Under the alternative hypothesis, the test statistic of equation 8.5 follows the noncentral t distribution with $2(n-1)$ degrees of freedom, and a noncentrality parameter δ defined as,

$$\delta = \sqrt{n}\frac{\Delta - \Delta_0}{s_p\sqrt{2}}$$

▶ **Unequal-variance t-test**

Under the alternative hypothesis, the test statistic of equation 8.7 follows the noncentral t distribution with a number of degrees of freedom $\nu = (n-1)(s_1^2 + s_2^2)^2/(s_1^4 + s_2^4)$, and a noncentrality parameter δ defined as follows:

$$\delta = \sqrt{n}\frac{\Delta - \Delta_0}{\sqrt{s_1^2 + s_2^2}}$$

In both situations of equal and unequal variances, you can compute the power of the test for various values of n, once you have your Minimum Detectable Difference $\Delta - \Delta_0$, and the standard deviations s_1^2 and s_2^2. You will need to use R to compute the power, since Excel and Calc do not implement Non-central t distribution.

8.5. Small Unpaired Samples with Symmetric Non-normal Distribution (Wilcoxon Rank-Sum Test)

The test statistic used with the Wilcoxon rank-sum test is quite different from those associated with the z and t tests seen so far. The z and t tests belong to the category of *Parametric Tests* in the sense that they both rely to some extent upon the Normal and the Student distributions. Both distributions are theoretical distributions not tied to the data at hand, which are used solely for the purpose of approximating the parameters (hence parametric) of those distributions. The Wilcoxon Rank-Sum test on the other hand is a nonparametric test, that does not rely on any theoretical distribution.

The Wilcoxon Rank-Sum test I will described in this section was developed by Wilcoxon (1949). Earlier however, Mann and Whitney (1947) developed a similar test, but used a different formulation for the test statistic. Consequently, the test to be described here is sometimes referred to as the Wilcoxon-Mann-Whitney test.

8.5.1 The Test Statistic

Note that the null hypothesis of equation 8.1 can be rewritten as H_0: $(\mu_1 - \Delta_0) - \mu_2 = 0$. Therefore testing this hypothesis amounts to comparing the shifted mean of population 1 $(\mu_1 - \Delta_0)$ to the population 2 mean μ_2. To understand how the Wilcoxon Rank-Sum test statistic W works, consider the following experiment:

(a) Suppose that you select a sample of 3 units from population 1 (named $x_{(1)1}, x_{(1)2}, x_{(1)3}$) and a sample of 4 units from population 2 (named $x_{(2)1}, x_{(2)2}, x_{(2)3}, x_{(2)4}$).

(b) You shift all 3 population 1 units to the left by Δ_0. That is the new population 1 measurements are now $x_{(1)1} - \Delta_0$, $x_{(1)2} - \Delta_0$, and $x_{(1)3} - \Delta_0$.

(c) You pool both samples together and sort the 7 measurements (i.e. the 4 original population 2 measurements, and the 3 shifted population 1 measurements) in ascending order.

Figure 8.3 shows a scenario where all (shifted) population 1 measurements (in shaded circles labeled as 1) exceed all population 2 measurements (in plain circles labeled as 2). This suggests that the null hypothesis is unlikely to be true. Figure 8.4 on the other hand depicts a scenario where the null hypothesis is likely to be true. You can see that measurements from both populations intermingle very well. Below the horizontal axis of each figure are the ranks associated with the measurements. *If you sum the ranks associated with population 1 measurements only, the resulting rank sum[7] will be much higher when the null hypothesis is false (Figure 8.3) than when the null hypothesis is true (Figure 8.4). It is this rank sum that will be the test hypothesis.*

In Figure 8.3 the rank-sum of the shifted population 1 measurements is $5 + 6 + 7 = 18$, and it reduces to $2 + 4 + 6 = 12$ under the null hypothesis.

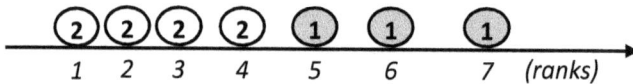

Figure 8.3. Units of the pooled sample sorted in ascending order when the null hypothesis is false

---

[7]Here is where the name Rank-Sum Test originated

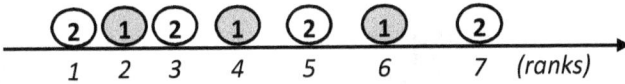

Figure 8.4. Units of the pooled sample sorted in ascending order when the null hypothesis is true

The Wilcoxon Rank-Sum test statistic W is defined as follows:

$$
W = \begin{array}{l} \text{Rank Sum of all Shifted} \\ \text{Measurements from Population 1} \end{array}
$$

(8.11)

Validity Conditions [8]

(1) The raw data prior to the transformation into ranks, is of ordinal, interval, or ratio type.

(2) Each sample is selected randomly from the population it represents.

(3) The two samples are independent.

(4) The two population variances are identical.

8.5.2 Description of the Wilcoxon's Rank-Sum Test

For any given sample sizes n_1 and n_2 the distribution of the test statistic is discrete and has been tabulated in Appendix A (see Tables A.6 through A.11. Tables A.6-A.11 are more complete than what is usually found in many statistic textbooks, and could be used for calculating P-values and critical values.

---

[8]These assumptions were made by several authors such as Conover (1980, 1999), Daniel (1990), or Marascuilo and McSweeney (1977). However, there are empirical evidence suggesting that conditions (1),and (4) may not be critical. That is, the test may well be used effectively with ordinal data, as well as with unequal variances

Under the hypothesis of equality of the means (the null hypothesis) the Wilcoxon test statistics W will be distributed symmetrically around a center M determined as the middle of the minimum and the maximum values of W. You may verify that $W_{min} = n_1(n_1 + 1)/2$, and that $W_{max} = n_1(n_1 + 2n_2 + 1)/2$. Consequently, the center of symmetry of the distribution of W is $M = n_1(n_1 + n_2 + 1)/2$.

The Fisher's P-value Approach

The calculation of the P-value of the Wilcoxon rank-sum test is done using the appropriate expression from Table 8.7 along with the probability distribution of W of Table A.6.

Table 8.7 : P-Value for testing 2 Population Means with the Wilcoxon Rank-Sum Test *(Hₐ is the research hypothesis)*

(H_a)	P-value
$\mu_1 - \mu_2 > \Delta_0$	$P(W \geq W_{obs})$
$\mu_1 - \mu_2 < \Delta_0$	$P\big(W \geq n_1(n_1 + n_2 + 1) - W_{obs}\big)$
$\mu_1 - \mu_2 \neq \Delta_0$	$P\big(\lvert W - M \rvert \geq \lvert W_{obs} - M \rvert\big)$ [a]

[a]Note that this P-value can be re-expressed as follows: $P(W \geq M + \lvert W_{obs} - M \rvert) + P(W \leq M - \lvert W_{obs} - M \rvert)$, and this expression should be used along with Table A.6

Example 8.4 _____

In order to evaluate the impact of Tobacco smoke on the infants exposed it, two independent samples of infants were selected. The first sample contains 7 infants who were not exposed to any smoke, while the second sample contains 8 infants exposed to tobacco smoke. The experimenter then measured the urinary concentration of cotanine (related to nicotine) from each infant and reported the following data:

Group	Scores							
Unexposed Infants	8	11	12	14	20	43	111	
Exposed Infants	35	56	83	92	128	150	176	208

Let μ_U and μ_E be the mean urinary concentration of cotanine of unexposed and exposed infants respectively. Since the purpose of this experiment is to demonstrate a higher concentration of cotanine among exposed infants, the research hypothesis could be formulated as follows: $\mu_E > \mu_U$ or $\mu_E - \mu_U > 0$ (which leads to a right-tailed test). It follows from Table 8.7 that the P-value is given by $P(W \geq W_{obs})$. I now need to evaluate W_{obs} the observed value of the Wilcoxon test statistic of equation 8.11. One of the many possible ways of computing W_{obs} is to MS Excel. Suppose that Exposed and Unexposed infants represent populations 1 and 2 respectively. You would then organize your data as shown in the "Unsorted Scores" portion of Table 8.8. The "Score" column (A) contains the input data. The "Pop" column (B) indicates whether the score is from population 1 or from population 2. The "Rank" column (C) contains sequential numbers from 1 through 15 (the total number of observations), and the "Rank1" column will be 0 for all population 2 infants and for population 1 infants, it will be the corresponding rank[9].

The next and last step is to select the 2 columns A and B and to sort them using column A as the sole sort key. This operation will yield the "Sorted Score" portion of Table 8.8. The sum of the last column (Rank1) is 87 and represents the observed test statistic. That is $W_{obs} = 87$.

In this problem, $n_1 = 8$ and $n_2 = 7$. Table A.9 provides upper-tailed probabilities for $(n_1 = 7, n = 8)$, *which by the way are identical to those associated with* $(n_1 = 8, n_2 = 7)$. It follows that $P(W \geq W_{obs}) = P(W \geq 87) < 0.001$ (since 87 exceeds 83, which already has a 0 probability). Therefore, the evidence in the sample in favor of the research hypothesis that stipulates

---

[9]In "Rank1" I actually put the following Excel formula =IF(B?=1,C?,0)

that exposed infants have a higher cotanine concentration.

Table 8.8 : Calculation of the Wilcoxon W-Statistic

Unsorted Scores				Sorted Scores			
Score (A)	Pop (B)	Rank (C)	Rank1 (D)	Score (A)	Pop (B)	Rank (C)	Rank1 (D)
35	1	1	1	8	2	1	0
56	1	2	2	11	2	2	0
83	1	3	3	12	2	3	0
92	1	4	4	14	2	4	0
128	1	5	5	20	2	5	0
150	1	6	6	35	1	6	6
176	1	7	7	43	2	7	0
208	1	8	8	56	1	8	8
8	2	9	0	83	1	9	9
11	2	10	0	92	1	10	10
12	2	11	0	111	2	11	0
14	2	12	0	128	1	12	12
20	2	13	0	150	1	13	13
43	2	14	0	176	1	14	14
111	2	15	0	208	1	15	15
Total			36	Total			87

The Neyman-Pearson Approach

To define the Neyman-Pearson method to the Wilcoxon's rank-sum test, let c_α and $c_{\alpha/2}$ be implicitly defined by $P(W > c_\alpha) = \alpha$ and $P(W > c_{\alpha/2}) = \alpha/2$ respectively. That is, c_α is the $100(1-\alpha)^{th}$ percentile of the W-distribution. The Neyman-Pearson's five-step hypothesis testing procedure is summarized as follows:

(1) The Hypotheses	$H_0: \mu_1 - \mu_2 = \Delta_0$ vs H_a: *(specify)*
(2) Significance Level	α *(to be specified)*
(3) Test Statistic	*Wilcoxon W-Statistic of equation 8.11*
(4) Decision Rule	
Alter. Hypothesis	Decision
$H_a: \mu_1 - \mu_2 \neq \Delta_0$	$\triangleright$ *Reject H_0 if $W_{obs} \geq c_{\alpha/2}$ or* $W_{obs} \leq 2M - c_{\alpha/2}$
$H_a: \mu_1 - \mu_2 < \Delta_0$	$\triangleright$ *Reject H_0 if $W_{obs} \leq 2M - c_\alpha$*
$H_a: \mu_1 - \mu_2 > \Delta_0$	$\triangleright$ *Reject H_0 if $W_{obs} \geq c_\alpha$*
(5) Final Decision	*Compute W_{obs} and apply the decision rule*

Because the law of probability of the Wilcoxon test statistic is discrete, the percentiles c_α or $c_{\alpha/2}$ are generally not accurate as will be seen in the next example. That is, it is often impossible to find a number c_α that fully satisfy the condition $P(W \geq c_\alpha) = \alpha$. You will then use the number that yields a probability as close to α as it can get.

Example 8.5 ————————————————————

I consider the same data of Example 8.6, and like to apply the Neyman-Pearson procedure for testing the hypothesis $\mu_E \leq \mu_U$. The 5 steps are described as follows:

▶ The Hypotheses
$$H_0: \quad \mu_E \leq \mu_U$$
$$H_a: \quad \mu_E > \mu_U$$

▶ The Significance Level

I am assuming here that the testing will be done at the $\alpha = 5\%$ significance level. Feel free to experiment with alternative significance levels such as 1% or 2%.

▶ The Test Statistic. The test statistic is W of equation 8.11.

▶ The Decision Rule

Before I can formulate the decision rule I need to obtain $c_{0.05}$ the 95th percentile of the W-distribution. It follows from Table A.9 - column (7,8) - that $P(W \geq 71) = 0.047$, which represents the best approximation of the 95th percentile you can get. The decision rule can now be formulated as follows:

$$\text{Reject } H_0 \text{ if } W_{obs} > 71.$$

▶ The Final Decision

It follows from Example 8.6 that $W_{obs} = 87$ which exceeds the critical value $c_{0.05} = 71$. Therefore the test hypothesis H_0 must be rejected.

Tables A.6 through A.11 can be very helpful in determining critical and P-values as the past two examples have shown. However, these tables do not extend beyond samples with sizes exceeding 10. Hence the question: what do you do if you must deal with large samples? The solution to this problem is to use the Z statistic defined as follows:

$$Z = \frac{W - n_1(n_1 + n_2 + 1)/2}{\sqrt{n_1 n_2 (n_1 + n_2 + 1)/12}}. \tag{8.12}$$

The useful result here is that the law of probability of the Z statistic is close to the standard Normal distribution when the sample sizes exceed 10 and the null hypothesis is true.

8.5.3 The Power Analysis

You must have noticed by now that I am a passionate advocate of the power analysis. I do believe that this analysis provides further and useful insight into the Neyman-Pearson testing procedure. However, performing the power analysis with the Wilcoxon test is a difficult task. The problem stems from the difficulty to describe the law of probability of the test statistic W, when the test hypothesis is false. This problem does not have a clear-cut solution. Some approximate solutions exist that are beyond the scope of this book.

8.6. Large Paired Samples (z-test for dependent samples)

Let $\overline{x}_1$ and $\overline{x}_2$ be the means associated with the samples selected from populations 1 and 2 respectively. Likewise, s_1^2 and s_2^2 are the variances associated with the samples selected from populations 1 and 2. The test statistic used for testing the equality of means when the samples are large and dependent is defined as follows:

$$z = \frac{\overline{x}_1 - \overline{x}_2}{\sqrt{s_1^2/n + s_2^2/n - \mathsf{cov}(\overline{x}_1, \overline{x}_2)}}, \qquad (8.13)$$

where $\mathsf{cov}(\overline{x}_1, \overline{x}_2)$ represents the covariance between the two sample means $\overline{x}_1$ and $\overline{x}_2$, and n is the common size of the 2 samples. This covariance can be expressed as follows:

$$\mathsf{cov}(\overline{x}_1, \overline{x}_2) = s_{1.2}^2/n, \qquad (8.14)$$

$s_{1.2}$ being a sample covariance between the variables x_1 and x_2. Since each sample has size n that exceeds 30, the law of probability of the z statistic is the standard Normal distribution.

Validity Conditions

(1) The data being analyzed is of interval or ratio type.

(2) The sample size n exceeds 30.

You may want to know that this test is not used very often in practice. The reason for this is that many researchers prefer to use the paired t-test discussed in the next section. Although the paired t-test was developed to deal with small samples, it remains valid for large samples and is implemented in Excel and in many other statistical or non-statistical packages.

8.7. Small Paired Samples with Approximate Normal Distribution (Paired t-test)

As previously mentioned, a paired experimental design involves a single sample of subjects is selected and two observations made on each subject on two occasions. Therefore, two series of sample scores are produced from the exact same sample following the experiment. Such a design is very common in studies that aim at demonstrating the effectiveness of a drug, a treatment, or an intervention program. Using the same individuals on both occasions has the advantage of eliminating the potential bias due to the interaction between subject and treatment.

In the paired designed, each subject i in the sample possesses two measurements x_{1i} and x_{2i}. The basic idea of the paired t-test is to compute the difference $d_i = x_{1i} - x_{2i}$ for each individual, and to apply the one-sample t-test on the differences as discussed in chapter 6. Table 8.9 shows a dataset that is suitable for analysis with the pairwise t-test. It contains the distribution of crimes by geographic area before and after the implementation of an anti-crime program. The objective of the study being to assess the program's effectiveness. The "Difference" row is what should be analyzed with the one-sample t-test of chapter 6.

The test statistic to be used is the ratio of the mean of the difference scores $(\bar{d})$ to its standard error (s_d). That is,

$$t = \frac{\bar{d}}{s_d}. \tag{8.15}$$

Table 8.9 : Impact of anti-crime program on the distribution of crimes by high-crime area

Reference	Geographic Area							
Period	A	B	C	D	E	F	G	H
Before	14	7	4	5	17	12	8	9
After	2	7	3	6	8	13	3	5
Difference	12	0	1	-1	9	-1	5	4

The law of probability associated with this t statistic is the Student t distribution with $n-1$ degrees of freedom. The data in Table 8.9 can be analyzed using the R as shown Figure 8.6. The whole testing procedure is implemented with the top 3 lines of instructions preceded with the ">" sign. The p-value of 0.03591 is an indication of strong evidence in the sample in favor of the effectiveness of the anti-crime program.

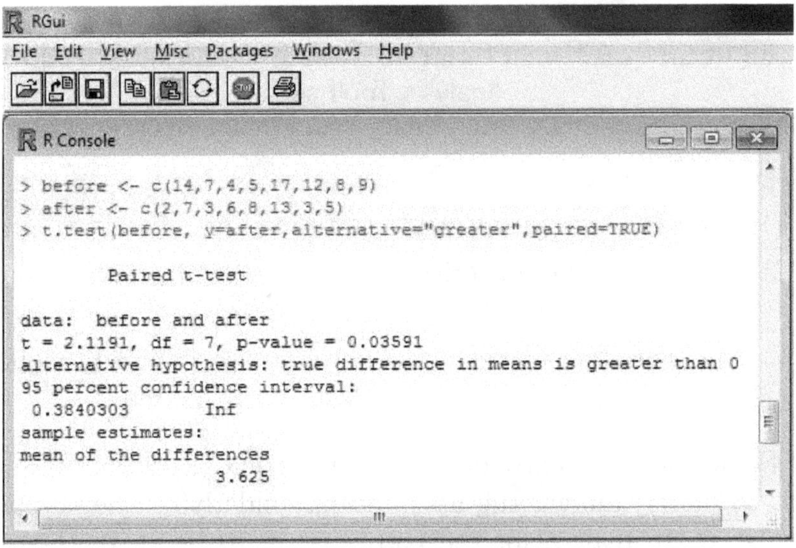

Figure 8.6. A Paired t-Test on Table 8.9 Data with the R Package

The Excel's Analysis ToolPak has a module for performing the paired *t*-test (see Section E.2 of Appendix E for further information regarding the ToolPak). I analyzed Table 8.9 data with this tool and the results (more detailed than R's results) are shown in Figure 8.7.

	L	M	N
t-Test: Paired Two Sample for Means			
		Before	After
Mean		9.5	5.875
Variance		20.286	12.696
Observations		8	8
Pearson Correlation		0.2982	
Hypothesized Mean Difference		0	
df		7	
t Stat		2.1191	
P(T<=t) one-tail		0.0359	
t Critical one-tail		1.8946	
P(T<=t) two-tail		0.0718	
t Critical two-tail		2.3646	

Figure 8.7. A Paired *t*-Test on Table 8.9 Data with Excel's Analysis ToolPak

Validity Conditions

(1) The data being analyzed is of interval or ratio type.

(2) Each of the 2 data series is a random sample of the population it represents.

(3) The scores obtained on each occasion are assumed to follow the Normal distribution, at least approximately.

Some statistics textbooks include the homogeneity of variances (i.e. $\sigma_1^2 = \sigma_2^2$) among the assumptions underlying the the paired *t*-test. In reality, for the paired *t*-test to be valid, the two variances need not be equal. However, a large variance in one population coupled with small sample sizes may lead to an artificially high difference in sample means, to the point of causing a false rejection of the null hypothesis. If you suspect a substantial difference between the variances, it may advised to use the Wilcoxon's signed-rank test discussed in the next section.

8.8. Small Paired Samples with Symmetric Non-normal Distribution (Wilcoxon's Signed-rank test for dependent samples)

When the sample is small ($n < 30$) and that the data are symmetric but non-normal, then the one-sample Wilcoxon Signed-Rank discussed in chapter 6 should be applied on the difference scores ($d_i = (x_{1i} - x_{2i}) - \Delta_0$). The corresponding test statistic is defined as follows:

$$S_+ = \text{Sum of Ranks Associated with the positive values of } d_i. \tag{8.16}$$

The law of probability of S_+ was discussed in chapter 6, and its percentiles tabulated (see Tables A.1 through A.5 in Appendix A).

Validity Conditions

(1) The measurements collected on both occasions (i.e. x_{1i} and x_{2i}) are interval or ratio data.

(2) The observations can be paired.

(3) The distribution of difference scores is symmetric (note that the distribution of the initial raw scores needs not be symmetric).

Refer to chapter 6 for a formal discussion on the procedures for calculating the P-value, and for implementing the Neyman-Pearson five step hypothesis procedure.

CHAPTER $\boxed{9}$

The Chi-Square Test

OBJECTIVE

This chapter discusses several versions of the chi-square test. The family of chi-square tests expands the statistical tests for proportions, which are based on dichotomous variables, to multiple-level categorical variables.

CONTENTS

9.1. Overview

This chapter generalizes the discussions presented in chapters 5 and 7 regarding the one-sample and two-sample tests for proportions. These two chapters have one thing in common, which is that they both deal with the analysis of dichotomous variables expressed in the form of proportions[1]. Therefore, one may consider extending the testing procedures from two-level dichotomous variables to more general multiple-level (or k-level) categorical variables, and from 2 populations to 3 populations or more.

Chapters 5 and 7 also have one main difference. While chapter 5 deals with a single sample, chapter 7 deals with two samples. The one-sample test of chapter 5 will have the following two extensions both of which are based on a single sample:

(*a*) The first extension generalizes the one-sample test involving one binary factor to the one-sample test involving one multiple-level factor, called the *Chi-Square Goodness-Of-Fit test*. A distribution of subjects among the k categories (or levels) of the categorical factor will be hypothesized, and the problem to resolve will be about testing how well the data fit the hypothetical distribution. Hence the name "Goodness-of-fit test."

(*b*) The second extension further generalizes the one-sample test involving one multiple-level factor to the one-sample test involving two multiple-level factors. This generalization will test whether the two factors are independent, and will lead to the *Chi-Square Test of Independence*.

---

[1]The proportion of males in a population of individuals is implicitly obtained by defining a 0-1 dichotomous variables where 1 represents the male and 0 the female, and by calculating the relative number of individuals with a 1.

The two-sample test of hypothesis of chapter 7 on the other hand, will also have two generalizations described as follows:

(c) The first extension generalizes the testing of two population proportions to the testing of 3 population proportions or more, resulting in a statistical procedure called the *"Chi-square homogeneity test for dichotomous factors."* The problem is indeed about testing whether the different populations are homogeneous with respect to the characteristic of interest.

(d) The second extension further generalizes the testing of 3 population proportions or more to the testing of the dependency between a multiple-level categorical variable and subjects' membership in one of 3 populations or more. This generalized dependency test is called the *"Chi-square homogeneity test for multiple-level categorical factors."* This problem amounts to testing homogeneity of the populations with respect to the categorical factor of interest.

To summarize, this chapter will focus on the following 4 hypothesis testing procedures:

▶ The Chi-Square Goodness-of-Fit Test.

It generalizes the one-sample t-test for proportions from a dichotomous factor to a more general categorical variable.

▶ The Chi-Square Test of Independence.

It generalizes the Chi-Square Goodness-of-Fit test to two general categorical factors to test their independence.

▶ The Chi-Square Homogeneity test for Dichotomous Factors.

It generalizes the two-sample t-test for proportions from two populations to 3 populations or more with a dichotomous factor.

▶ The Chi-Square Homogeneity test for Multiple-Level Categorical Factors.

It generalizes the two-sample t-test for proportions from two samples to 3 samples or more and a more general categorical factor.

Figure 9.1 summarizes the different hypothesis testing procedures discussed in chapters 5 and 7, as well as those to be discussed in this chapter, along with a broad description of the situation in which they are used. The shaded boxes contain the 4 testing procedures that will be discussed in this chapter.

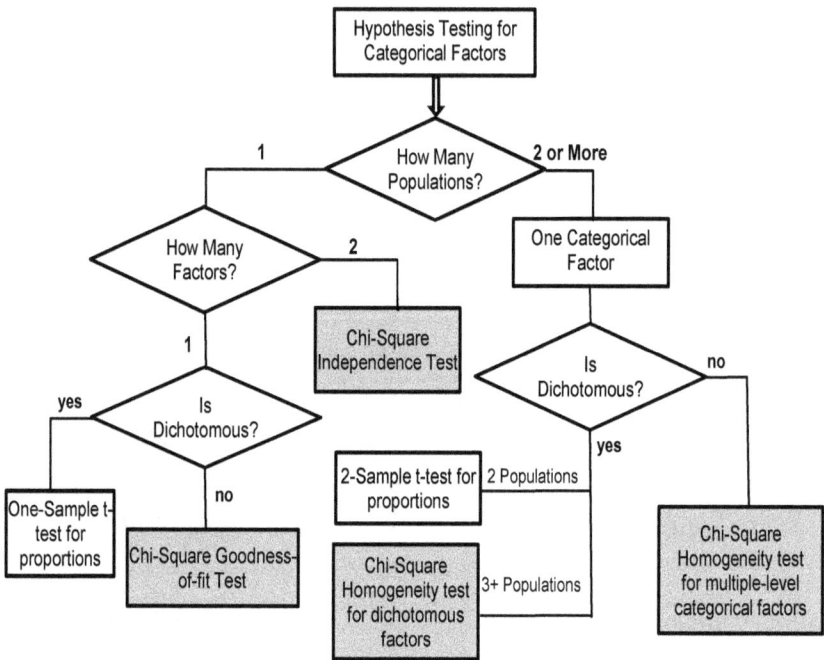

Figure 9.1. Hypothesis Testing Involving Categorical Factors

9.2. The Chi-Square Goodness-of-Fit Test

In chapter 5, I presented a number of techniques for comparing the population proportion to a hypothesized value.

For example, you could test the claim that fewer than 25% of all work-related accidents at ABC Inc. take place between 10AM and 11AM. In this section, I am concerned with a broader problem. I want to know whether work-related accidents are evenly distributed during a typical workday.

One option for addressing this problem is to divide a typical workday into 8 one-hour long time slots and to tally the number of work-related accidents as shown in Table 9.1, to select a sample of 80 accidents, and distribute them among the 8 categories.

Table 9.1 : Distribution of 80 Work-Related Accidents by Time of the Day

Time	Number of Accidents	Time	Number of Accidents
8 to 9 A.M.	6	1 to 2 P.M.	7
9 to 10 A.M.	6	2 to 3 P.M.	8
10 to 11 A.M.	20	3 to 4 P.M.	19
11 to 12 P.M.	8	4 to 5 P.M.	6

From Table 9.1, you can obtain the series of observed counts n_i, for $i = 1, \cdots, 8$, where $n_1 = n_2 = 6$, $n_3 = 20$, and so on. Before you can define the hypotheses, you need to define the series of expected counts. If the accidents are *Uniformly* distributed, as claimed in the research question, then the expected counts will be m_i, for $i = 1, \cdots, 8$, where $m_1 = m_2 = \cdots = m_8 = 10$ (obtained as $10 = 80/8$). The value of the expected cells is determined from the research question.

The Hypotheses

To formulate the hypotheses rigourously, I need to be a little careful about the notations I use. Remember that hypotheses are always stated with respect to population parameters. It is the population parameter that is unknown. What comes from the sample will always be known, at least after the data is collected.

Let N_i be the population count in cell i, and n_i the sample count in cell i. Furthermore, M_i will be the expected population count under the null hypothesis, and m_i the expected sample count in cell i based on the number n of units in the sample. The null and alternative hypotheses are formulated as follows:

$$\begin{cases} \mathsf{H_0}\colon N_i = M_i \text{ for all cells } i, \\ \mathsf{H_a}\colon N_i \neq M_i \text{ for at least one cells } i \end{cases} \tag{9.1}$$

There are situations in practice where it is not convenient to formulate the hypotheses with respect to population counts. Population counts are not always well defined, and attempting to define them well is sometimes irrelevant. In this case, the hypotheses should be defined in terms of population proportions. Consider Table 9.1 data for example representing the distribution of a sample of 80 accidents by time of the day. All you really want to known is whether the next accident is as likely to occur between 8 and 9AM as it is in any other time slot. That is, you want to know whether the probability of occurrence of an accident is the same in all time slots. What is the population of accidents? The answer in this case is that a clear-cut definition of the population of accidents is downright irrelevant. Then it becomes convenient to define the hypotheses as follows:

$$\begin{cases} \mathsf{H_0}\colon \pi_i = \pi_{0i} \text{ for all cells } i, \\ \mathsf{H_a}\colon \pi_i \neq \pi_{0i} \text{ for at least one cells } i \end{cases} \tag{9.2}$$

where π is the "true" probability of membership into cell i, and π_{0i} the hypothetical (or expected) probability of membership into category i.

The Test Statistic

I assume that you will have a number k of categories in your study. For Table 9.1 data, $k = 8$. After you collect your sample data, you will be in the position to compute the test statistic denoted as χ^2 (read chi square), which is defined as follows:

$$\chi^2 = \sum_{i=1}^{k} \frac{(n_i - m_i)^2}{m_i} = \sum_{All\ cells} \frac{(Observed - Expected)^2}{Expected}. \quad (9.3)$$

If the null hypothesis is true, then the law of probability associated with the χ^2 statistic is approximated reasonably well with the Chi-square distribution with $k - 1$ degree of freedom, under the general validity conditions of chi-square tests stated below. If the hypotheses are formulated with respect to expected probabilities as in equation 9.2, the expected sample counts m_i of the test statistic are obtained as $m_i = n\pi_{0i}$.

Decision Rule

The null hypothesis is tested at a nominal significance level α to be specified by you the researcher. The null hypothesis H_0 will then be rejected if the observed test hypothesis χ^2_{obs} exceeds the critical value c_α, which represents the $100(1 - \alpha)^{th}$ of the chi-square distribution with $k - 1$ degrees of freedom.

Using Excel 2010, you could calculate the critical value as c_α =CHISQ.INV.RT($\alpha,k - 1$). If you are using Excel 2007, you may replace the CHISQ.INV.RT with CHIINV and use the same parameters.

Validity Conditions of the Chi-Square Test

The following conditions need to be met to ensure the validity of the Chi-square goodness-of-fit test:

(a) The data to be analyzed are nominal. *If the type of your data is ordinal, you may still use the chi-square test. However, alternative and more effective statistical tests that exploits the ordinal nature of the data would be more more appropriate (see chapter 10 for more on this)*

(b) The data being analyzed represent a random sample of *independent* observations.

(c) $m_i \geq 5$ for all cells. That is, the expected cell sample count m_i must exceed 5 for all cells i. *The primary objective of this condition is to ensure the validity of the chi-square distribution with $k - 1$ degrees of freedom as a reasonable approximation to the law of probability of the test statistic. Note that there is no agreement in the statistical community regarding this third condition.*

Cochran (1952) has proposed another condition, which essentially requires all expected cell counts to equal or exceed 1, and the percent of small cells with an expected count of 4 or less to remain below 20%.

However, if some of your cells are very small (e.g.1,2, or 3), my recommendation is to collapse the smallest cells in a logical manner so as to create bigger cells for analysis.

If your interest in the validity conditions of the chi-square test goes beyond what is offered here, you may want to read Zar (1999, P. 470) for further discussion.

Example 9.1

Using Table 9.1 data, I want to test the hypothesis that work-related accidents are equally likely to occur in any of the 8 time slots of the day, at the 2% significance level. Let π_i be the probability that an accident occurs in time interval i ($i = 1, \cdots, 8$) where time interval 1 is 8 to 9AM, and time interval 8 is 4 to 5 PM.

(a) *The Hypotheses*

H_0: $\pi_1 = \pi_2 = \cdots = \pi_8 = 0.125$ versus

H_a: $\pi_i \neq 0.125,$ *for some time interval i*

(b) *The Significance Level.* It is considered to be $\alpha = 0.02$.

(c) *The Test Statistic.* The chi-square test statistic of equation 9.3 with expected counts expressed in the form of $m_i = n\pi_{0i} = 80 \times 0.125 = 8$.

(d) *The Decision Rule.* The critical value $c_\alpha = c_{0.02}$ is calculated from Excel 2010 as 16.62 =CHISQ.INV.RT(0.02,8 − 1). The decision rule is then formulated as follows:

Reject H_0 if $\chi^2_{obs} \geq 16.62$

(e) *The Final Decision.* Table 9.2 shows all the steps for calculating the observed chi-square test statistic $\chi^2_{obs} = 24.6$, which exceeds the critical values 16.62. This leads to the rejection of the null hypothesis. Consequently work-related accidents do not have an uniform frequency of occurrence during a typical workday.

Table 9.2 : Chi-Square Test of Table 9.1 Data

$(i)^a$	$(n_i)^b$	$(m_i)^c$	$(n_i - m_i)^2$	$(n_i - m_i)^2/m_i$
8 to 9 A.M.	6	10	16	1.6
9 to 10 A.M.	6	10	16	1.6
10 to 11 A.M.	20	10	100	10.0
11 to 12 P.M.	8	10	4	0.4
1 to 2 P.M.	7	10	9	0.9
2 to 3 P.M.	8	10	4	0.4
3 to 4 P.M.	19	10	81	8.1
4 to 5 P.M.	6	10	16	1.6
Total	80	80	.	**24.6**

[a]Time interval
[b]Observed Accidents
[c]Expected Accidents

Example 9.2

Suppose you want to study the causes of missing goods in a men's clothing store. To this end, you decide to select a sample of 100 boxes that had been tampered with and ascertained that, for 60 of the boxes, the missing pants, and shoes were attributed to shoplifting. For 30 other boxes employees had stolen the goods, and for the remaining 10 boxes the missing items are due to poor inventory control. *Can I claim that shoplifting is twice as likely to be the cause of the loss as either employee theft or poor inventory control and that employee theft and poor inventory control are equally likely? Use the 0.05 significance level.*

Solution

Let π_S, π_E, and π_I be the probability that a missing good is due to shoplifting, employee theft, and poor inventory respectively. The exact values of these probabilities are unknown to me. It follows from the research question that the hypothetical values π_{0S}, π_{0E}, and π_{0I} of these probabilities must satisfy the following conditions:

$$\begin{cases} \pi_{0S} = 2(\pi_{0E} + \pi_{0I}), \\ \pi_{0E} = \pi_{0I}. \end{cases} \tag{9.4}$$

Using equation 9.4, and the fact that the 3 hypothetical probabilities must sum to (i.e. $\pi_{0S} + \pi_{0E} + \pi_{0I} = 1$), you can conclude that $4\pi_{0E} = 1$. Consequently, $\pi_{0E} = \pi_{0I} = 0.25$, and $\pi_{0S} = 0.50$. Now I can formulate the hypotheses, and carry out the remaining hypothesis testing steps.

(a) *The Hypotheses*

H_0: $\pi_S = 0.50$, *and* $\pi_E = \pi_I = 0.25$ versus

H_a: *One of the 3 probabilities π_S, π_E, and π_I is different from its hypothetical value.*

(b) *The Significance Level.* It is considered to be $\alpha = 0.05$.

(c) *The Test Statistic.* It is the chi-square test statistic of equation 9.3 with expected counts expressed in the form of $m_i = n\pi_{0i}$. This leads to the following expected values $m_S = 100 \times 0.5 = 50$, $m_E = m_I = 100 \times 0.25 = 25$.

(d) *The Decision Rule.* The critical value $c_\alpha = c_{0.05}$ is calculated with Excel 2010 as 5.992 =CHISQ.INV.RT(0.05,3 − 1). The decision rule is then formulated as follows:

Reject H_0 if $\chi^2_{obs} \geq 5.992$.

(e) *The Final Decision.* Table 9.3 shows all the steps for calculating the observed chi-square test statistic $\chi^2_{obs} = 12$, which exceeds the critical values 5.992. This leads to the rejection of the null hypothesis. Consequently, the distribution of missing items does not match its hypothesized description.

Table 9.3: Chi-Square Test of Missing Good Data

Causes	(n_i)	(m_i)	$(n_i - m_i)^2$	$(n_i - m_i)^2/m_i$
Shoplifting	60	50	100	2
Employee theft	30	25	25	1
Poor inventory	10	25	225	9
Total	100	100	.	12

Example 9.2 has demonstrated that the availability of expected counts should not always be taken for granted. At times, you will be required to do some preparatory work before obtaining these expected counts.

The P-value

The P-value associated with the chi-square test of homogeneity can be obtained using Excel 2010 for example in the same way the critical value was obtained.

$$\text{P-value} = \text{CHISQ.DIST.RT}(\chi^2_{obs}, k - 1).$$

In Example 9.2, I obtained $\chi^2_{obs} = 12$. Since the test statistic follows the chi-square distribution with 2 degrees of freedom, the P-value is obtained as 0.00248 =CHISQ.DIST.RT(12,2) . This too indicates a strong evidence in the sample in favor of the research hypothesis H_a.

Using the R Package

The `chisq.test` function provided by the R package is very convenient for performing the chi-square goodness-of-fit test. You need to define the vector of observed counts `x.observed`, and the vector of hypothetical (or expected) probabilities `p.expected`, the name of which is assigned to parameter `p`. This function would be used as follows:

$$\text{chisq.test}(x.observed, \ p = p.expected)$$

Example 9.3 ──────────────────────────

The missing good problem of Example 9.2 can be resolved in a straightforward manner suing the following R instructions:

```
> x.observed <- c(60,30,10)
> p.expected <- c(0.50,0.25,0.25)
> chisq.test(x.observed, p = p.expected)

        Chi-squared test for given probabilities

data: x.observed
X-squared = 12, df = 2, p-value = 0.002479
>
```

9.3. The Chi-Square Test of Independence

In section 9.2, I discussed the chi-square goodness-of-fit test where the "true" distribution of subjects across different levels of a single factor is tested against a hypothetical distribution. This technique is built upon the empirical distribution of a single sample of subjects across the same factor levels. The question I want to answer in this section is whether or not two factors I am interested are connected somehow. For example, when reading a newspaper, some readers will focus on the news, sports, or the comics sections depending on their interests. An interesting question is whether the section that the readers read first is somehow related to the type of community (City, Suburb, Rural) in which they reside. The two factors of interest in this example are Factor 1 = "Newspaper Section" with 3 categorical levels (News, Sports, Comics), and Factor 2 = "Community of Residence" with 3 categorical levels (City, Suburb, Rural). Because I will be testing independence between the 2 factors, this test is called the test of independence. The name Chi-square associated with the test stems from the chi-square distribution that approximates the law of probability of the test statistic reasonably well under mild regularity conditions.

The empirical evidence that feeds this statistical technique is provided by a single sample of n individuals or objects that will be cross-classified by factors 1 and 2. Table 9.4 shows the distribution of $n=1,024$ newspaper subscribers by their community of residence, and the section of the newspaper that they read first.

The problem to be addressed is whether based on Table 9.4, one may conclude that the community of residence affects the section of the newspaper that is read first. The strategy for answering this question consists of finding the configuration of the contingency table that is expected under the assumption of independence of the two factors, and then measure the gap that separates the observed and the expected configurations using the appropriate measure that will later be called the test statistic. A gap that exceeds your comfort zone defined by the critical region will force you to reject the independence hypothesis.

Table 9.4 : Distribution of 1,204 newspaper subscribers by community and the section of the newspaper they read first

	Newspaper Section			
Community	News	Sports	Comics	Total
City	170	124	90	384
Suburb	120	112	100	332
Rural	130	90	88	308
Total	420	326	278	1,024

How do you figure out the contingency table configuration under the hypothesis of independence ? This task is accomplished using the 2 "Total" margins of Table 9.4. The vertical margin for example gives you for each community, the likelihood for it to be the residence of a newspaper subscriber. The ratio $384/1024 = 0.375$ represents the probability for a subscriber to reside in a city. If you now know that 420 subscribers read the news section first, you would normally expect $420 \times 0.375 = 157.5$ of them (ignore the decimal point for now) to reside in a city. That is how expected cell counts are calculated. You may now create the entire expected table as shown Table 9.5. You could see each expected cell count as following the Binomial distribution where the probability of success is given by the vertical marginal proportion.

In general you are going to have a contingency table with **r** rows and **c** columns. A typical row will be labeled as i, while a column is labeled as j, which defines a table cell (i, j). An arbitrary subject will be classified in row i with probability $\pi_{i\cdot}$, in column j with probability $\pi_{.j}$, and into cell (i,j) with probability π_{ij}. In the mathematical language, independence means that all 3 probabilities are linked as $\pi_{ij} = \pi_{i.}\pi_{.j}$.

Table 9.5 : Expected Distribution of 1,204 newspaper subscribers

	Newspaper Section			
Community	News	Sports	Comics	Total
City	157.50	122.25	104.25	384
Suburb	136.17	105.70	90.13	332
Rural	126.33	98.05	83.62	308
Total	420	326	278	1024

The Hypotheses

The null and alternative hypotheses for the test of independence are formulated as follows:

$$\begin{cases} \text{H}_0: & \pi_{ij} = \pi_{i.}\pi_{.j} \text{ for all table cells } (i,j), \\ \text{H}_a: & \pi_{ij} \neq \pi_{i.}\pi_{.j} \text{ for a particular cell } (i,j). \end{cases} \qquad (9.5)$$

The specific formulation of these hypotheses is a technical issue. What is essential is the understanding that the null hypothesis represents the independence of factors, while the alternative hypothesis represents some form of dependency of factors.

The Test Statistic

In a typical contingency table with r rows, and c column, n_{ij} represents the observed count in cell (i,j), $n_{i.}$ the marginal total of row i, and $n_{.j}$ the marginal total of column j. As explained earlier, the expected count associated with cell (i,j) is defined as follows:

$$m_{ij} = \frac{n_{i.}n_{.j}}{n}. \qquad (9.6)$$

The test statistic of the chi-square independence test, denoted by χ^2 is defined as follows:

$$\chi^2 = \sum_{All\ cells} \frac{(Observed - Expected)^2}{Expected} \tag{9.7}$$

This test statistic can be rewritten in a more mathematical form as follows:

$$\chi^2 = \sum_{i=1}^{r}\sum_{j=1}^{c} \frac{(n_{ij} - m_{ij})^2}{m_{ij}}. \tag{9.8}$$

The law of probability of the test statistic associated with the chi-square test of independence is the Chi-square distribution with $(r-1)(c-1)$ degrees of freedom.

Be aware that a number of statistical packages do not use the test statistic of equation 9.8 on 2×2 tables. Instead, they implement a slightly modified version called the *Continuity-corrected chi-square statistic* initially proposed by Yates (1934), and defined as follows:

$$\chi^2 = \sum_{i=1}^{r}\sum_{j=1}^{c} \frac{(|n_{ij} - m_{ij}| - 0.5)^2}{m_{ij}}, \tag{9.9}$$

where $|\cdot|$ represents the absolute value (the R package by the way, uses equation 9.9 by default on 2 by 2 tables, and equation 9.8 with no correction in all other situations). Note that the chi-square distribution is not the exact law of probability of χ^2 of equation 9.8. Instead, it is only an approximation, and the continuity correction aims at improving that approximation. The term "continuity correction" reflects the fact that the χ^2 statistic is discrete, and must be smoothed a little further so that the law of probability of the corrected statistic is closer to the chi-square distribution.

The Decision Rule

As previously indicated, the test statistic of equation 9.8 measures the gap between observed and expected distributions of subjects by factors 1 and 2. You now need to decide what threshold it must exceed before the hypothesis of independence must be rejected. It is you the researcher who decides what threshold will be used. The common guideline has always been to formulate the rule so that the probability of wrongly rejecting H_0 remain below a significance level α of your choice.

The threshold that accomplishes this goal is denoted by $\chi^2_{\alpha,(r-1)(c-1)}$, and is implicitly defined by,

$$P(\chi^2 \geq \chi^2_{\alpha,(r-1)(c-1)}) = \alpha. \qquad (9.10)$$

This critical value can be calculated using Excel 2010 as follows:

$$\chi^2_{\alpha,(r-1)(c-1)} \quad \text{=CHISQ.INV.RT}(\alpha,(r-1)(c-1))$$

Hence, the decision rule:

> Reject H_0 if $\chi^2_{obs} \geq \chi^2_{\alpha,(r-1)(c-1)}$.

The P-value is defined as,

> P-value $= P(\chi^2 \geq \chi^2_{obs})$, $\qquad (9.11)$

and can be calculated with Excel 2010 in a straightforward manner as,

$$\text{P-value} = \text{CHISQ.DIST.RT}(\chi^2_{obs}, (r-1)(c-1)). \qquad (9.12)$$

Validity Assumptions

The chi-square test of independence should be used when the following conditions are all satisfied:

(a) The two factors being analyzed must be nominal. *That is, these factors must represent mutually exclusive categories.*

(b) The subjects must represent a random sample, and be independent. *That is the sample is not supposed to contain duplicates for example.*

(c) The cell expected counts must equal or exceed 5. *Again, the alternative conditions of Cohran (1952) could be used. They stipulate that (i) $m_i \geq 1$ and (ii) 20% of the cells or fewer can have expected counts of 5 or less.*

Example 9.4 _____

I now want to complete the test of independence between the community of residence and the newspaper section as described in Table 9.4. The table of expected counts (Table 9.5) has already been produced. Table 9.6 contains all 9 squared corrected Pearson residuals[2] $(n_{ij} - m_{ij})^2/m_{ij}$ going into the calculation of the test statistic of equation 9.8.

The summation of all 9 cells led to the observed test statistic $\chi^2_{obs} = 7.3399$. The number of degrees of freedom associated with the chi-square distribution of the test statistic is $df =$

---

[2]Note that the Pearson residual associated with cell (i,j) is $(n_{ij} - m_{ij})/\sqrt{m_{ij}}$

$(3-1)(3-1) = 4$. Assuming that the test is being done at the 5% significance level, the corresponding critical value is obtained from Excel 2010 as

$$\chi^2_{0.05,4} = 9.488 \;\; =\text{CHISQ.INV.RT}(0.05,4)$$

Table 9.6 : $(n_{ij} - m_{ij})^2/m_{ij}$ for all Table Cells (i,j)

| | Newspaper Section | | |
Community	News	Sports	Comics
City	0.9921	0.0251	1.9478
Suburb	1.9206	0.3761	1.0802
Rural	0.1067	0.6617	0.2297

Observed chi-square statistic χ^2_{obs} $=$ 7.3399

The observed test statistic of 7.3399 is smaller than the critical value 9.488. Therefore, the null hypothesis cannot be rejected. There is insufficient evidence in the sample to support any impact that the community of residence may have on the newspaper section that subscribers read first.

If you are interested in the p-value, you may obtain it from Excel 2010 as,

$$\text{P-value} = 0.119 \;\; =\text{CHISQ.DIST.RT}(7.3399,4) \;,$$

which is too large, indicating weak evidence in the sample in favor of the alternative hypothesis.

Using the R Package

The `chisq.test` function provided by the R package is very convenient for performing the chi-square test of independence. All you need to do is define the contingency table in a matrix `table.observed`, and use it as single parameter in calling chisq.test. This function would be used as follows :

$$\texttt{chisq.test}(\textit{table.observed})$$

The R package can actually produce other interesting things such as the expected counts,

> chisq.test(*table.observed*) $expected,

or the Pearson residuals, which you can obtained as,

> chisq.test (*table.observed*)$residuals

Example 9.5 _____

The test of independence between the community of residence and the newspaper section as described in Table 9.4 can be implemented in a straightforward manner suing the following R instructions:

```
> table.observed <- matrix(c(170, 124, 90,
+ 120, 112, 100,
+ 130, 90, 88), nrow = 3,byrow=TRUE)
> chisq.test(table.observed)

        Pearson's Chi-squared test

data: table.observed
X-squared = 7.3399, df = 4, p-value = 0.1190
>
```

9.4. Chi-Square Homogeneity Test for Binary Factors

Table 9.7 shows the distribution of 400 plant seeds by seed type, and by their germination status indicating whether the seed germinated within 5 weeks of planting. It follows from this table that 87 of the 147 type 3 seed germinated during the observation period. Chapter 7 entitled "Two-sample test of hypothesis for proportions" describes the techniques for comparing two seed types (Type 1 and 2 for example) with respect to their probabilities to germinate within 5 weeks. The 2 populations under investigation in this case are the population of type 1 seeds, and the population of type 2 seeds. In this section, I am dealing not with 2 populations, but with 5 populations. I want to be able to answer a more general question, which is whether all 5 populations of plant seeds are similar (or homogeneous[3]) with respect to their probabilities to germinate with 5 weeks.

As previously discussed, the strategy for solving this problem is to create a copy of Table 9.7 showing the expected distribution of 400 plant seeds under the hypothesis of population homogeneity. The gap separating the observed and expected counts will be evaluated with the test statistic before making the final determination.

Table 9.7 : Distribution of 400 Plant Seeds by Type and Germination Status

Seed Type	Germinated	Failed to Germinate	Total
1	31	7	38
2	57	33	90
3	87	60	147
4	52	44	96
5	10	19	29
Total	237	163	400

The observed and the expected tables will share the same vertical and horizontal Total margins (also called row totals, and column totals), and only the inside of the tables will differ. Consi-

---

[3]Here is where the name "test of homogeneity" originates.

der for example the 38 Type 1 seeds, and the 237 seeds that germinated. If all seed types are equally likely to germinate, then of all 400 experimental seeds, you would normally expect that $38 \times 237/400 = 22.5$ will germinate[4].

Let n be the total number of sample subjects from all populations under investigation ($n = 400$ in Table 9.7), n_{i+} the number of subjects in sample i (e.g. $n_{2+} = 90$ in Table 9.7), n_{+1}, and n_{+2} the number of subjects in the first and second categories respectively (e.g. $n_{+1} = 237$ and $n_{+2} = 163$ in Table 9.7). For an arbitrary population i, the expected counts m_{i1} and m_{i2} in categories 1 and 2 respectively are calculated as follows:

$$m_{i1} = n_{i+} \frac{n_{+1}}{n}, \;\; and \;\; m_{i2} = n_{i+} - m_{i1}. \qquad (9.13)$$

That is, once you calculate the first column expected count m_{i1}, the second column count is derived from it. Using equation 9.13, I calculated the expected version (see Table 9.8) of Table 9.7 under the assumption of homogeneity

Table 9.8 : Expected Distribution of 400 Plant Seeds by Type and Germination Status Under Homogeneity Hypothesis

Seed Type	Germinated	Failed to Germinate	Total
1	22.52	15.49	38
2	53.33	36.68	90
3	87.10	59.90	147
4	56.88	39.12	96
5	17.18	11.82	29
Total	237	163	400

---

[4]The basic idea here is that all 5 seed types will share the exact sample germination probability, which is $237/400 = 0.5925$

Hypotheses and Test Statistic

Let P be the number of populations under investigation, and π_{i1} the probability that a randomly selected subject from population i be classified into category 1. The null hypothesis is that of homogeneity of the populations with respect to the likelihood of the subjects to belong to category 1. The null and alternative hypotheses are formulated as follows:

$$\left\{ \begin{array}{ll} \text{H}_0: & \pi_{11} = \pi_{21} = \cdots = \pi_{\text{P}1}, \\ \text{H}_a: & \textit{2 populations have unequal category 1} \\ & \textit{membership propensities} \end{array} \right. \qquad (9.14)$$

The *Test Statistic*, denoted by χ^2 is expressed as,

$$\begin{aligned} \chi^2 &= \sum_{\substack{All\ cells}} \frac{(Observed - Expected)^2}{Expected}, \\ &= \sum_{i=1}^{\text{P}} \sum_{j=1}^{2} \frac{(n_{i1} - m_{i1})^2}{m_{i1}} \end{aligned} \qquad (9.15)$$

The law of probability governing the variation of this test statistic is the chi-square distribution with P $-$ 1 degrees of freedom.

The Decision Rule and Its Use

Suppose that you will be testing homogeneity in the populations at a significance level α. The null hypothesis should be rejected if the observed test statistic χ^2_{obs} exceeds the critical value denoted as $\chi^2_{\alpha,\text{P}-1}$, and which represents the $100(1-\alpha)^{th}$ percentile of the chi-square distribution with P $-$ 1 degrees of freedom. If you are an Excel user, then the critical value could be calculated with Excel 2010 as follows:

$$\chi^2_{\alpha,\text{P}-1} = \text{CHISQ.INV.RT}(\alpha,\text{P-1}) \qquad (9.16)$$

The decision rule can now be formulated as follows:

$$\text{Reject } H_0 \text{ if } \chi^2_{obs} \geq \chi^2_{\alpha,\text{P}-1}. \qquad (9.17)$$

Validity of the Homogeneity Test

The chi-square test of homogeneity described above will be valid only if the following conditions are all met:

(a) The 2 categories being studied are mutually exclusive.

(b) The samples are selected randomly and independently from each of the P populations.

(c) The expected count in each of the cells is 5 or greater (*you may use Cochran's set of conditions here as well (Cochran, 1952, and as discussed in the previous section*).

Example 9.6 _____

I now want to perform a test of homogeneity of the 5 plant seed populations, whose sample data are displayed in Table 9.7. I have already calculated the expected counts in Table 9.8. My next step is to compute the test statistic.

Table 9.9: Squared Pearson residuals $(n_{ij} - m_{ij})^2/m_{ij}$

Seed Type	Germinated	Failed to Germinate
1	3.1977	4.6494
2	0.2533	0.3683
3	0.0001	0.0002
4	0.4187	0.6088
5	3.0024	4.3654
χ^2_{obs} (Observed test statistic):		16.8640

Assuming that you want to test homogeneity at the 5% significance level, the critical value will be $\chi^2_{0.05,4} = 9.4877$ obtained with Excel 2010 as =CHISQ.INV.RT(0.05,4) , where $4 = 5 - 1 = P - 1$. It appears that the observed test statistic exceeds the critical value. Consequently, the null hypothesis must be rejected. There is sufficient evidence in the sample against the homogeneity assumption. The seed types do not have the same propensity for germination.

On the use of R for testing homogeneity

If you are an R user, you will want to know that the chisq.test function can be used for testing populations homogeneity, in the same way it was used for testing independence in Example 9.5.

9.5. Chi-Square Homogeneity Test for Multiple-Level Factors

In section 9.4, I presented the chi-square test of homogeneity for binary factors, where several populations are compared to each other with respect to the propensity for their members to belong to either one of the 2 categories under investigation. The same procedure can be extended from binary factors to multiple-level factors with 3 levels or more in a straightforward manner. For multiple levels, you will be testing the homogeneity of several populations with respect to membership propensity of belonging to any of numerous categories under investigation.

The procedure for testing homogeneity of several populations for multiple-level factor is identical to the procedure for testing 2 multiple-level factors for independence (see section 9.3), only the interpretation of results differ. You could see the different populations being compared as the levels of a second factor, and test for independence using the exact same procedure of section 9.3. When interpreting your results, you will need to remember that you are testing homogeneity of the populations, and not independence of two factors.

CHAPTER 10

Analysis of Variance

OBJECTIVE

In this chapter, you will learn one of the most popular statistical techniques among practitioners. The Analysis of variance. It allows you to perform a global test of several population means, and is an extension of the two-sample t-test for means to the more general situation of 3 populations or more.

CONTENTS

10.1. Overview

Table 10.1 shows the productivity of 3 employees at a sporting goods store. Productivity is measured by the number of customers served on different days that were randomly selected independently for each employee. It appears for example that Jennifer served 74 customers on day 2. You as manager would like to know whether Peter, Jennifer, and Kosta all have the same productivity level. Table 10.1 provides a limited amount of information on the employees' productivity based a few days that you randomly selected. This will result in a measure of productivity that is subject to sampling error. That is why I must use a statistical test that accounts for this error, before I can determine whether all 3 employees are equally productive or not.

Table 10.1 : Number of customers served by employee

Peter	Jennifer	Kosta
57	65	45
53	74	52
49	69	47
56	70	50
	68	49
	71	

In chapter 8, I presented the two-sample test of hypothesis for means. Can I not use that technique to perform 3 pairwise t-test between Peter and Jennifer, between Jennifer and Kosta, and between Peter and Kosta? And why would I need a new testing procedure called the Analysis of Variance (**ANOVA**) to do this job? The answer is yes, you can very perform 3 pairwise t-tests to do this job. But there is a problem with that solution that I am now going to explain. If you can live with that problem you will certainly not need ANOVA.

The fundamental question I am trying to answer is whether all 3 employees have the same productivity level or not. This will be the global null hypothesis. You may decide to perform each of the 3 pairwise t-tests at the 5% significance level, and reject the global null hypothesis if any of the 3 pairwise null hypotheses of equal performance between 2 employees is rejected. With this strategy, the probability of wrongly rejecting the global null hypothesis (i.e. the global significance level) will likely not be what you want it to be. You will therefore lose the most appealing aspect of the theory of hypothesis testing, which is to have complete control over the significance level.

The second problem posed by the pairwise t-test approach is that even if you set the 3 pairwise significance levels so that the global significance level is what you want, the result will be a weak test of the global hypothesis. This procedure will be weak in the sense of being unable to reject the global null hypothesis because none of the pairwise was rejected. A special procedure optimized on the testing the global null hypothesis will be more powerful for detecting small differences than 3 independent pairwise tests. The Analysis of Variance (or ANOVA) is one such global procedure.

The ANOVA will take different forms depending on the type of data you want to analyze. Table 10.1 describes data from 3 samples selected independently for each of the 3 employees. The Employee is the single factor with 3 levels being analyzed here, each factor level having its own independent data. This will lead to the *Single-Factor ANOVA* (also called *One-Way ANOVA*) for independent samples to be discussed in section 10.2. There are times when a researcher wants to monitor change over time, or measure progress. Table 10.2 shows performance data of a sample of 9 employees observed on the first Saturday of the months of January, February, March, and April. The Month is a single factor with 4 levels that is being investigated, and performance measurements are taken from the same employee sample, and repeated for each of the 4 levels of the study factor. The resulting analysis will be called *Single-Factor Repeated-Measures ANOVA* or *One-Way Repeated-Measures ANOVA* discussed in section 10.4. Different techniques must be used for different study designs to ensure the validity of the law of probability associated with the test statistics.

Table 10.2 : Number of customers served by month and by employee

Employee	January	February	March	April
John	57	57	59	62
Paul	53	55	59	58
Peter	49	45	58	57
Mary	56	60	65	67
Gabriel	45	60	49	56
Kosta	55	59	66	68
Jennifer	47	66	62	65
Marquez	77	79	82	80
Donald	67	69	68	69

The ANOVA generally requires the Normal distribution to be a reasonable approximation of the law of probability underlying the sample data, whether the samples are dependent or independent. When satisfied, this assumption often leads to powerful statistical procedures. Unfortunately, the Normality assumption will not always hold for your data. In this case, you will use the Kruskal-Wallis test for independent samples as discussed in section 10.3 , and the Friedman test discussed in section 10.5. The use of these two tests will result in some loss of power in the ability to reject the null hypothesis of equality when some differences indeed exist. Figure 10.1 depicts the different ANOVA techniques that I will discuss in this chapter along with the conditions under which they will be used.

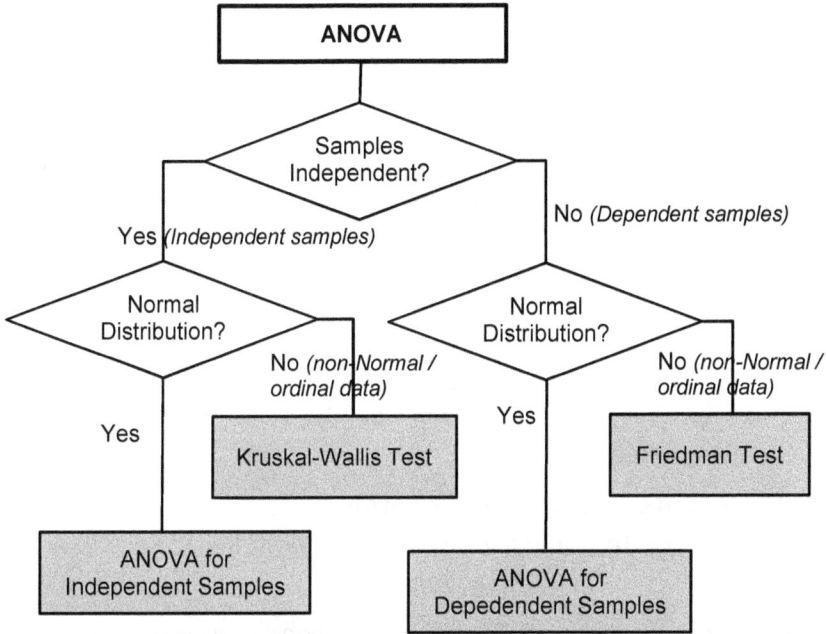

Figure 10.1. ANOVA Procedures and their Conditions of Applicability

10.2. ANOVA for Independent Samples

In this section, I will discuss the single-factor Analysis of Variance for independent samples, which is the technique you will need to analyze Table 10.1 data. In this table, you are dealing with 3 populations, which are the population of Peter's performance scores, the population of Jennifer's performance scores, and the population of Kosta's performance scores. Each population may contain a very large number of scores, with Table 10.1 only providing a snapshot of it with sample scores. The null hypothesis you want to test is whether the population means μ_P,

μ_J, and μ_K are all equal, using the sample data of Table 10.1. The null and alternative hypotheses are formulated as follows:

$$\begin{cases} \text{H}_0: \ \mu_P = \mu_J = \mu_K, \\ \text{H}_a: \ \textit{Two of the 3 means } \mu_P, \mu_J, \mu_K, \ \textit{are unequal} \end{cases} \quad (10.1)$$

The general strategy for solving this problem is to use the 3 samples of Table 10.1 as follows:

(a) Compute a measure of variation of the 3 sample means (i.e. each sample mean is taken as the unit of analysis), known as the *Between-Groups Variability*. This will evaluate the extent to which, Peter, Jennifer, and Kosta differ. If they all have the same means, the between-groups variability will be small.

(b) Compute a measure of variation within the 3 samples (this is essentially some average of the 3 sample variances), known as the *Within-Groups Variability*.

(c) Compute the ratio F of the Between-Groups Variability to the Within-Groups Variability.

$$\text{F} = \frac{\textit{Between-Groups Variability}}{\textit{Within-Groups Variability}}. \quad (10.2)$$

The reasoning here is that, if $\mu_P = \mu_J = \mu_K$, then you would normally have about the same variation between samples as within samples, and the F should be close to 1. If it exceeds 1 by a "sufficiently" large margin (beyond a critical value), then you would reject the null hypothesis.

You can now see that the ANOVA procedure described here amounts to a series of variance calculations, hence the name *Analysis of Variance*. The F ratio, which will later be referred to as the ANOVA test statistic, is the ratio of the variance of means to mean of variances.

Validity Conditions

When implementing ANOVA in general, you will be comparing k population means (in Table 10.1 data, $k = 3$). Before deciding to use the single-factor ANOVA for independent samples, you need to ensure that the following conditions are all satisfied:

(a) The data being analyzed is interval or ratio.

(b) Each sample has been randomly selected from the population it represents. *For all practical purposes, this condition will ensure that the samples are independent, and that the values are not repeated*

(c) The law of probability underlying the sample data can be reasonably well approximated by the Normal distribution. *This condition will ensure the validity of the law of probability associated with the test statistic.*

(d) *Homogeneity of Variance.* The variances of the underlying k populations represented by k samples must be equal to one another. *Although there are statistical procedures for testing variance homogeneity, such as the Hartley's F_{max} test (see Hartley (1940, 1950)), I advise against using them as part of your ANOVA analysis. Using them will make your whole testing procedure unwieldy with no guarantee whatsoever of improving things. My advice is to use your knowledge of the experimental processes that generated the data (this subject-matter knowledge could be valuable), and determine if there are reasons to believe that the k populations may have different variances. Do not hesitate to discuss about this issue among peers to get different opinions. If you do not find such a reason, then assume the variance homogeneity condition satisfied.*

Why does one need the homogeneity of variance condition? I indicated earlier that if the between-groups variability exceeds the within-groups variability by a substantial margin, you would reject the null hypothesis of equality of population means. This logic will hold, only if between-groups variability can be primarily attributed to the difference in means. Whether this logic can still hold if the population variances are heterogenous is uncertain.

10.2.1   The ANOVA Computations

In this section, I will present the mechanics of ANOVA is a more formal way. Although, the actual computations will often be done by the computer, it is in the best interest of the researcher to have some grasp of the processing activities being performed with the input data.

You have k populations under investigation, each of which is represented by a sample of data. The number of observations in each sample may differ from sample to sample, and sample i will contain n_i observations (e.g. in Table 10.1, sample 2 is the Jennifer sample containing $n_2 = 6$ observations). The total number of observations collected for the entire experiment is n_T. That is $n_T = n_1 + n_2 + \cdots + n_k$ (in Table 10.2, $n_T = 4 + 6 + 5 = 15$).

Let s_i^2 be the sample variance associated with sample i. The within-group variability, which is often referred to in many statistics textbooks as the Mean Square Error (**MSE**) is given by:

$$\text{MSE} = \frac{1}{n_T - k} \sum_{i=1}^{k} (n_i - 1) s_i^2. \tag{10.3}$$

The **MSE** is also called the Error Mean of Squares, and represents the weighted sum of the individual sample variances (note that $n_T - k$ is the summation of all values of $n_i - 1$ over the k samples). The term error is used here to reflect the fact that the variation within the sample cannot be attributed to the factor being investigated, and is seen as coming from random errors that you cannot control.

Let $\bar{x}_i$ be the mean of the n_i observations associated with sample i, and $\bar{x}$, the mean of all n_T observations in the experiment. The between-groups variability, also referred to in statistics textbook as the Mean Square Treatment[1] (MSTR) or the Treatment Mean of Squares, is given by:

$$\text{MSTR} = \frac{1}{k-1} \sum_{i=1}^{k} n_i(\bar{x}_i - \bar{x})^2, \qquad (10.4)$$

and measures the variance of the sample means.

The Hypotheses

$$\left\{ \begin{array}{l} \text{H}_0: \mu_1 = \mu_2 = \cdots = \mu_k, \\ \text{H}_a: \textit{Two of the } k \textit{ means are unequal} \end{array} \right. \qquad (10.5)$$

The objective is to decide whether or not the null hypothesis must be rejected at a prescribed significance level α specified by the researcher.

The Test Statistic

The test statistic for testing H_0 is the F-ratio defined as,

$$F = \frac{\text{MSTR}}{\text{MSE}}. \qquad (10.6)$$

When the null hypothesis is true, the law of probability associated with the F statistic is well approximated by the F distribution with 2 numbers of degrees of freedom given by $df_1 = k - 1$ and $df_2 = n_T - k$.

---

[1]The use of the term "Treatment" here stems from the origin of ANOVA in agricultural field trials to evaluate the effectiveness of treatments by pesticides

The Decision Rule

The null hypothesis will be rejected if the observed value F_{obs} of the test statistic exceeds the $100(1-\alpha)^{th}$ percentile of the $F(k-1, n_T - k)$ distribution that I will denote by $F_{\alpha, k-1, n_T}$.

The critical value $F_{\alpha, k-1, n_T}$ can be obtained in several ways depending on the tool you are using.

Excel 2010: $F_{\alpha, k-1, n_T}$ =F.INV.RT$(\alpha, k - 1, n_T - k)$

Excel 2007: $F_{\alpha, k-1, n_T}$ =FINV$(\alpha, k - 1, n_T - k)$

Calc: $F_{\alpha, k-1, n_T}$ =FINV$(\alpha; k - 1; n_T - k)$

R: $F_{\alpha, k-1, n_T}$ <-qf$(\alpha, k - 1, n_T - k,$ lower.tail=FALSE)

Decision Rule: Reject H_0 if $F_{obs} \geq F_{\alpha, k-1, n_T}$.

Unless you are using OpenOffice Calc, you will seldom need these functions. Excel (both versions 2007 & 2010 considered in this book) through the Analysis ToolPak, and the R package have modules and function for performing ANOVA, which will do all the calculations for you.

Example 10.1

I now want to use the ANOVA technique and the sample data of Table 10.1 in order determine whether Peter, Jennifer, and Kosta have the same work performance. The null and alternative hypotheses for this problem are defined in expression (10.1), and will be tested at the significance level $\alpha = 0.05$.

Using Excel's Analysis ToolPak[2], I was able to produce the ANOVA table whose content is shown in Table 10.3.

---

[2]Section E.4 of Appendix shows the steps for using Excel to produce the ANOVA Table 10.3

Table 10.3: ANOVA Table for Table 10.1 Performance Data

ANOVA

Source of Variation	SS	df	MS	F	P-value	F crit
Between Groups	785.8	2	392.9	7.44	0.0079	3.89
Within Groups	633.53	12	52.79			
Total	1419.33	14				

The 3 most important numbers in this table are (i) the observed F statistic $F_{obs} = 7.44$, the P-value $= 0.0079$, and the critical value F crit $= 3.89$ (this corresponds to $F_{0.05,2,12}$). It appears that F_{obs} exceeds the critical value. Consequently the null hypothesis must be rejected. Therefore, Peter, Jennifer, and Kosta do not perform at the same level.

As for the remaining numbers in the ANOVA table, MS stands for Mean Square, and MSTR $= 392.9$, the Mean Square Treatment that measures the between-groups variability, while MSE $= 52.79$, the Mean Square Error that measures the within-groups variability.

ANOVA Using R

Using the R package to perform the single-factor ANOVA requires the input data to contain the Factor column (i.e. Employee column in Example 10.1), and the Count column (i.e. Score in Example 10.1)). I decided to create a `csv` file (my preferred option for creating data-frames for R) named,

`chapter10_example10_1.csv,`

and organized as shown in Table 10.4.

The ANOVA analysis can then be performed as shown in Figure 10.2. The `read.csv` function reads your `csv` file to create the data-frame perf.data. The `aov` function performs the ANOVA itself, while summary produces the ANOVA table you need. This

ANOVA is similar to that of Excel, with the exception that the critical value is not provided. This should not be a problem since this critical value can be obtained from R as shown earlier in this section.

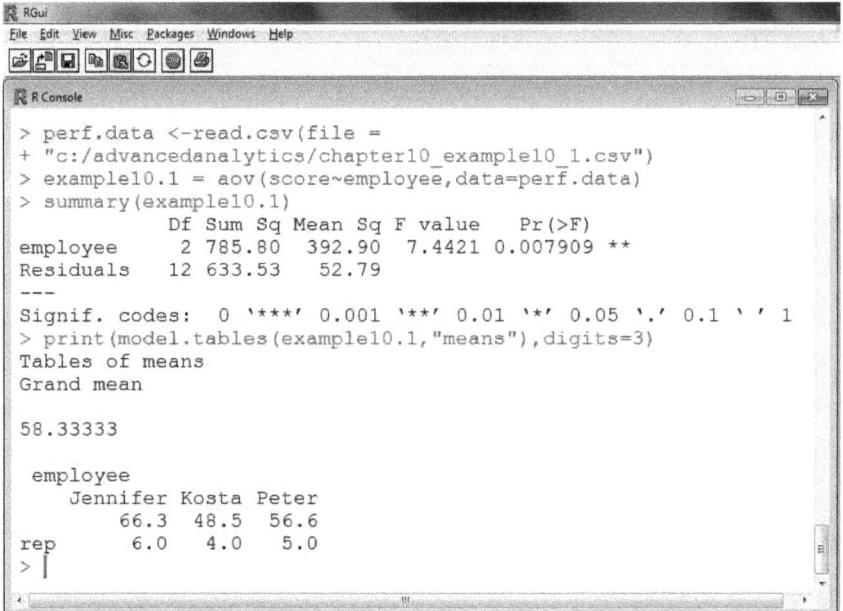

```
> perf.data <-read.csv(file =
+ "c:/advancedanalytics/chapter10_example10_1.csv")
> example10.1 = aov(score~employee,data=perf.data)
> summary(example10.1)
            Df Sum Sq Mean Sq F value    Pr(>F)
employee     2 785.80  392.90  7.4421  0.007909 **
Residuals   12 633.53   52.79
---
Signif. codes:  0 '***' 0.001 '**' 0.01 '*' 0.05 '.' 0.1 ' ' 1
> print(model.tables(example10.1,"means"),digits=3)
Tables of means
Grand mean

58.33333

 employee
    Jennifer Kosta Peter
        66.3  48.5  56.6
rep      6.0   4.0   5.0
>
```

Figure 10.2. ANOVA Analysis of Example 10.1 using R

You also use the print() function in order to generate summary statistics similar to those produced by Excel.

Table 10.4 : The `chapter10_example10_1.csv` File

```
employee,score
Peter,57
Peter,53
Peter,49
Peter,56
Peter,68
```

```
Jennifer,65
Jennifer,74
Jennifer,69
Jennifer,70
Jennifer,49
Jennifer,71
Kosta,45
Kosta,52
Kosta,47
Kosta,50
```

10.2.2 The Pairwise Comparisons

The Problem

In Example 10.1, I rejected the null hypothesis of equality of means, meaning that Peter, Jennifer, and Kosta do not perform at the same level. This conclusion suggests that there are high and low performers in the group of 3 employees. I will want to know who performs better than who? Achieving this goal requires that I conduct a series of pairwise comparisons. These often unplanned tests, which are conducted after the data have been collected are known under the term *Post hoc* analysis.

You do not have to conduct all pairwise comparisons possible. You have the luxury of deciding whether to conduct 2, 3 or any arbitrary number of pairwise comparisons. The statistical problem to be concerned about here stems from the fact that each additional pairwise comparison you perform increases the probability of committing one type I error within your predetermined set (or family) of comparisons. Therefore, you want to exercise some control over this probability.

You will commit a *Familywise type I error* on the family of comparisons if you commit a type I error on anyone of the pairwise comparisons within the family. The associated probability will be denoted by α_{FW}. The type I error you commit on any specific comparison is referred to as the *Per Comparison type I error*, and the associated probability denoted by α_{PC}.

Suppose that the number of pairwise comparisons is c. Then α_{FW} and α_{PC} will be linked as follows:

$$\alpha_{FW} = c \cdot \alpha_{PC}. \tag{10.7}$$

Example 10.1 offers a maximum of 3 possible pairwise comparisons (Peter-Jennifer, Jennifer-Kosta, Peter-Kosta). If each of them is done at the 0.05 significance (i.e. $\alpha_{PC} = 0.05$) then equation 10.7 leads to a familywise type I error probability of $\alpha_{FW} = 3 \times 0.05 = 0.15$. This value may be too high for some researchers.

The dilemma that the statistical community has faced for years with this problem is that reducing the familywise type I error probability will necessarily increase the type II error probability. Reducing the familywise type II error probability will necessarily increase the type I error probability. A wide variety of methods have been proposed for addressing this problem, ranging from Fisher's LSD[3] test, which controls the familywise type II error, and not the type I error, to the Bonferroni-Dunn test (see Dunn (1961)) that controls the familywise type I error, but not the type II error. Between these two extremes some compromises, which I will not discussed in this book, were proposed such as the Tukey's HSD[4] test by Tukey (1953), or the Scheffé test by Scheffé (1953).

---

[3]LSD stands for Least Significant Difference
[4]HSD stands for Honestly Significant Difference

The Hypotheses and the Test Statistic

Suppose that a pairwise comparison you are interested consists of testing two population means μ_a and μ_b. Depending on the type of comparison you are interested in, the hypotheses could be formulated as follows:

Two-tailed test: H_0: $\mu_a = \mu_b$ *versus* H_a: $\mu_a \neq \mu_b$,
Right-tailed test: H_0: $\mu_a \leq \mu_b$ *versus* H_a: $\mu_a > \mu_b$,
Left-tailed test: H_0: $\mu_a \geq \mu_b$ *versus* H_a: $\mu_a < \mu_b$

The test statistic t for testing these hypotheses is defined as follows:

$$t = \frac{\overline{x}_a - \overline{x}_b}{\sqrt{\mathsf{MSE}(1/n_a + 1/n_b)}}, \qquad (10.8)$$

where MSE is the Mean Square Error from the ANOVA table, n_a is the number of observations associated with factor level a, and n_b the number of observations associated with factor level b.

The law of probability of the t statistic of equation 10.8 is well approximated by the Student t-distribution $t(n_T - k)$ with $n_T - k$ degrees of freedom.

Fisher's LSD Test

The Fisher's LSD test is sometimes referred to as the multiple t tests, and will prescribe a per-comparison type I error probability $\alpha_{PC} = \alpha$ to be applied to all pairwise comparisons of interest. Because the number of pairwise comparisons is not an integral part of the testing procedure, you will want to limit the number of such comparisons to a minimum[5] in order to prevent the familywise type I error from getting out of control (see equation

---

[5]Some sources recommend the number of comparisons using Fisher's LSD method not to exceed the number of degrees of freedom associated with the MSTR (the between-groups variability measure)

10.7). If you perform c comparisons, the familywise type I error probability will be $\alpha_{FW} = c \cdot \alpha$. The LSD test is based on the following decision rules:

Two-tailed test: Reject H_0 if $|t_{obs}| \geq t_{\alpha/2, n_T - k}$,
Right-tailed test: Reject H_0 if $t_{obs} \geq t_{\alpha, n_T - k}$,
Left-tailed test: Reject H_0 if $t_{obs} \leq -t_{\alpha, n_T - k}$,

Bonferroni-Dunn Test

The Bonferroni-Dunn prescribes a familywise type I error probability α_{FW}, which will be achieved by using a per-comparison type I error probability $\alpha_{PC} = \alpha_{FW}/c$, where c is the number of pairwise comparisons you have planned. The Bonferroni-Dunn test is based on the following decision rules:

Two-tailed test: Reject H_0 if $|t_{obs}| \geq t_{\alpha_{PC}/2, n_T - k}$,
Right-tailed test: Reject H_0 if $t_{obs} \geq t_{\alpha_{PC}, n_T - k}$,
Left-tailed test: Reject H_0 if $t_{obs} \leq -t_{\alpha_{PC}, n_T - k}$,

I indicated in previous chapters that the critical value $t_{\alpha, df}$ associated with the t-statistic can be obtained with Excel 2010 as $=\text{T.INV}(1 - \alpha, df)$.

Example 10.2

Consider once again the work performance data of Example 10.1. Since the null hypothesis of equality of means was rejected, I now like to perform pairwise comparison to determine what employees have differences in performance levels that are statistically significant. All pairwise tests are two-tailed, and Table 10.5 summarizes these analyzes according to the Fisher's LSD and Bonferroni-Dunn methods.

It follows that according to the Fisher LSD approach, the Peter-Jennifer and Jennifer-Kosta differences are statistically significant at the per-comparison significance level of 0.05, unlike the Peter-Kosta difference, which is not. However, all 3 comparisons carry a familywise significance level of 0.15.

According to the Bonferroni-Dunn approach, only the Jennifer-Kosta difference is statistically significant at the per-comparison significance level of 0.0167 (=0.05/3), the other differences are not. However, all 3 Bonferroni-Dunn pairwise comparisons now carry an overall familywise significance level of only 0.05.

Table 10.5 : Pairwise Comparisons of Table 10.1 Performance Data

Comparison	t-stat	Fisher's LSD		Bonferroni-Dunn	
		Crit Val[a]	type 1[b]	Crit Val[a]	type 1[b]
Peter-Jennifer	-2.21	2.179	0.05	2.780	0.0167
Peter-Kosta	1.66	2.179	0.05	2.780	0.0167
Jennifer-Kosta	3.80	2.179	0.05	2.780	0.0167

[a] *Crit Val stands for Critical Value*
[b] *type 1 refers to the per-comparison type I error probability*

It appears from Table 10.5 that Fisher's LSD test will have good power in terms of its ability to reject the null hypothesis when it is false, but may yield a much higher type I error probability than the Bonferroni-Dunn test. Note that the number of degrees of freedom associated with the MSTR is 2, which is the maximum number of comparisons recommended for Fisher's LSD test. Two comparisons will yield a familywise type I error probability of 0.10, which is reasonable. The Bonferroni-Dun test controls the type I error probability very well, but is very conservative. Unless you have a strong interest in conduction many pairwise comparisons, my recommendations would be to stick with Fisher's LSD test, and to limit the number of comparisons to 2 or 3.

10.3. Kruskal-Wallis Test

The single-factor ANOVA discussed in section 10.2 requires the law of probability underlying the data to be reasonably close to the Normal distribution, and the population variances to show some homogeneity. You may not feel comfortable with one or both of these assumptions. An alternative approach widely used by researchers is the Kruskal-Wallis test. It was suggested by Kruskal (1952) and, Kruskal and Wallis (1952), and requires the data to be ordinal. This test was designed primary to compare population medians, and not population means. Should this be a problem? Not really. In fact, if the underlying populations are symmetric, population means and medians become identical. Otherwise, the median is what you should be interested in.

The Mann-Whitney test was discussed in chapter 8 for testing two population means when the t-test assumptions were violated. The Kruskal-Wallis test generalizes the Mann-Whitney test to 3 populations or more. Both tests are based on ranks instead of raw data, which makes them applicable to a variety of data types. In that sense, the Kruskal-Wallis test is another nonparametric test, which does not require the data to follow a particular law of probability. The downside being the loss of power due to the use of ranks in place of actual measurements.

10.3.1 Kruskal-Wallis Test Statistic

Assuming that you want to compare k population means, the hypotheses would be formulated as follows:

$$
\begin{cases}
\text{H}_0\text{: } \mu_1 = \mu_2 = \cdots = \mu_k, \\
\text{H}_a\text{: } \textit{Not all } \mu_i \textit{ are equal.}
\end{cases}
\tag{10.9}
$$

Whether the μ_i represents the population mean or the population median is irrelevant when defining the Kruskal-Wallis test. You will not be estimating these parameters. Instead, you will merely be comparing them.

Test Statistic

The Kruskal-Wallis test statistic will be denoted by K. It is calculated by first ranking all n_T observations (all samples combined) in ascending order from 1 to n_T. Let $\overline{R}_{i.}$ be the average of the ranks associated with population i, and $\overline{R}$ the average of all the ranks. The K statistic is calculated as follows:

$$\mathsf{K} = \frac{12}{n_T(n_T+1)} \sum_{i=1}^{k} n_i \left(\overline{R}_{i.} - \overline{R}\right)^2. \tag{10.10}$$

The law of probability associated with this test statistic is well approximated by the Chi-square distribution with $k - 1$ degrees of freedom, when the null hypothesis is true. If the null hypothesis is true, then all population distributions would be the same. Any random sample taken from any population would yield approximately the same mean rank $\overline{R}_{i.}$. This would lead to a small value for the K statistic, which by the way tells you how far you expect the mean rank from any given sample to be away from the overall mean rank.

Let $\chi^2_{\alpha,k-1}$ be the $100(1 - \alpha)^{th}$ percentile of the chi-square distribution. The decision rule for the Kruskal-Wallis test if formulated as follows:

$$\text{Reject } H_0 \text{ if } \mathsf{K}_{obs} \geq \chi^2_{\alpha,k-1}. \tag{10.11}$$

Validity Conditions

Several authors including Conover (1980, 1999), Daniel (1990), or Marascuilo and McSweeney (1977) mentioned a series of assumptions upon which they indicated the Kruskal-Wallis was based. These are:

(a) Each of the k samples is randomly selected from the population it represents. *(A key aspect of this assumption is the need to avoid a large number of duplicates in the observations)*

(b) The k samples under study are independent. *(A key aspect of this assumption is to avoid using the Kruskal-Wallis test on repeated measurements taken on several occasions. Repeated-measure procedures are discussed in the next section should be used.)*

(c) The analytic variable used for ranking is continuous. *(This assumption is not critical to ensuring the validity of the Kruskal-Wallis test, and is often ignored by practitioners. The concern here is that if the null hypothesis of true, you want all k populations to be homogeneous, which may not quite be the case with discrete variables. If you are dealing with an unusual dataset, you may want to pay attention to this issue.)*

(d) The probability distributions underlying the sample data are identical in their shape. *(The purpose of this assumption is to ensure that a rejection of the null hypothesis can only be attributed to the difference in means.)*

(e) Each independent sample has a size of 5 or more. *(This ensures the validity of the chi-square approximation)*

Example 10.3

Suppose that you want to analyze Table 10.1 data using the Kruskal-Wallis test, because of your suspicion that the Normality assumption necessary for a valid single-factor ANOVA may not be satisfied. The test of equality of means will be conducted at the 0.05 significance level. To compute the test statistic, I created Table 10.6 using Excel, where you can see the observed Kruskal-Wallis test statistic $K_{obs} = 7.2413$. To obtain the rank column, I used this very convenient Excel function named RANK.AVG(?,?,1) (see section E.9 of Appendix E for a description of this function under the title "Rank Tests").

Table 10.6 : Computation of the Observed K Statistic

Name	Score	Rank	$\overline{R}_i$	$\overline{R}$	n_i	$n_i(\overline{R}_i - \overline{R})^2$
Peter	57	9	7.7	8	5	0.45
Peter	53	7				
Peter	49	3.5				
Peter	56	8				
Peter	68	11				
Jennifer	65	10	11.25	8	6	63.375
Jennifer	74	15				
Jennifer	69	12				
Jennifer	70	13				
Jennifer	49	3.5				
Jennifer	71	14				
Kosta	45	1	3.5	8	4	81
Kosta	52	6				
Kosta	47	2				
Kosta	50	5				
K_{obs} *(Observed K statistic)*$\longrightarrow$						7.2413

The critical value is 5.99, obtained using Excel 2010 as follows:

=CHISQ.INV.RT(0.05,2) .

Because $K_{obs} = 7.2413$ exceeds 5.99, you must reject the null hypothesis of mean equality.

Kruskal-Wallis Test with the R Package

The R package offers the `kruskal.test()` function for implementing the Kruskal-Wallis test. Its use is straightforward as shown in Figure 10.3. Data for Peter, Jennifer, and Kosta are defined as vectors, which are then supplied as arguments to the `kruskal.test()` function. "Kruskal-Wallis chi-squared" is the observed value of the test statistic K_{obs}, `df` the number of degrees of freedom $(k - 1)$ associated with the chi-square distribution, and the P-value.

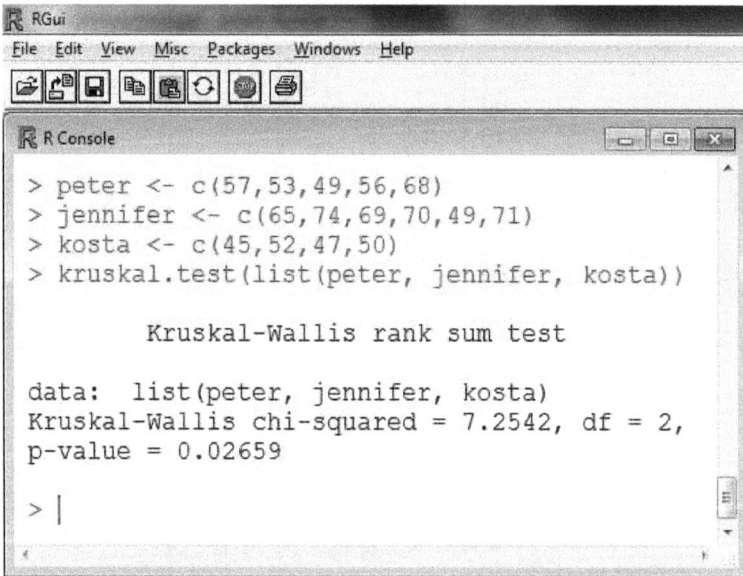

```
> peter <- c(57,53,49,56,68)
> jennifer <- c(65,74,69,70,49,71)
> kosta <- c(45,52,47,50)
> kruskal.test(list(peter, jennifer, kosta))

        Kruskal-Wallis rank sum test

data:  list(peter, jennifer, kosta)
Kruskal-Wallis chi-squared = 7.2542, df = 2,
p-value = 0.02659

>|
```

Figure 10.3. Kruskal-Wallis Test of Table 10.1 with the R Package

The critical value is not provided by the `kruskal.test()` function, but can be obtained as `qchisq(1-0.05,2)`, 0.05 being the significance level, and 2 the number of degrees of freedom.

You may have noticed that the observed Kruskal-Wallis statistic given by R in Figure 10.3, which is 7.2542, is slightly different from the Excel result of Table 10.6, which is 7.2413. This discrepancy stems from the fact that R's Kruskal-Wallis statistic is tie-corrected.

Tie Correction for the Kruskal-Wallis Test

Several sources in the literature recommend that the Kruskal-Wallis test statistic be divided by an adjustment factor C to correct it for the presence of tied scores. Tie correction will increase the test statistic slightly making the test more powerful. The correction factor C is defined as follows:

$$C = 1 - \frac{\sum_{i=1}^{s}(t_i^3 - t_i)}{n_T^3 - n_T}, \qquad (10.12)$$

where s is the number of different series of ties, and t_i the number of tied scores within the ith series. The resulting tie-corrected Kruskal-Wallis statistic H_C is defined as follows:

$$H_C = H/C. \qquad (10.13)$$

It follows from Table 10.6 that the number of series of ties in Example 10.3 is 1 (i.e. $s = 1$), and the number of tied scores within that series is 2 (Peter and Jennifer have the same score of 49). Therefore the tie correction factor is $C = 1 - (2^3 - 2)/(15^3 - 15) = 1 - 6/3360 = 0.998214$. The tie-corrected Kruskal-Wallis statistic becomes $H_C = 7.2413/0.998214 = 7.2542$. The law of probability of the tie-corrected statistic is expected to be closer to the chi-square distribution with $k-1$ degrees of freedom, than the law of probability of the uncorrected statistic.

10.3.2 Pairwise Comparisons

After rejecting the null hypothesis of equality of means with the Kruskal-Wallis test, it becomes important to determine which means are different from one another. This goal is achieved with the pairwise comparisons. For the sake of comparing two means μ_a and μ_b, the pairwise test will be similar to that of section 10.2, with the following two exceptions:

(i) The test statistic is,

$$z = \frac{\bar{x}_a - \bar{x}_b}{\sqrt{n_{\mathrm{T}}(n_{\mathrm{T}} + 1)(1/n_a + 1/n_b)/12}} \qquad (10.14)$$

(ii) The law of the probability associated with the z test statistic is approximated with the Standard normal distribution.

Fisher LSD test, or the Bonferroni-Dunn test can be used with the above test statistic as previously done in section 10.2.

10.4. ANOVA for Dependent Samples

The one-factor ANOVA for dependent samples is a technique for comparing means or medians of 3 populations or more, which are created by repeated measurements taken from the same initial group of subjects or objects. Several names have been used in the literature to designate this technique. It is also known as single-factor within-subjects ANOVA, single-factor repeated-measures ANOVA, or randomized-blocks one-way ANOVA. Each of these names is justified by the type of data being analyzed, and the type of statistical technique used, as I will show later.

Table 10.7 contains return rate data in a men's clothing store by salesperson and day of the week. The problem of interest here is to test if the return rate depends on the day of the week, using the daily return rates associated with the same group of 9 salespersons.

Table 10.7 : Return Rates of clothing in a men's clothing store by salesperson and day of the week

Day	Salesperson								
	1	2	3	4	5	6	7	8	9
Monday	2.8	5.9	3.3	4.4	1.7	3.8	6.6	3.1	0.0
Tuesday	3.6	1.7	5.1	2.2	2.1	4.1	4.7	2.7	1.3
Wednesday	1.4	0.9	1.1	3.2	0.8	1.5	2.8	1.4	0.5
Tuesday	2.0	2.2	0.9	1.1	0.5	1.5	1.4	3.5	1.2

The group of 9 employees used in Table 10.7 is not required to be a random sample of the employee population. The purpose of this study is not to compare employees with respect to the return rates of purchased items. Instead, the problem is to compare the 4 days with respect to the return rate ($k = 4$ represents the number of populations being compared, "$x = return\ rate$" is the analytic variable, and "$n = 9$" is the number of subjects used in the experiment). Had you selected 9 employees randomly and independently for each of the days that you could have used the single-factor ANOVA for independent samples discussed in the previous section. However, the repeated-measures design may be more indicated in this case if you believe that the employee may have an impact on the return rate.

The randomization of the experiment that produced Table 10.7 data comes from the days, which you may select randomly from the set of 7 days. Alternatively, you could preselect the 4 days you are interested in, then for each employee randomly assign the order in which you will observe the return rates[6]. The

---

[6]Random assignment of days means that for one employee, the Tuesday rate will be observed first, followed by the Monday's (the following week), while the order of observation may differ for another employee with the first

objective is to avoid a situation where the Friday return rate affects the Saturday rate for a particular employee.

The Hypotheses

For k arbitrary populations to compare, the hypotheses to be tested would be formulated as follows:

$$\begin{cases} \mathsf{H}_0: \ \mu_1 = \mu_2 = \cdots = \mu_k, \\ \mathsf{H}_a: \ Note \ all \ \mu_i \text{'s} \ are \ equal \end{cases} \qquad (10.15)$$

For the return rate problem of Table 10.7, the null hypothesis becomes $\mathsf{H}_0: \ \mu_{\mathrm{MO}} = \mu_{\mathrm{TU}} = \mu_{\mathrm{WE}} = \mu_{\mathrm{TH}}$, and the strategy for addressing this problem revolves around (i) quantifying the the between-day variability using the Mean Square Treatment (MSTR) as measure, (ii) quantifying the within-employee variability using the Mean Square Error (MSE)[7], (iii) If the MSTR exceeds the MSE by a wide margin, you can conclude that the observed day effect is not merely due to random errors, but actually reflects a real change of the return rate by day. The null hypothesis must then be rejected.

Table 10.7 gives the impression that there are 2 factors being investigated here. That is untrue. "Day of the week" is the only factor of interest. The repeated-measures design uses Employee to separate the Employee effect, which could be large, from the error effect so that the latter can be minimized. The minimized error effect will make the day-of-week effect stand out when both effects are compared so that the null hypothesis can be rejected when it is wrong.

In Table 10.7, return rates are not simply organized by day of the week, they are also organized by blocks of data defined by

---

rate being taken on a Wednesday for example.

[7] *This actually measures the effect of random errors beyond the experimenter's control. Therefore, the MSE will be adjusted to remove the effect that the between-day variability has on the within-employee variability*

the Employee who is not a subject in the study. It is this blocking role played by the Employee factor to reduce experimental error that earned this ANOVA technique the name Randomized-Blocks One-Way ANOVA.

The Test Statistic and the Decision Rule

Although the repeated-measures ANOVA will typically be implemented using a computer program, I like to briefly discuss here how the test statistic is obtained, and how it should be interpreted. This discussion will allow you to have a deeper understanding of the software output.

Let x_{ij} be the score associated with the ith block, and j^{th} treatment. The Mean Square Treatment (MSTR), which measures of between-treatment variability is given by:

$$\text{MSTR} = \frac{n}{k-1} \sum_{j=1}^{k} (\overline{x}_{.j} - \overline{x})^2, \qquad (10.16)$$

where $\overline{x}_{.j}$ is treatment j's mean, n is the number of measurements per treatment, and $\overline{x}$ the overall mean. The Mean Square Error (MSE) for measuring within-block variability, is given by:

$$\text{MSE} = \frac{1}{n-1} \sum_{i=1}^{n} s_i^2 - n\text{MSTR}/(n-1), \qquad (10.17)$$

where s_i^2 is block i's sample variance.

The test statistic for the repeated-measures single-factor Analysis of Variance is defined as follows:

$$F = \frac{\text{MSTR}}{\text{MSE}}. \qquad (10.18)$$

The law of probability associated with this test statistic is reasonably well approximated by Fisher's F distribution with $k - 1$ and $(k - 1)(n - 1)$ degrees of freedom.

The decision rule for controlling the Type I error at a significance level α is:

$$\text{Reject } H_0 \text{ if } F_{obs} \geq F_{\alpha, k-1, (n-1)(k-1)}, \qquad (10.19)$$

where $F_{\alpha, k-1, (n-1)(k-1)}$ is the $100(1 - \alpha)^{th}$ percentile of the F distribution $F\big[k - 1, (k - 1)(n - 1)\big]$.

Example 10.4

Using the return rate data of Table 10.7 and the repeated-measures ANOVA, I like to test the hypothesis of equality of return rates across days of the week at the 5% significance level.

Response

The null and alternative hypotheses are defined as follows:

$$\begin{cases} \text{H}_0: \; \mu_{\text{MO}} = \mu_{\text{TU}} = \mu_{\text{WE}} = \mu_{\text{TH}}, \\ \text{H}_a: \; \mu_{\text{MO}}, \mu_{\text{TU}}, \mu_{\text{WE}}, \; and \; \mu_{\text{TH}} \; are \; not \; all \; equal. \end{cases}$$

I used Excel (either version 2007 or 2010 will do the job) and Table 10.7 data to produce the ANOVA table displayed in Table 10.8. The detailed steps for using Excel to do this can be found in section E.5. It is crucial at this stage to interpret the ANOVA table properly by using only relevant numbers. It follows from from Table 10.7 that the data pertaining to the study factor (i.e. "Day of the week") is displayed horizontally, row-wise. Therefore, the ANOVA table row labeled as "Rows" contains the numbers most relevant to our analysis. The row labeled as "Columns" contains information pertaining to the blocking factor Salesperson, which is irrelevant for our analysis.

Table 10.8: Repeated-Measures ANOVA for the Return Rate Data of Table 10.7

ANOVA						
Source of Variation	SS[a]	df[b]	MS[c]	F	P-value	F crit
Rows	28.001	3	9.334	6.580	0.0021	3.009
Columns	26.265	8	3.283	2.315	0.0535	2.355
Error	34.044	24	1.418			
Total	88.310	35				

[a] *SS stands for Sum of Squares*
[b] *df = degrees of freedom*
[c] *MS stands for Mean Square, and is the ratio of SS to df*

It follows from Table 10.8 that for the purpose of our analysis, $MSE = 1.418$, $MSTR = 9.334$, the observed test statistic $F_{obs} = 6.580$, and the 5% critical value is $F_{0.05,3,24} = 3.009$. Since the observed test statistic exceeds the critical value, the null hypothesis must be rejected. You may also use the P-value of 0.0021, which is small, as an indication of strong evidence in the sample in favor of the alternative hypothesis.

Using R for Implementing Repeated-Measures ANOVA

The R package offers the possibility to conduct the repeated-measures single-factor ANOVA. Like in almost all statistical packages, contingency tables must use a list format where a series of variables display the information vertically. To analyze Table 10.7 data, I created a CSV file named **chapter10_example10_4.csv**, an extract of which is shown in Table 10.9. This table only shows the data pertaining to the first 2 salespersons. Because variable names in R are case-sensitive, I systematically lowercase them.

Table 10.9 : Extract of `chapter10_example10_4.csv`

```
employee,day,rate
P1,MON,2.8
P1,TUE,3.6
P1,WED,1.4
P1,THU,2
P2,MON,5.9
P2,TUE,1.7
P2,WED,0.9
P2,THU,2.2
```

To produce the ANOVA table, you need to execute the commands shown in Figure 10.4.

(i) The first line reads the input dataset.

(ii) The second line is the most important and most delicate to handle. It uses the **aov** function with 2 arguments. The second argument is the dataset created in line 1. In the first argument `rate~day+Error(employee/day)`, `rate` represents the analytic variable, `day` is the factor of interest, and `employee` the blocking factor. Note that in the repeated-measures design, randomization occurs within the blocking factor, which represents the main source of random errors. Therefore, the proper evaluation of the error mean of squares (**MSE**) requires the calculation of the within-blocks variation from which the variation between main-factor levels must be subtracted to account for the crossing of both factors (see equation 10.14). The term `Error(employee/day)` is used to instruct R how the **MSE** should be obtained.

You will notice that R does not automatically produce the critical value as Excel does. This critical value can be obtained from R as follows:

$$qf(0.05,3,24,lower.tail=FALSE),$$

where 0.05 is the significance level, $3 = k - 1$, and $24 = (n - 1)(k-1) = (9-1)(4-1)$ are the 2 numbers of degrees of freedom associated with the **F** statistic.

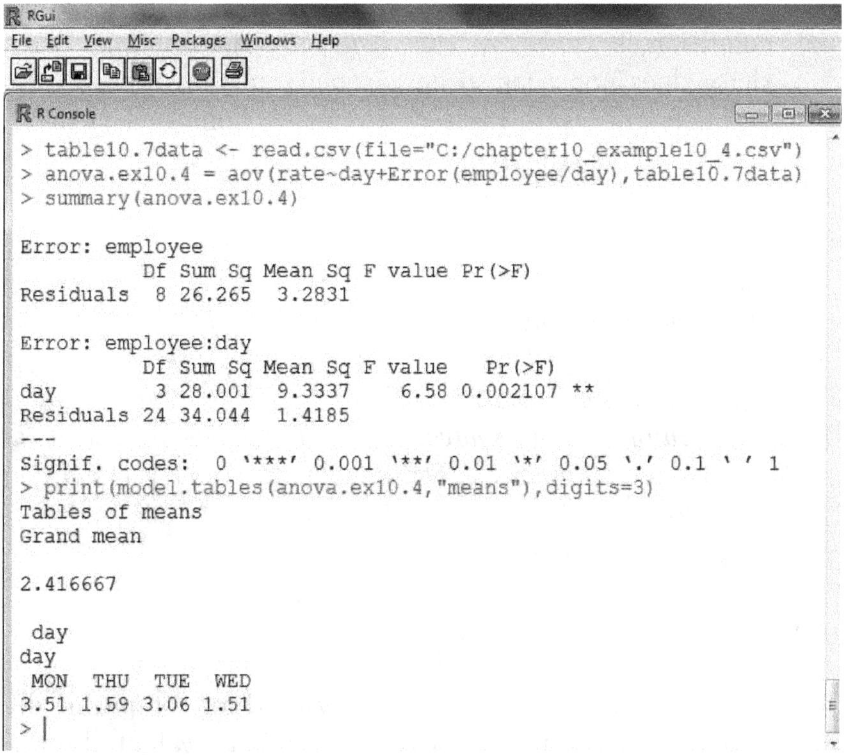

```
R RGui
File Edit View Misc Packages Windows Help

R R Console

> table10.7data <- read.csv(file="C:/chapter10_example10_4.csv")
> anova.ex10.4 = aov(rate~day+Error(employee/day),table10.7data)
> summary(anova.ex10.4)

Error: employee
          Df Sum Sq Mean Sq F value Pr(>F)
Residuals  8 26.265  3.2831

Error: employee:day
          Df Sum Sq Mean Sq F value   Pr(>F)
day        3 28.001  9.3337    6.58 0.002107 **
Residuals 24 34.044  1.4185
---
Signif. codes:  0 '***' 0.001 '**' 0.01 '*' 0.05 '.' 0.1 ' ' 1
> print(model.tables(anova.ex10.4,"means"),digits=3)
Tables of means
Grand mean

2.416667

 day
day
 MON  THU  TUE  WED
3.51 1.59 3.06 1.51
> |
```

Figure 10.4. Repeated-Measures ANOVA with R

Validity Conditions

(a) ANOVA for dependent samples is employed with interval or ratio data.

(b) The law of probability of the analytic variable x for each level of the study factor is well approximated by the Normal distribution.

(c) The sample of k observations within each of the n blocks has been randomly selected from the population it represents.

(d) The change in x-values between two arbitrary levels of the study factor will have a constant variation. This intriguing condition is known as the *Sphericity Assumption*. Sphericity does not refer to homogeneity in the variance of x across factor levels. Instead, it refers to homogeneity in the variance of the difference $x_a - x_b$ between two factor levels a and b, across all pairs of factor levels.

(My recommendation is to proceed with a visual examination of these differences to see if these variances are strikingly different. Weinfurt (2000, p. 331) and Howell (2002, p. 519) believe that a violation of this assumption will primarily affect the post hoc analysis, and not the main test. Many references have suggested methods for testing the sphericity assumptions. These methods should be avoided by practitioners, since the statistical community itself is not in agreement on their merits.)

10.5. Friedman's Test

The Friedman test is typically used for comparing different values of a population mean (or median) evaluated under different conditions, when the number of conditions is 3 or more. That is the measurements will be taken on the same sample of subjects on 3 occasions or more. For example, if you want to evaluate the impact of surface type (e.g. cement, clay, grass) on the performance of racing horses, you may want to select a sample of horses and test that same group on all 3 surfaces.

Since the samples of data are dependent (because based on the same subjects), neither the ANOVA for independent samples nor the Kruskal-Wallis technique can be used. The ANOVA for dependent samples of section 10.4 might not be applicable either if you are concerned about some of its validity conditions not being satisfied. The only option left will be the Friedman's test.

In chapters 6 and 8, I presented the Wilcoxon matched-pairs signed-rank test for comparing 2 population means based on 2 dependent samples. The Friedman test is an extension of the Wilcoxon test to a design involving 3 dependent samples or more. This statistical test, originally proposed Friedman (1937), has been given several names in the literature. Some authors refer to it as the "Friedman Two-Way Analysis of Variance by Ranks" due to the fact that it is based on ranks rather than on the original raw data, and that the *subject* and the *measurement period* are two factors (hence two-way) that affect the magnitude of the measurements. Others have called this technique, "Friedman's test for randomized block experiment." This terminology is borrowed from more advanced concepts in the study of ANOVA techniques that are beyond the scope of this book. This test is also sometimes called the "Repeated Measures ANOVA."

The conditions that must be satisfied to ensure the validity of the Friedman test are very general and met most the times. Several authors such as Conover (1980, 1999), Daniel (1990) have listed the following validity conditions:

► The sample of subjects being analyzed was randomly selected from the population it represents. *In practice, the subjects are not always selected following a rigorous random selection protocol. The key aspect here is to have a sample that is a reasonably good representation of the population.*

► The analytic variable is a continuous variable. *Again, the Friedman test is often used successfully in practice with dis-*

crete variables that take numerous values.

Consider once again the Table 10.7 data on return rates in a men's clothing store by salesperson and day of the week. I like to analyze this dataset again in this section with the Friedman test.

10.5.1 The Hypotheses

The Friedman test should normally be interpreted as the test of equality of the population medians. If you are willing to assume that the underlying probability distributions are symmetric, then testing population medians amounts to testing the population means. Even if the symmetric nature of the underlying distribution is uncertain, the test remains valid but only for testing population medians.

The primary goal in comparing three populations or more is to prove that they are different. Therefore, you will naturally want to protect yourself against the possibility of rejecting the hypothesis of homogeneity among the populations when in reality they are all the same. Consequently, if k dependent populations (or one population observed on k occasions) are being analyzed, the test hypothesis[8] will be defined as follows:

$$H_0: \mu_1 = \mu_2 = \cdots \mu_k, \qquad (10.20)$$

where the symbol μ represents the mean or the median depending on whether the underlying distribution is symmetric or not. The alternative hypothesis is defined as,

$$H_a: \textit{Not all medians } \{\mu_1, \cdots, \mu_k\} \textit{ are equal}^{[9]}. \qquad (10.21)$$

---

[8]This is will be the "null" hypothesis in Fisher's terminology for P-values.

[9]Note that this alternative hypothesis should not be formaule as $H_a: \mu_1 \neq \mu_2 \neq \cdots \neq \mu_k$. This formulation is wrong because the null hypothesis must be rejected even if some means are equal.

10.5.2   The Test Statistic

I am assuming here that you have collected k observations for each of the n subjects (or experimental conditions) in the study. In Table 10.7 for example, you would have $k = 4$ and $n = 9$. The Friedman's test statistic, which I will denote[10] by F_r can be obtained as follows:

(a) Rank all k observations in ascending order from 1 to k separately for each of the n subjects (or experimental conditions). If there are ties, then the average all ranks involved is assigned to each score. Let r_{ij} be the jth rank associated with subject i. That is r_{ij} may take any value from 1 to k.

(b) Sum the ranks separately for each of the k measurement periods (or experimental conditions). *This operation will lead to k rank sums $r_{+1}, \cdots, r_{+k}$ for all k measurement periods.*

(c) The test statistic is calculated as follows:

$$F_r = \frac{12}{nk(k+1)} \sum_{j=1}^{k} r_{+j}^2 - 3n(k+1). \qquad (10.22)$$

The law of probability of this test statistic is approximated by the chi-square distribution with $k-1$ degrees of freedom. This approximation is known to hold even for moderate values of n.

(d) If your dataset contains ties, then it is widely-accepted in the statistical community that the above test statistic must

---

[10]This notation is formed with letter F for Friedman and r to indicate that the statistic is based on ranks, and not on raw measurement scores.

be divided by a tie-correction factor[11] C defined as follows:

$$C = 1 - \frac{\sum_{i=1}^{T}(t_i^3 - t_i)}{n(k^3 - k)}, \qquad (10.23)$$

where T is the total number of tie series, and t_i is the number of tied scores in the ith series of ties. Table 10.7 for example contains a single series of ties associated with salesperson 6. That series is $\{1.5, 1.5\}$ and contains only 2 tied scores. Consequently, $T = 1$, $t_1 = 2$. Since $n = 9$, and $k = 4$, the tie-correction factor C is given by $C = (2^3 - 2)/(9(4^3 - 4)) = 0.98889$.

Once the tie-correction factor C is calculated, it can be used to compute the tie-corrected Friedman test statistic as follows:

$$\mathsf{F}_r^* = \mathsf{F}_r/C. \qquad (10.24)$$

Although tie correction is recommended primarily when the number of ties is excessive, it is nevertheless systematically implemented in several statistical packages.

If you are among those who want to understand why this test statistic works, I am going to provide a simple justification. From the k rank sums $r_{+1}, \cdots, r_{+k}$ you can compute the rank means $\bar{r}_{+1}, \cdots, \bar{r}_{+k}$. When the null hypothesis H_0 is true, you would normally expect these means to be close to one another. The Friedman test statistic is a statistical measure that tells about how far you would expect any given rank mean to stray from the overall average value (which by the way is given by $(k+1)/2$), and can

---

[11]The expression used for computing C (see equation 10.23) is based on a methodology described in Daniel (1990) and Marascuilo and McSweeney (1977)

be rewritten as follows:

$$F_r = \frac{12n}{k(k+1)} \sum_{j=1}^{k} \left[\bar{r}_{+j} - (k+1)/2 \right]^2. \qquad (10.25)$$

10.5.3 The P-value

To compute Fisher's P-value, you must compute the observed value $F_{r(obs)}$ of the Friedman statistic, then evaluate the P-value as the probability for F_r to exceed its observed value. That is,

$$\text{P-value} = P\left(F_r \geq F_{r(obs)}\right). \qquad (10.26)$$

Tables C.1 and C.2 of Appendix C show various ways that this probability may be calculated with Excel, OpenOffice Calc, or the R package. It follows from Table C.1 that $\mathsf{CHISQ.DIST}(F_{r(obs)}, k - 1, \mathsf{TRUE})$ calculates the probability $P\left(F_r \leq F_{r(obs)}\right)$. Consequently, the P-value would be obtained as,

$$= 1\text{- } \mathsf{CHISQ.DIST}(F_{r(obs)}, k - 1, \mathsf{TRUE}).$$

Example 10.5 _____

I now want to use the data in Table 10.7 to test the hypothesis that the clothing return rate varies by day of the week. To this end, I created Table 10.10 that is a version of Table 10.7 where raw return rates are replaced with their ranks within each salesperson's series of return rates. Where there are ties such as for salesperson 6, the rank average is used. Rank sums are then obtained in the column labeled as "Rank Sum." These sums are then squared before the observed test statistic $F_{r(obs)}$ can be calculated following equation 10.22 using $n = 9$ (number of salespersons) and $k = 4$ (the number of days).

Table 10.10 : Rankings of return rates pertaining to clothing
in a men's clothing store by salesperson

Month	\multicolumn{9}{c}{*Salesperson*}	Rank Sum[a]	Squared Sum[b]
	1	2	3	4	5	6	7	8	9		
Monday	2	1	2	1	2	2	1	2	4	17	289.00
Tuesday	1	3	1	3	1	1	2	3	1	16	256.00
Wednesday	4	4	3	2	3	3.5	3	4	3	29.5	870.25
Thursday	3	2	4	4	4	3.5	4	1	2	27.5	756.25
$F_{r(obs)}$ - Observed test statistic											9.767
$F^*_{r(obs)}$ - Tie-corrected test statistic											9.8776

[a]This column represents row-wise rank sums r_{+j}, $j = 1, \cdots, 4$
[b]This column represents squared rank sums r^2_{+j}, $j = 1, \cdots, 4$

To produce the above table from Table 10.7, I used Excel to
duplicate the first table replacing each cell of the duplicate table
with an Excel formula that looks like =RANK.AVG(B2,B$2:B$5)
that could easily be copied throughout the table.

As for the P-value, I obtained it as follows:
0.01965 =1- CHISQ.DIST(9.876,4-1,TRUE) , which represents a
strong evidence in favor of the research hypothesis that return
rates change by day of the week.

In Example 10.5, I used Excel 2010 to obtain the P-value
associated with the tie-corrected Friedman test statistic. If you
are an Excel user, it is one way to resolve this problem. If you
are an R user, you may even have an easier time. In fact, Figure
10.5 shows what I did with R to analyze Table 10.7 data. The
results are the same. All you need to do with R is define the
return.rates matrix properly, and to call the friedman.test
function to obtain both the test statistic and the associated P-
value.

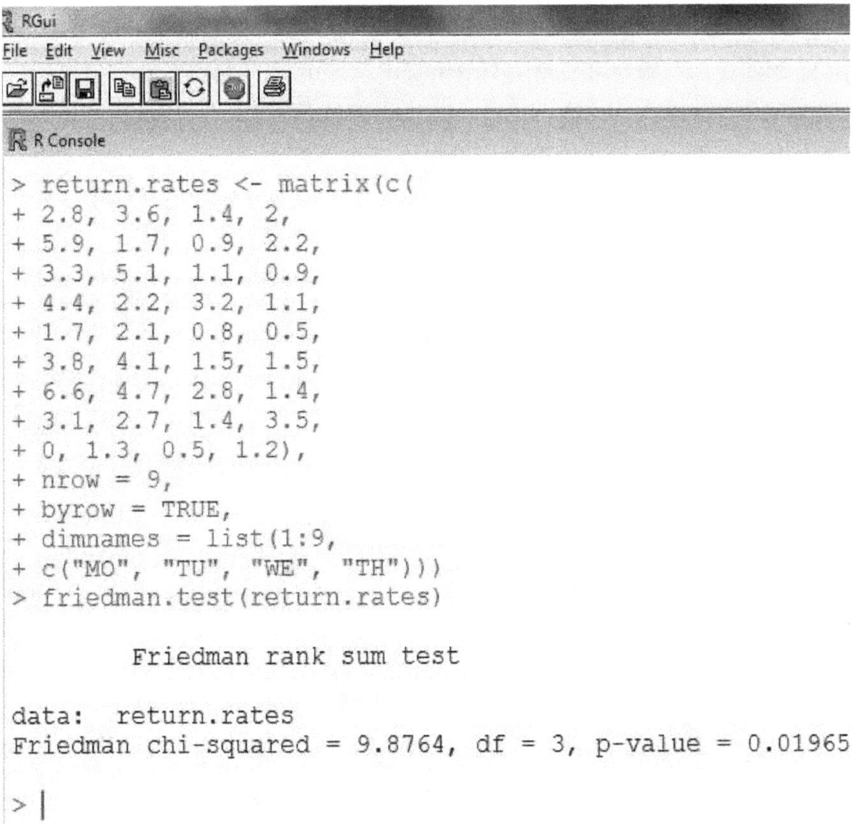

```
R RGui
File  Edit  View  Misc  Packages  Windows  Help

R R Console
> return.rates <- matrix(c(
+ 2.8,  3.6,  1.4,  2,
+ 5.9,  1.7,  0.9,  2.2,
+ 3.3,  5.1,  1.1,  0.9,
+ 4.4,  2.2,  3.2,  1.1,
+ 1.7,  2.1,  0.8,  0.5,
+ 3.8,  4.1,  1.5,  1.5,
+ 6.6,  4.7,  2.8,  1.4,
+ 3.1,  2.7,  1.4,  3.5,
+ 0,  1.3,  0.5,  1.2),
+ nrow = 9,
+ byrow = TRUE,
+ dimnames = list(1:9,
+ c("MO", "TU", "WE", "TH")))
> friedman.test(return.rates)

        Friedman rank sum test

data:  return.rates
Friedman chi-squared = 9.8764, df = 3, p-value = 0.01965

> |
```

Figure 10.5. Analysis of Table 10.1 Data Using R

10.5.4 The Neyman-Pearson's Critical Region

As you must have learned by now, the Neyman-Pearson approach to hypothesis testing requires the calculation of a critical value that determines the critical region, which must include the observed test statistic before the null hypothesis can be rejected.

For the Friedman test, there is only one possible alternative hypothesis that is formulated as shown in 10.27. Assuming that

α is the test's significance level, and that c_α is the $100(1 - \alpha)^{th}$ percentile of the Chi-square distribution with $k - 1$ degrees of freedom (c_α is also known as the critical value), the decision rule is formulated as follows:

$$\boxed{\text{Reject } H_0 \text{ if the } F^*_{r(obs)} \geq c_\alpha} \qquad (10.27)$$

Using Excel 2010, the critical value c_α is calculated as =CHISQ.INV($1 - \alpha, k$-1). See Tables C.3 and C.4 in Appendix C for alternative methods for computing this critical values with Excel 2007, Calc 3.2.0 or R.

If you want to test the null hypothesis of Example 10.5 at a significance level of $\alpha = 5\%$, then the critical value needed to accomplish this is 7.815 calculated as =CHISQ.INV(1-0.05,4-1) .

10.5.5 Pairwise Comparison

In the previous example I rejected the null hypothesis, suggesting that the return rates for clothing is not uniform across days without indicating which days are different and which are similar. You may want to know whether the Wednesday return rate exceeds that of Tuesday. These are the follow-up analyzes (or post-hoc analyzes) that I will discuss in this section.

As previously indicates, you as researcher needs to first determine the number of pairwise analyzes that you are interested in. In the clothing return rate example, you may decide that you only want to know whether the Wednesday return rate exceeds that of Tuesday (i.e. H_a: $\mu_{WE} > \mu_{TU}$). In this case you will have a single pairwise comparison of interest. If you want to compare Monday to Tuesday, Tuesday to Thursday, and Monday to Thursday then you will have 3 pairwise comparisons. Let me assume that you are interested in c comparisons. With c comparisons come c decisions to be made regarding the rejection or non-rejection of the different null hypotheses. In the context of pairwise comparison, you commit a type I error if one of the c null hypotheses is erroneously rejected. Your problem will then be to ensure that the probability of this special type I error does not exceed a specified threshold α.

Example 10.6

Consider the return rates of Example 10.7, and suppose you want to know whether the pairwise differences between Tuesday, Wednesday and Thursday are statistically significant. More specifically, you want to check the truthfulness of the following claims:

(a) H_a: $\mu_{TU} < \mu_{WE}$, *(b)* H_a: $\mu_{TH} < \mu_{WE}$, and *(c)* H_a: $\mu_{TU} < \mu_{TH}$. This leads to the following corresponding test hypotheses:

(a) H_0: $\mu_{TU} \geq \mu_{WE}$, *(b)* H_0: $\mu_{TH} \geq \mu_{WE}$, and *(c)* H_0: $\mu_{TU} \geq \mu_{TH}$.

Solution

The number of pairwise comparisons I am interested in here is c = 3. This set of 3 comparisons forms your family of pairwise comparisons. I then choose the *Familywise Type I Error Rate* of $\alpha_{FW} = 0.05$, which ensures that the overall probability of wrongly rejecting any of the 3 null hypotheses will not exceed 5%.

However, the pairwise comparisons will be tested one at a time. For each test, I apply a special significance level called the *Per Comparison Type I Error Rate*, and given by:

$$\alpha_{PC} = \alpha_{FW}/c. \qquad (10.28)$$

Although I have not shown interest in the other possible pairwise comparisons, you should know that there are a total of six[12] pairwise comparisons possible with four days.

---

[12]The number 6 here represents the number of combinations of 4 days

Going back to the 3 pairwise comparisons, the per comparison type I error rate to be used will be $\alpha_{PC} = 0.05/3 = 0.0167$. For the sake of testing *(a)* H_0: $\mu_{TU} \geq \mu_{WE}$ versus H_a: $\mu_{TU} < \mu_{WE}$, the null hypothesis will be rejected if the difference $RS_{WE} - RS_{TU}$ between the Wednesday and the Tuesday rank sums exceeds the critical difference CD_{tu} given by:

$$CD_{TU} = d_\alpha \sqrt{nk(k+1)/6}, \qquad (10.29)$$

where d_α is the $100(1-\alpha_{PC})^{th}$ percentile of the standard Normal distribution for the one-sided test of hypothesis, and is $100(1 - \alpha_{PC}/2)^{th}$ percentile of the standard Normal distribution for the two-sided test. Since the tests in this example are all one-sided, d_α is the 98.33^{th} percentile of the Normal distribution. That is $d_\alpha = 2.127$ (obtained from Excel as =NORM.S.INV(0.9833)). Therefore, $CD_{TU} = 2.127\sqrt{9 \times 4(4+1)/6} = 11.65$. Table 10.11 summarizes all 3 pairwise comparisons.

Table 10.11 : Pairwise Comparisons

Months	*Month Difference*	*Statistical Significance*[a]
Wednesday − Tuesday	$29.5 - 16 = 13.5$	Statistically significant
Wednesday − Thursday	$29.5 - 27.5 = 2.0$	Not significant
Thursday − Tuesday	$27.5 - 16 = 11.5$	Statistically significant

[a]Difference will be statistically significant only if it exceeds $d_\alpha = 2.127$

taken 2 at a time.

CHAPTER $\boxed{11}$

Regression Analysis

OBJECTIVE

This chapter introduces the important topic of regression analysis. Among other things, you will learn to describe mathematically the relationship between a measurement variable of interest, and a number of explanatory variables. The resulting regression equation may be used for forecasting and for weighting the importance of some explanatory variables.

CONTENTS

11.1.   Introduction

Regression analysis encompasses a large collection of statistical techniques for predicting the magnitude of one variable (the response variable) based on your knowledge of other variables (the explanatory variables). The field of regression analysis is immense, and I decided to confine myself to a limited number of regression models that are commonly used, and which I found particularly useful in practice.

For the purpose of predicting a measurement response variable such sales revenue, or costs of having a Merchant account at a credit card processing company, I will discuss the *Linear Regression Model*.

For the purpose of predicting a binary response variable such as the response of the solicited individual (e.g. 0 = did not respond, 1=responded) to a credit-card pre-approved solicitation campaign, I will discuss the *Logistic Regression Model*.

For the purpose of predicting counts of events[1] such as the number of car accidents, I will discuss the *Log-linear Regression* also called *Poisson Regression*.

Linear Regression

Consider for example a credit card processing company that charges a fixed monthly merchant account fee of $25 and 1.5% of the price of each transaction. Then the total monthly cost to you can be expressed as follows:

$$\text{Cost} = \$25 + 0.015 \times (\text{Number of Transactions}). \qquad (11.1)$$

In this case, I will say that there is a relationship between **Cost**

---

[1]This response variable is neither measurable nor dichotomous, it is countable.

and **Number of Transactions** that is linear and deterministic (or systematic). In its abstract form, this relationship can be expressed as $y = a + bx$ where $y =$ **Cost**, and $x =$ **Number of Transactions** are the variables, and $a = 25$ and $b = 0.015$ are the constants that must be known for the relationship to be useful.

The real value of an expression such as equation 11.1 lies in the capability it gives you to predict future costs based on your anticipated number of transactions. It also allows you to weight the importance of the number of transactions on the total cost, by looking at the magnitude of its coefficient. It creates the opportunity for a more effective management of your resources, and more streamlined evaluation of different strategies.

Note that the study of deterministic relationships such as the one described by equation 11.1 requires algebra, not statistics. This becomes a statistical problem when a deterministic link is what you expect, and not necessarily what you will observe. It is this new uncertainty and the need to study the gap between the observed and the expected that will make the analysis of the relationship among variables a statistical problem.

Consider for example Table 11.1 data showing the number of rooms and number of kilowatt-hours (**KWH**) of 10 residences in which live households of the same size. You would normally *expect* the number of kilowatt-hours to increase with the number of rooms. Is this relationship going to be deterministic? It is very unlikely, primarily because the actual kilowatt-hours consumed will depend on the number of rooms, and probably on a number other unpredictable small factors such as TV usage.

Table 11.1: Number of rooms and number of kilowatt-hours used in 10 single-family residences

Number of Rooms	12 9 14 6 10 8 10 10 5 7
Kilowatt-Hours (thousands)	9 7 10 5 8 6 8 10 4 7

You will express your expected linear relationship between the number of rooms and the KWH as $E(KWH) = a + b \times$ Rooms where a and b are 2 constants to be determined. That relationship will certainly not be formulated as $KWH = a + b \times$ Rooms. Because it would be downright inaccurate. In its abstract form, this relation will be rewritten as:

$$E(y) = a + bx, \tag{11.2}$$

where y is the response or the dependent variable, x is the predictor or the independent variable, a and b are the regression coefficients to be calculated using the data collected (I will show in section 11.2 how this is accomplished). I like to stress out that the statistical model can only model an existing pattern. That is, if there is no empirical evidence on the existence of a linear structure between x and y, such a link should not be expected. Figure 11.1 displays a scatter plot of the number of kilowatt-hours as a function of the number of rooms.

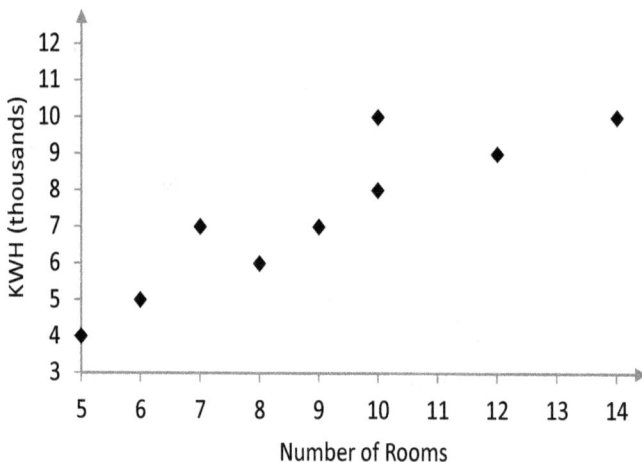

Figure 11.1. Scatter plot of Table 11.1 Data

The independent variable is always on the horizontal axis also known as the x-axis, while the response variable is on the vertical axis also known as the y-axis. Although the relationship between the 2 variables is not neat, one can clearly see an upward linear trend that supports the expectation of a linear relationship.

Logistic Regression

Suppose that you want to launch a fund-raising campaign to collect funds for an education program. Since the fund-raising kit is expensive you want to distribute it to those individuals who are likely to respond to your solicitation campaign. The response variable in this case is binary and defined as follows: $y = 1$, *if the solicited individual responds with a pledge, and $y = 0$, if the individual does not respond.*. Your problem could be to determine whether educational attainment (x) for example, measured in terms of number of years spent in school, is a good predictor for response. This will allow you to target groups of individuals with a certain level of education.

The real problem here is to determine how the dichotomous response variable y could be linked to a predictor such as educational attainment in a way that makes sense. The solution to this problem is to consider π, the probability for an arbitrary solicited individual to respond to the campaign *(note: π is also the expected value of y)*, and to link it to education as follows:

$$\pi = \frac{\exp(a + bx)}{1 + \exp(a + bx)}, \tag{11.3}$$

where exp is the exponent function, a and b being the parameters of the logistic regression model that will be calculated using the data. Cornfield (1962) has provided a theoretical justification for using equation 11.3. Another equivalent form of equation 11.3 commonly found in textbooks is the following:

$$\log\left(\frac{\pi}{1 - \pi}\right) = a + bx. \tag{11.4}$$

In equation 11.4, the ratio $\pi/(1 - \pi)$ is called the *Odds* (e.g. the odds that someone will respond to a solicitation campaign as opposed to not responding). The log odds $\log(\pi/(1 - \pi))$ is generally referred to as the *Logit*. Its relationship with x is linear, a characteristic that will often be used to check the validity of the logistic model.

To use logistic regression in practice, you will typically need an input dataset containing a number of sample units and their values on a dichotomous 0-1 variable y and an independent variable x. The outcome of the analysis, will be the list of sample units with an estimated probability of response attached to each. You as user will use these probabilities to discriminate the more interesting units from the least interesting.

Log-linear Regression

Suppose that you as researcher want to investigate the claim that the frequency of births by caesarean section is higher in private hospitals where future mothers must pay a fee, than in public hospitals where there is no fee. To this end, you decide to collect the data displayed in Table 11.2.

Table 11.2 : Frequency of C-Section Births by Hospital Type[a]

Births	236	739	970	2,371	309	679	26	1,272	3,246	1,904
Hospital Type	0	1	1	1	1	1	0	1	1	1
Caesareans	8	16	15	23	5	13	4	19	33	19

[a]0=Private Hospital, 1=Public Hospital

To answer the research question stated above, one option would be to perform a straight comparison between the number of Caesareans in private and public hospital using the Wilcoxon rank-sum test[2]. The problem with this approach is that the

[2]The *t*-test may not be applicable since the number of c-sections is a

number of c-sections is certainly more affected by the number of births in the hospital than by the hospital type. You may want to get around this issue by taking the proportion of c-sections rather than the raw counts. There is still a major problem, which is that 1 c-section in a hospital where only 2 births took place, represents an unduly high 50% c-section rate, which does not provide an accurate picture of the practices in that particular hospital.

When the data being analyzed represent counts (or estimated proportions), the recommended approach is the log-linear regression. The response variable in this case would be,

$$y = Number\ of\ caesarean\ sections$$

Let m be the expected value of y (i.e. the expected number of c-sections), $x_b =$ *the number of births*, and $x_t =$ *the hospital type*. The log-linear model for the caesarean section problem is the following:

$$\log(m) = \beta_0 + \beta_b x_b + \beta_t x_t, \tag{11.5}$$

where β_0, β_t, and β_b are the 3 log-linear model parameters that will be calculated using Table 11.2 data. I will discuss later in this chapter how these calculations can be done. For the sake of being concrete, let us assume that these parameters are estimated as follows: $\beta_0 = 1.23$, $\beta_b = 0.00028$, and $\beta_t = 1.047$ (these values are fictitious, and do not come from Table 11.2). Then model 11.5 can be written as, $\log(m) = 1.23 + 0.00028x_b + 1.047x_t$. Consequently, the expected number of c-sections m could be rewritten as[3] $m = e^{0.00028}e^{0.00028x_b}e^{1.047x_t}$. Now for a public hospital (i.e. $x_t = 1$), the expected number of c-sections is $m_{PU} = e^{0.00028}e^{0.00028x_b}e^{1.047}$, and

---

discrete variable and is probably not normally distributed (not even approximately)

[3]Remember the following properties of the logarithm and exponent functions, which state that $e^{\log(x)} = x$, $\log(e^x) = x$, and $e^{a+b} = e^a e^b$.

will be $m_{\mathrm{PR}} = e^{0.00028} e^{0.00028 x_b}$ for a private hospital. The ratio of these two numbers is,

$$m_{\mathrm{PU}}/m_{\mathrm{PU}} = e^{1.047} = 2.85,$$

which indicates that you would normally expect the number of caesarean sections to be almost 3 times higher in public hospitals than in private hospitals.

Note that, even if you did not want to compare c-section rates in public and private hospitals, you could still have used the log-linear model in order to predict the number of c-sections based on the number of births. I will discuss other applications of the log-linear regression later in this chapter.

11.2. Linear Regression

You can quantify the association between several measurement variables using one of the many correlation coefficients available in the literature. Regression analysis is not just about quantifying association. It provides the methodology for *describing* and quantifying these associations. You will typically have one variable to study, and an interest in understanding the specific way in which it is related to other variables. The study variable is called the *Dependent Variable*, or the *Response Variable* and is denoted by y. The variables used to explain the y are known as the *Independent Variables* or the *Explanatory Variables*, or the *Predictors*. You may want to understand for example how the amount of life insurance individuals carry is affected by the their income, and risk aversion score. In this example, "amount of life insurance" is the dependent variable, while income, and risk score are the two independent variables.

If only one independent variable is used, it will generally be denoted by x. Sometimes, you will use several independent variables, which will be denoted by x_1, x_2, x_3 for example. For now, I assume for simplicity that you want to study the effect of a single independent variable x on the dependent variable y. I will discuss the more complex case of multiple independent variables later on.

Let μ be the expected value of the independent variable y (i.e. $\mu = E(y)$). The whole linear regression theory is based upon the premise that the link between the expected value of y and x is deterministic, and that the observed value y differs from its expected value by an uncontrolled and unpredictable amount e called the *random error*. The discrepancy between the observed and the expected value of y is an error in the sense that the expectation is often seen as a prediction. This discrepancy is considered random because of the admission that its magnitude is unpredictable. Since μ is a function of x, the observed independent variable can then be represented as follows:

$$y = \mu(x) + e \qquad (11.6)$$

The statistical model 11.6 has two components, the deterministic component $\mu(x)$ also called the regression term, and the random component e. Each form that the regression term $\mu(x)$ takes will define a special regression model. If $y = $ *life insurance amount* and $x = $ *income*, then I may postulate that the expected life insurance amount and the individual income level are linked as $\mu(x) = \beta_0 + \beta x$, which leads to the statistical model, $y = \beta_0 + \beta_1 x + e$. This is a linear model, since the regression term in this case is a linear function of x. Moreover, β_0, and β_1 are called the *regression coefficients*. In practice, these regression coefficients are calculated using a sample of n subjects from which the series of data points $(x_1, y_1), (x_2, y_2), \cdots, (x_n, y_n)$ will be collected. To justify the procedures used for computing the regression coefficients, and lay ground for statistical inference, I will need the following framework:

*For each of these n subjects, the linear regression mo-
del relating x and y values is,*

$$y_i = \beta_0 + \beta_1 x_i + \varepsilon_i, \qquad (11.7)$$

*where y_i, x_i, and ε_i are variables associated with a
specific subject i, the n random error terms ε_i are
independent, and they all share the same law of probability, which is the Normal distribution with mean
0, and variance σ^2.*

Equation 11.7 expresses the way I see things, but will not
itself be useful in practice. What you will need in practice is the
estimated regression function, also called the *regression equation*.

The Regression Equation

Based on your sample data, you will calculate β_0 and β_1.
Because your calculations will be based on your own personal
data, what you will get are not the exact values of β_0 and β_1.
Instead, you will get their estimated values that I will denote by
b_0 and b_1. These estimated values are subject to some sampling
variability that I will briefly discuss later in this chapter. Using
the estimated regression coefficients b_0 and b_1 you can predict
the value of y associated with a particular individual i with the
following *regression equation*:

$$\widehat{y}_i = b_0 + b_1 x_i. \qquad (11.8)$$

$\widehat{y}_i$ (read y i hat) the predicted response of individual i will differ
from the observed value, and the difference $e_i = y_i - \widehat{y}_i$ is called
the residual.

Throughout this section on linear regression, y (the dependent
variable) will always be quantitative (i.e. continuous). However,

the independent variable could be either categorical or quantitative. The independent variable type will define a special type of linear regression model.

11.2.1 Types of Linear Regression Models

When a regression model involves a quantitative dependent variable, and quantitative independent variables, it is often referred to as *ordinary regression model*. The use of a single independent variable will result in a *simple linear regression* model. In this section, I will briefly mention a few special linear models that are based on categorical independent variables, or a mixture of categorical and quantitative independent variables.

Use of Dummy Independent Variables

If you want to relate the amount of life insurance to individual income, you may also want to study the effect of living in urban area or not. The way to do it is to create an indicator variable for urbanicity before including it in the regression model as follows:

$$y_i = \beta_0 + \beta_1 x_i + \beta_2 u_i + \varepsilon_i, \tag{11.9}$$

where $u_i = 1$ if i resides in an urban area, and $u_i = 0$ if not. Indicator variables are sometimes called dummy, binary, or dichotomous variables. This model suggests that you expect the amount of life insurance to be $\mu(x) = \beta_0 + \beta_1 x$ for non-urban residents, and $\mu(x) = (\beta_0 + \beta_2) + \beta_1 x$ for urban residents. That is in urban and non-urban areas, the amount of life insurance increases the same way by income level, but have different minimum amounts.

Application to Time Series

Suppose that you want to model monthly car sales revenue, but believe that it might be affected by advertising expenditures and season (Winter, Spring, Summer, and Fall). Let y be the monthly car sales, and x the monthly advertising expenditures. You would resolve this regression problem by defining 3 dummy variables for the season (not 4). These dummy variables would be WIN, SPR, and SUM, and the resulting regression model would be,

$$y_i = \beta_0 + \beta_1 x_i + \beta_2 \text{WIN}_i + \beta_3 \text{SPR}_i + \beta_2 \text{SUM}_i + \varepsilon_i, \quad (11.10)$$

where $\text{WIN}_i = 1$ if the sales revenue reported by car dealer i represents cars sold in Winter, and $\text{WIN}_i = 0$ otherwise. Why did I decide not to define a dummy variable for the Fall season? It is because that was unnecessary, and may even have created some computational issues. In fact, if $\text{WIN} = \text{SPR} = \text{SUM} = 0$, that tells me that it was the Fall season. Therefore, model 11.10 will use all the information available. Was it possible to pick any 3 out of 4 seasons? The answer is yes, it would not have made any difference.

Analysis of Variance Models

At times, you will encounter the term *Analysis of variance model*. These are regression models where all independent variables are categorical. If you want to study monthly car sales as a function of the season, you could formulate the following model:

$$y_i = \beta_0 + \beta_1 \text{WIN}_i + \beta_2 \text{SPR}_i + \beta_3 \text{SUM}_i + \varepsilon_i, \quad (11.11)$$

which is similar to model 11.10. Although, analysis of variance models can be studied within the framework of linear regression models, they are often based on a distinct statistical methodology, using their own terminology. If the treatment of analysis of variance is generally separated from that of ordinary regression models, it is because their structure permits numerous simplifications, and a different interpretation of results.

Model 11.11 is normally formulated in the analysis of variance jargon as,

$$y_{ij} = \mu + \alpha_i + \varepsilon_{ij}, \tag{11.12}$$

where i is a particular season (Winter, Spring, Summer, or Fall), $\alpha_i = $ *the effect of season i on the sale amount y_{ij} reported by car dealer j in season i.* Season is not an explicit independent variable in the model. It is now called a factor, and its values (Winter, Spring, Summer, or Fall) are the factor levels. What is in the model is the factor effect α.

Analysis of Covariance Models

You will be dealing with an Analysis of covariance (ANCOVA) model, when the list of independent variables includes a mixture of categorical as well as quantitative variables. If a quantitative independent variable x, and a multiple-level factor are used, a typical ANCOVA model will take the form,

$$y_{ij} = \mu + \alpha_i + \beta x_{ij} + \varepsilon_{ij}, \tag{11.13}$$

where α_i is the effect of factor's i^{th} level of the factor, and x_{ij} the x value associated with factor level i and subject j. ANCOVA models also are often treated with a special methodology for convenience like ANOVA models, although they could well be treated as special linear models.

11.2.2   Regression Coefficients in Linear Regression

In this section, I will consider the simple linear model of equation 11.7 and show how the regression coefficients β_0 and β_1 should be estimated using sample data, in order to obtain the more useful regression equation 11.8. To fix ideas, consider the power consumption data of Table 11.1. The linear trend that exists in this data is depicted in figure 11.1. My problem is to find the regression line that "best" fits that data. Figure 11.2 displays the same data as figure 11.1, and 2 possible regression lines along with their respective regression equations (Linear and Linear 2).

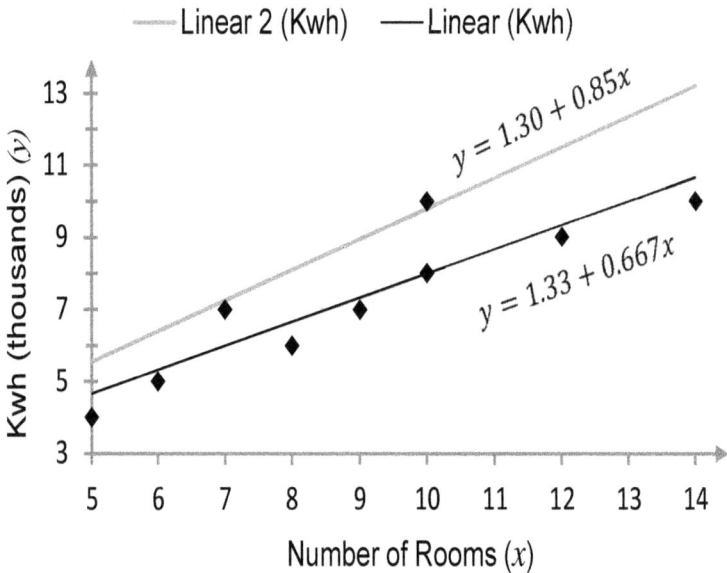

Figure 11.2. The regression line of the Power Consumption Data

A fundamental question here is which of the two regression lines is better? A simple look at Figure 11.2 reveals that "Linear" appears to stay consistently closer to the bulk of the data than

"Linear 2." Intuitively, you may want to favor "Linear" as the best regression line. How does this crude intuition translate into a formal statistical procedure ? There are actually several possible approaches that have been proposed in the literature. The most popular one by far is the *Least Squares Method* that I will now discuss.

If you are using a sample of n observations, then each pair of coefficient (b_0, b_1) will generate a series of n residuals $e_i = y_i - (b_0 + b_1 x_i)$. The distance between the regression line and the observations is measured by the sum of squared residuals SS_e defined as follows:

$$SS_e = \sum_{i=1}^{n} \left(y_i - (b_0 + b_1 x_i)\right)^2. \qquad (11.14)$$

The objective of the least squares method is to find the 2 coefficients b_0 and b_1 that will give the sum of squares SS_e its smallest value possible. Mathematical expressions for b_0 and b_1 that satisfy this condition can be derived using special calculus methods that I will not detail in this book. This same least squares method is used for a larger number of independent variables, and regression coefficients.

11.2.3 Regression Calculations

There are numerous software options for obtaining the regression equation. The purpose of this section is to mention some of them. Mathematical expressions for computing the regression coefficients exist. Unless you are a software developer or a mathematical statistician, it is unclear to me why you would want to mess up with those. When regression analysis involves 2 or more independent variables, you may even need to master matrix algebra in order to manipulate these equations effectively. These technicalities are not essential for an effective use of statistics in practice.

In the following example, I will show you how to obtain the equation of a simple linear regression, with a single independent variable. I will show how this equation is obtained with OpenOffice Calc, Excel, and R.

Example 11.1 ──────────────────────────────

Table 11.3 shows the college and high school grade point averages (GPA) of 6 students selected randomly from college's files. The objective is to predict college performance from high school GPAs using a regression equation.

Table 11.3 : College and High School GPAs of 6 Students

Student	GPA.HS(x)	GPA.CO(y)
1	2.0	1.7
2	2.6	2.8
3	2.8	2.1
4	3.2	2.3
5	3.3	2.6
6	3.6	3.0

The dependent variable is college GPA (i.e. $y = $ GPA.CO) and the independent variable, the high school GPA (i.e. $x = $ GPA.HS). The regression equation will have the form $y = b_0 + b_1 x$. The coefficient b_0, which is not associated with any independent is called the *Intercept*[4], and the coefficient associated with the unique independent variable x of a simple linear regression model is called the *Slope*. The slope measures the extent to which you expect the college GPA to increase as a result of an increase of 1 point of the high school GPA.

───────────────────────

[4]The *intercept* in this example, is the college GPA you expect from a student who had a 0 GPA in high school

▶ **OpenOffice Calc 3.2.0.** Assume that the 6 high school GPAs are in cells A2:A7, and the 6 college GPAs in cells B2:B7.

Calculate intercept as b_0 =INTERCEPT(B2:B7;A2:A7)

Calculate the slope as b_1 =SLOPE(B2:B7;A2:A7) .

This yields the following regression equation:

GPA.CO $= 0.6325 + 0.6117 \times$ GPA.HS

▶ **Excel 2007/2010.** The Excel procedures for obtaining the slope and the interecept are very similar to those of Calc, with a few minor exceptions. The function names are identical, and only the semicolon (;) should be replaced with the comma (,).

In addition to providing the SLOPE and INTERCEPT functions, Excel allows you to create the scatterplot of Figure 11.3 where the regression line and the regression equation are displayed. This option is useful in that it allows you to visualize the linear trend in your data, and produce the regression at the same time. See section E.6 of Appendix E to see the details for creating Figure 11.3.

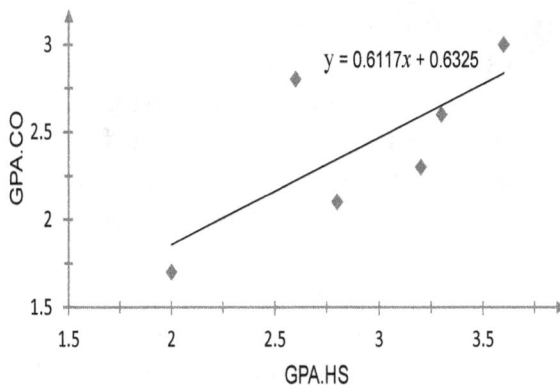

Figure 11.3. Regression Equation GPA Data of Table 11.3

▶ **R Package**. Obtaining the slope and the intercept with the R package is straightforward as shown in Figure 11.4.

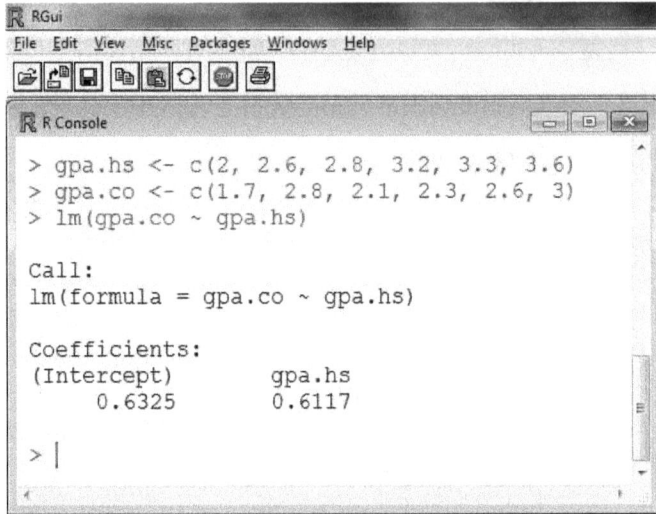

```
R RGui
File  Edit  View  Misc  Packages  Windows  Help

R R Console

> gpa.hs <- c(2, 2.6, 2.8, 3.2, 3.3, 3.6)
> gpa.co <- c(1.7, 2.8, 2.1, 2.3, 2.6, 3)
> lm(gpa.co ~ gpa.hs)

Call:
lm(formula = gpa.co ~ gpa.hs)

Coefficients:
(Intercept)         gpa.hs
     0.6325         0.6117

> |
```

Figure 11.4. Regression Equation for Table 11.3 Data Using R

In this section, I have shown how the 2 coefficients of a simple linear regression model can be calculated. However, if your model involves 2 independent variables or more, then it becomes a multiple linear regression with 3 regression coefficients or more to estimate. The Intercept and Slope functions of OpenOffice Calc 3.2.0, and Excel can no longer be used to compute the coefficients of a multiple regression model. The only two options you will have among those considered in this book, are the Excel Analysis ToolPak, and the R package. I will show later in this section 11.2 how multiple regression model coefficients can be calculated, along with the different tests of hypothesis.

Although your decision to develop a simple linear regression model is often based on a visual inspection of the scatterplot, and your suspicion of the existence of a linear pattern in the data, you still need a formal statistical procedure for testing the goodness of your model. Although there exists a wide variety of procedures for testing the quality of the model, I will confine myself to presenting 3 of them, which I found useful in practice. These are,

(a) *The Test of Regression Relation.* This procedure is about testing the null hypothesis that all regression coefficients associated with independent variables are 0, versus the alternative that not all of the coefficients are 0. This test allows you to determine whether the independent variables help explain the independent or not, and if developing the regression even helps at all.

For the simple regression model $y = \beta_0 + \beta_1 x + \varepsilon$, this amounts to testing H_0: $\beta_1 = 0$ versus H_a: $\beta_1 \neq 0$. For the regression model with 2 independent variables $y = \beta_0 + \beta_1 x_1 + \beta_2 x_2 + \varepsilon$, this amounts to testing H_0: $\beta_1 = 0$ *and* $\beta_2 = 0$ versus H_a: *Not both* β_1 *and* β_2 *equal 0.*

(b) *The Coefficient of Determination.* The coefficient of determination is a useful statistical measure that quantifies the percentage of total variation in the data that is explained by the regression model. It is the ratio of the regression sum of squares (denoted by SSR) to the total sum of squares (denoted by SSTO). While SSR sums the deviations of all predicted y values from the overall mean of observed values, the SSTO sums the deviations of all observed y values from their overall mean.

(c) *The Test of Individual Coefficients.* Testing individual regression coefficients for statistical significance, amounts to testing the null hypothesis that a given coefficient is 0 versus the alternative that it is different from 0. This procedure is useful in practice, since it allows you to determine whether the independent variable associated with the coefficient being tested has any role at all to play in the model.

These 3 types of statistical procedures will be further discussed in the next section in the context of multiple linear regression models with two independent variables.

11.2.4 Statistical Inference in Regression Analysis

In this section, I will consider regression models with two independent variables, and will occasionally refer to the more general situation of 3 independent variables or more. I assume that y is the dependent variable, x_1 and x_2 the two independent variables, and the analysis based on a sample of n subjects labeled as $i = 1, \cdots, n$. Moreover, y_i, x_{1i}, and x_{2i} represent values taken by the ith subject on the y, x_1, and x_2 variables respectively. The general formulation of the regression model with two independent variables that will ensure the validity of the statistical procedure is the following:

$$y_i = \beta_0 + \beta_1 x_{1i} + \beta_2 x_{2i} + \varepsilon_i, \qquad (11.15)$$

where β_0, β_1, and β_2 are the regression coefficients, and the ε_i's (for $i = 1, \cdots, n$) are n random and independent error variables, each of which following the Normal distribution with mean 0 and variance σ^2.

Throughout this section, I will illustrate the statistical methods using the data of Table 11.4. Table 11.4 contains sales data of a new body lotion (in hundreds of dollars), by city, as well as by the city population (in thousands) and the per capita income. I like to model the sales revenue using Population and Per Capita Income (PCI) as independent variables. My intention is to consider equation 11.15 as my regression model, and use Table 11.4 to compute the 3 regression coefficients β_0, β_1, and β_2.

I now want to explain how the quality of the regression analysis will be evaluated. After you estimate the regression coefficients, you will obtain the regression equation $\widehat{y}_i = b_0 + b_1 x_{1i} + b_2 x_{2i}$, where b_0, b_1, and b_2 are the estimated values of the regression parameters based on the sample data. Consider the following 3 statistical measures:

$$\text{SSTO} = \sum_{i=1}^{n}(y_i - \bar{y})^2, \quad \text{SSR} = \sum_{i=1}^{n}(\widehat{y}_i - \bar{y})^2,$$
$$\text{SSE} = \sum_{i=1}^{n}(y_i - \widehat{y}_i)^2, \tag{11.16}$$

Table 11.4 : Lotion Sales, Population, and PCI of 15 Cities

Sales[a] Revenue (y)	Population (thousands) (x_1)	Per Capita Income (x_2)
126.36	274	2,450
93.60	180	3,254
173.94	375	3,802
102.18	205	2,838
52.26	86	2,347
131.82	265	3,782
63.18	98	3,008
149.76	330	2,450
90.48	195	2,137
42.90	53	2,560
196.56	430	4,020
180.96	372	4,427
112.32	236	2,660
80.34	157	2,088
165.36	370	2,605

[a]This represents the sales revenue in hundreds of dollars

SSTO is the total sum of squares and measures the total variation in the data. SSR is the regression sum of squares, and represents the variation in data that is caused by regression. Note that if the regression equation is perfect, in the sense of producing predicted values $\widehat{y}_i$ that match the corresponding observed one y_i then both sums of squares will be identical. Therefore, the proximity of these 2 quantities will help evaluate the quality of the regression model. SSE is the error sum of squares, which one will expect to be small if the regression model is good. The reasoning used here is similar to that previously used in the Analysis of Variance (ANOVA) context, where regression is seen as a treatment applied to the sample of subjects to produce the predicted values. The SSR here plays the same role that the treatment sum of squares SSTR played in the ANOVA context.

F test for regression relation

For the sum of squares of equation 11.16, you may derived two useful mean of squares, which are the Regression Mean of Squares (MSR), and the Error Mean of Squares (MSE). These 2 mean squares are defined as follows:

$$\text{MSR} = \frac{\text{SSR}}{p - 1} \ and \ \text{MSE} = \frac{\text{SSE}}{n - p}, \qquad (11.17)$$

where p is the number of regression coefficients ($p = 3$ for equation 11.15), and n is the number of subjects in the sample.

Hypotheses

For the purpose of testing the regression relation, the hypotheses are formulated as follows:

$$\left\{ \begin{array}{l} \text{H}_0\text{: } \beta_1 = \beta_2 = 0, \\ \text{H}_a\text{: } \beta_1 \neq 0 \ or \ \beta_1 \neq 0 \end{array} \right. \qquad (11.18)$$

You may have noted that parameter β_0 is not part of neither the null nor the alternative hypothesis. It is because the objective here is to test the relevance of including the two independent variables x_1 and x_2 into the regression model. Therefore, only the coefficients associated to the 2 variables are tested for statistical significance.

Test Statistic

The statistic for testing the hypotheses 11.18 is defined as,

$$F = \frac{\mathrm{MSR}}{\mathrm{MSE}}. \tag{11.19}$$

If the null hypothesis is true, then the law of probability associated with the F statistic of equation 11.19 is well approximated by the $F(p-1, n-p)$ distribution. This results in the following decision rule:

$$\boxed{\text{Reject } H_0 \text{ is } F_{obs} > F_{\alpha,p-1,n-p,}} \tag{11.20}$$

where α is the test significance level, and $F_{\alpha,p-1,n-p}$ the $100(1 - \alpha)^{th}$ percentile of the F distribution. Note that rejecting the null hypothesis only tells you that one coefficient or perhaps more are non null, but will not tell you which coefficient is null if any. To answer the question, you need to test some of the coefficients individually.

t-Test of Individual Coefficient

Let β_k be any given regression coefficient ($k = 0, 1, 2$), b_k its sample-based estimated value, and $s(b_k)$ the standard error associated with b_k. Testing H_0: $\beta_k = 0$ versus H_a: $\beta_k \neq 0$ is based on the following t statistic:

$$t = \frac{b_k - \beta_k}{s(b_k)}, \tag{11.21}$$

which follows the Student's t distribution with $n - p$ degrees of freedom. This leads to the following decision rule:

$$\text{Reject } H_0 \text{ if } t_{obs} > t_{\alpha/2,n-p}, \qquad (11.22)$$

where $t_{\alpha/2,n-p}$ is the $100(1 - \alpha/2)^{th}$ percentile of the Student's distribution with $n - p$ degrees of freedom.

The F test for regression relation, and the t-test of individual coefficients can only tell you whether or not the independent variables x_1 and x_2 have some explanatory power, which justifies their use in a regression model. Once the importance of the independent variables is established, what is often missing is a measure of the strength of the relationship between y, x_1, and x_2 as described by the regression model. This measure is given by the coefficient of determination.

Coefficient of Determination

The coefficient of determination (often labeled as "R-Square" in most statistical package)[5], and denoted by R^2 is one of the most popular measures for evaluating the quality of a regression model. It is defined as,

$$R^2 = \frac{\text{SSR}}{\text{SSTO}}, \qquad (11.23)$$

where SSR and SSTO are the regression and error sums of squares respectively. It measures the proportion of total variation of the data that can be attributed to the regression model. This measures varies from 0 to 1 taking low values when the regression

---

[5] *Oftentimes the term Coefficient of Multiple Determination is used when dealing with multiple linear regression, while reserving the term Coefficient of Determination for simple linear regression models.*

is a poor fit, and higher values when the regression model is a good. Note that a mere addition of independent variables to the model may artificially increase the magnitude of the coefficient of determination. To correct this problem, it is often recommended to use the adjusted coefficient of determination defined as,

$$R_a^2 = 1 - \left(\frac{n-1}{n-p}\right)\frac{\text{SSE}}{\text{SSTO}}. \tag{11.24}$$

Note that $R_a^2 = 1 + (n-1)R^2/(n-p)$. In most applications both coefficients will be similar. However, in case there is noticeable difference, I would recommend using the adjusted value R_a^2.

Example 11.2 _____

In this example, I like to perform regression analysis on the data of Table 11.4, where the dependent variable is $y = $ *Sales Revenue*, $x_1 = $ *City Population*, and $x_2 = $ *Per Capita Income*. Using the Regression module of Excel's Analysis ToolPak, I produced the Summary, the ANOVA, and the Coefficients' tables shown below. Details for doing this can be found in section E.7 of Appendix E.

Multiple R represents the *Coefficient of Multiple Correlation*, which is the square root of the coefficient of multiple determination R^2. It appears that Multiple R, R^2, and R_a^2 are all very high and close to 1. This is an indication that the regression model is good at explaining the variation in the data. The "Standard Error" of 1.6982, also known as the *Standard Error of Estimate* represents the standard error associated with the prediction $\widehat{y}_i$ pertaining to subject i, that you would make using the regression equation[6].

_____

[6]You may want to know that this standard error equals the square root of the error mean of squares MSE

SUMMARY OUTPUT	
Regression Statistics	
Multiple R	0.9995
R Square	0.9989
Adjusted R Square	0.9988
Standard Error	1.6982
Observations	15

As for the ANOVA table, the main role it plays is to test the regression relation (see equations 11.18 through 11.20). This table shows the sums of squares in the SS column, the associated number of degrees of freedom (df), the means of squares (MS), the observed F statistic and associated P-value labeled as "significance F." That is, $MSR = 16,379.56$, $MSE = 2.88$, $F = 5,679.47$, and P-value $= 1.38E - 18$ (Note that $1.38E - 18$ is the scientific notation $1.38 \times 10^{-18} = 1.38/10^{18}$. The very tiny value of the P-value indicates that the null hypotheses 11.18 must be rejected.

ANOVA					
	df	SS	MS	F	Significance F
Regression	2	32,759.13	16,379.56	5,679.47	1.381E-18
Residual	12	34.61	2.88		
Total	14	32,793.73			

The last table provides all the statistical information pertaining to the regression coefficients. It follows from this table that the estimated values of the parameters β_0, β_1, and β_2 are respectively $b_0 = 2.693$, $b_1 = 0.387$, and $b_2 = 0.0072$. The t tests for individual coefficients described with equations 11.21 and 11.22 can be carried out using the test statistics "t stat," associated p-values, and the confidence interval boundaries LB95% and UB95%.

	Coefficients	S.E.	t Stat	P-value	LB95%	UB95%
Intercept	2.693	1.896	1.42	0.18	-1.4378	6.82
Population	0.387	0.0047	81.92	7.3E-18	0.3766	0.397
PCI	0.0072	7.6E-04	9.50	6.2E-07	5.5E-03	8.8E-03

Typically, when the P-value is smaller than 0.05, one may conclude that the regression coefficient is statistically significant, and can be considered non-zero. When 0 is included in the confidence interval, it is an indication that 0 could well be the "true" value of the regression coefficient. Therefore, the estimated coefficient is not statistically significant, and its inclusion in the regression model unjustified.

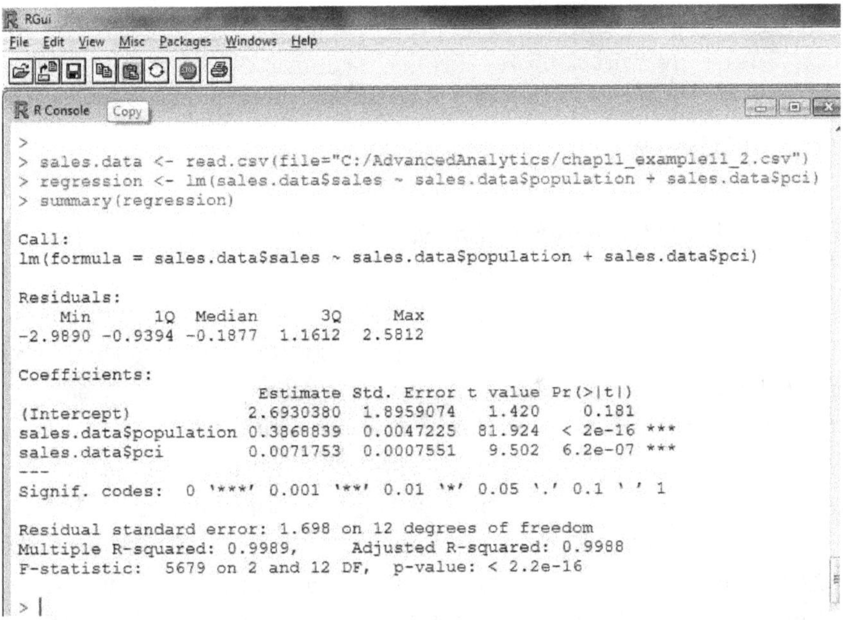

Figure 11.5. Multiple Regression of Table 11.4 Data with R

11.3.   Logistic Regression

In this section, I like to give you a glimpse of this important topic that is logistic regression. In my opinion, logistic regression is too important in practice to be entirely neglected even in an introductory statistics textbook, as is often the case.

Credit card companies for example use logistic regression extensively to streamline their solicitation campaigns by targeting prospects most likely to respond. These models are known in the credit card industry as *Credit Scorecards* or *Scoring models*. The primary purpose of these models is to assign to each individual of interest, a score from 0 to 1 representing the likelihood of that person to meet certain criteria.

The limited goal that I have set for this section is to show you the types of analyzes that are possible with logistic regression. You may find a more detailed account of this technique in Hosmer and Lemeshow (2000), my preferred reference on logistic regression models.

11.3.1 Logistic Regression with 1 Independent Variable

Suppose you want to know whether individuals in the higher income bracket are more likely to hold an American Express (AMEX) travel card. If you can select a random sample of American Express travel card holders, and another sample of non-holders then you may apply the two-sample t-test for means, and compare the mean income of the two populations of AMEX card holders and non-holders. This would be a *Prospective study* where you select subjects then observe their characteristics. Now, what do you do if all you have is a random sample of individuals initially selected for other purposes, and you want to use it to study AMEX card holders and non-holders with respect to their income? That is, you want to compare groups in a *Retrospective study*.

The group of AMEX card holders who turn up in the existing study cannot be considered as a representative sample of the entire population of AMEX card holders. Therefore, using the t-test is not feasible. The solution here is to use logistic regression.

To illustrate this approach, I will consider the data contained in Table 11.5. This data is about 100 individuals with their annual income in US dollars, the number of individuals (i.e. **Nb of cases**) at the specified income level, and the number of AMEX card holders.

Let y_i be a dichotomous variable defined as follows:

$$y_i = \begin{cases} 1, & \text{if individual } i \text{ possesses an AMEX card,} \\ 0, & \text{otherwise.} \end{cases} \qquad (11.25)$$

Let x_i be individual i's annual income, and π_i the probability of her having an AMEX card[7]. The logistic regression model stipulates that π_i and x_i are linked as follows:

$$\log\left(\frac{\pi_i}{1 - \pi_i}\right) = \alpha + \beta x_i, \qquad (11.26)$$

and Table 11.5 data will be used to estimate the values of α and β. The calculated values of these parameters will be denoted by a and b for α and β respectively. Note that π_i is a function of x_i and will also be denoted by $\pi_i(x_i)$. After presenting the method for calculating the regression coefficients, I will explain how they should be interpreted.

---

[7]Note that $\pi_i = P(y_i = 1)$

Table 11.5 : Distribution of American Express Card Holders by Income

Annual Income	Nb of Cases	AMEX Holders	Annual Income	Nb of Cases	AMEX Holders
$14K	2	0	$31K	1	0
$16K	2	0	$33K	2	0
$17K	6	2	$38K	5	2
$20K	10	2	$42K	5	4
$21K	8	1	$43K	3	3
$22K	9	2	$45K	4	3
$24K	5	1	$48K	1	0
$26K	8	0	$50K	2	0
$27K	2	0	$52K	1	0
$28K	1	1	$60K	1	0
$29K	2	0	$65K	6	5
$30K	12	2	$94K	2	2

Calculating the Regression Coefficients

Although the logistic regression coefficients will generally be obtained using a software, I like to briefly present the underlying approach. When fitting a logistic regression model, the input data cannot be condensed as it is in Table 11.5. Instead, individual-level information must be explicitly listed. For example, instead of having $17K for income, 6 for Number of cases, and 2 for AMEX card Holders, you need to have 6 records with the following data:

```
     Income  $17  $17  $17  $17  $17  $17
AMEX.Holder    0    0    0    0    1    1
```

Therefore, to fit a logistic regression model, you will need a series of n values $x_1, \cdots, x_n$ for the independent variable x and a series of n dichotomous 0-1 values $y_1, y_2, \cdots, y_n$ for the response variable. Table 11.6 has the format required for fitting logistic regression, and contains the same information as Table 11.15.

The 2 coefficients α and β of equation 11.26 are approximated by the 2 numbers $\widehat{\alpha}$ and $\widehat{\beta}$ that maximize the following function known as the *log likelihood function*:

$$L(\alpha, \beta) = \alpha \left(\sum_{i=1}^{n} y_i \right) + \beta \left(\sum_{i=1}^{n} y_i x_i \right) - \sum_{i=1}^{n} \log \left(1 + e^{\alpha + \beta x} \right) \quad (11.27)$$

Note that maximizing $L(\alpha, \beta)$ is equivalent to maximizing the probability to obtain the same outcome of Table 11.6 if the experiment that produced that data was to be repeated. This estimation method is known as the *Maximum Likelihood Method*.

Table 11.6 : American Express Card Holders by Income

INC	14,14,16,16,17,17,17,17,17,17,20,20,20,20,20,20,20, 20,20,20,21,21,21,21,21,21,21,21,22,22,22,22,22,22, 22,22,22,24,24,24,24,24,26,26,26,26,26,26,26,26,27, 27,28,29,29,30,30,30,30,30,30,30,30,30,30,30,30,31, 33,33,38,38,38,38,38,42,42,42,42,42,43,43,43,45,45, 45,45,48,50,50,52,60,65,65,65,65,65,65,94,94
AMEX	0,0,0,0,0,0,0,0,1,1,0,0,0,0,0,0,0,0,1,1,0,0,0,0,0,0,0,1, 0,0,0,0,0,0,0,1,1,0,0,0,0,1,0,0,0,0,0,0,0,0,0,0,0,1,0,0,0, 0,0,0,0,0,0,0,0,0,1,1,0,0,0,0,0,0,1,1,1,1,1,1,0,1,1,1,1, 1,1,0,0,0,0,0,0,1,1,1,1,1,0,1,1

Fitting the Logistic Regression Using R

The R package offers a simple way to fitting a logistic regression model. To fit model 11.26 to Table 11.6 data, I started by creating a `csv` text file named `chap11_tab11.6.csv` containing 2 columns of data delimited by commas, and representing the INC and AMEX variables (I actually used `inc` and `amex` in lower case since variable names in R are case-sensitive). I then submitted the R instructions of Program 11.1 listed below. These instructions were actually stored in a script file named `chapter11.logit1.r` and called from within R with the source() function as shown in Figure 11.6

```
01  input.data <- read.csv("chap11_tab11.6.csv",
header=T)
02  model <- "factor(amex) ~ inc"
03  print(model)
04  logitfit <- glm(model, family = binomial(link
= "logit"), data=input.data)
05  print(summary(logitfit))
```

Program 11.1. R Script for fitting logistic regression to Table 11.6 data

It follows from Figure 11.6 that the estimated coefficients are $\hat{\alpha} = -3.0580$ and $\hat{\beta} = 0.0663$ (these numbers come from the "Estimate" column of the "Coefficients" table shown on the figure)[8]. What can you do with these coefficients? Using them, you can state that an individual i with income x_i has a probability of carrying an AMEX card estimated at,

$$\hat{\pi}_i(x_i) = \frac{e^{-3.0580+0.0663x_i}}{1 + e^{-3.0580+0.0663x_i}}, \tag{11.28}$$

---

[8]R outputs several other statistics that I will not discuss in this introductory books. My focus will be on the coefficients only.

```
R R Console
> source("chapter11.logit1.r")
[1] "factor(amex) ~ inc"

Call:
glm(formula = model, family = binomial(link = "logit"),

Deviance Residuals:
    Min       1Q    Median       3Q      Max
-1.7330   -0.7462  -0.5883    0.7099   2.0331

Coefficients:
             Estimate Std. Error z value Pr(>|z|)
(Intercept) -3.05801    0.64620  -4.732 2.22e-06 ***
inc          0.06627    0.01790   3.701 0.000214 ***
---
Signif. codes:  0 '***' 0.001 '**' 0.01 '*' 0.05 '.' 0.

(Dispersion parameter for binomial family taken to be 1

    Null deviance: 122.17  on 99  degrees of freedom
Residual deviance: 103.43  on 98  degrees of freedom
AIC: 107.43

Number of Fisher Scoring iterations: 4

> |
```

Figure 11.6. Logistic Regression with R on Table 11.6 Data

Interpreting Regression Coefficients

▶ The fact that $\widehat{\beta} = 0.0663$ is positive, is an indication that the likelihood to carry an AMEX card increases with income. If you are an AMEX card issuer, and you have a file containing individuals along with their annual income, you may use equation 11.28 to score that file assigning a probability to carry a card. You may then use these scores to decide who will be contacted during a solicitation campaign.

▶ Equation 11.28 also indicates that the odds of carrying an AMEX over not carrying it is given by: $\widehat{\pi}_i(x_i)/\left[1-\widehat{\pi}_i(x_i)\right] = e^{-3.0580}e^{0.0663x_i}$. Consequently, if individual i's income goes up by \$1,000 from x to $x+1$, then the odds of carrying an AMEX card with change by a quantity called *Odds Ratio* (OR) obtained as follows:

$$\text{OR} = \frac{\widehat{\pi}_i(x_i+1)/\left[1-\widehat{\pi}_i(x_i+1)\right]}{\widehat{\pi}_i(x_i)/\left[1-\widehat{\pi}_i(x_i)\right]} = e^{0.0663} = 1.0686.$$

That is any increase of \$1,000 in income will result in an increase of 6.86% in the likelihood to carry an AMEX card.

Fitting the Logistic Regression Using Excel

Can you calculate the logistic regression coefficients using Excel? The answer is yes, and I will show you how. But first of all, if all you want is get the logistic regression coefficients, which is the case for many practitioners, then using Excel could be a good option. However, if you want to conduct more advanced analyzes such as testing the coefficients for statistical significance then I would advise to use the R package or another dedicated statistical package.

Logistic regression models can be fitted with Excel, provided you first install an Excel Add-In called **Solver**. It comes with Excel, and should be installed using the same procedures described for installing the **Analysis ToolPak** (see section E.2 of Appendix E)[9]. After installing Solver, you should see the Solver icon on the right side (beneath the ToolPak) after selecting "Data" from Excel's main menu.

---

[9]In Figure E.2, you can see "Solver Add-In" listed. You would select this option instead of Analysis ToolPak.

Now, note that equation 11.27 can be rewritten as,

$$L(\alpha, \beta) = \sum_{i=1}^{n} u_i, \tag{11.29}$$
$$where\ u_i = \alpha y_i + \beta y_i x_i - \log\big(1 + \exp(\alpha + \beta x)\big)$$

u_i is individual's i contribution to the likelihood function. To fit the logistic regression to Table 11.6 data, proceed as follows:

(a) Organize your Table 11.6 data as shown in Figure 11.7. The "INC" column contains income data for each individual in your sample, AMEX column flags all AMEX card owners with 1, and non-owners are labeled as 0. In the column labeled as u_i, type in an Excel formula for computing u_i of equation 11.29. Figure 11.7 shows in the formula bar, the Excel formula used in cell **C4**. Cells **D2** and **E2** will contain the final regression coefficients you are seeking.

(b) Sum all u_i values into cell **C102** (see Figure 11.8). Then Launch Excel's solver, and fill out the Solver's form as shown in Figure 11.8. The "Set Object" text box should contain cell number containing the sum of the u_i's (in this case **C102**). The "Max" radio button must be selected, because you want top maximize the likelihood function. The "By Changing Variable Cells" text box specifies the cells that will contain the regression coefficients (in this case, **D2** and **E2**). *Ensure that the checkbox labeled as "Make Unconstrained Variables Non-Negative" is unchecked.* Click the Solve button, and look at the estimated parameters in cells **D2** and **E2**.

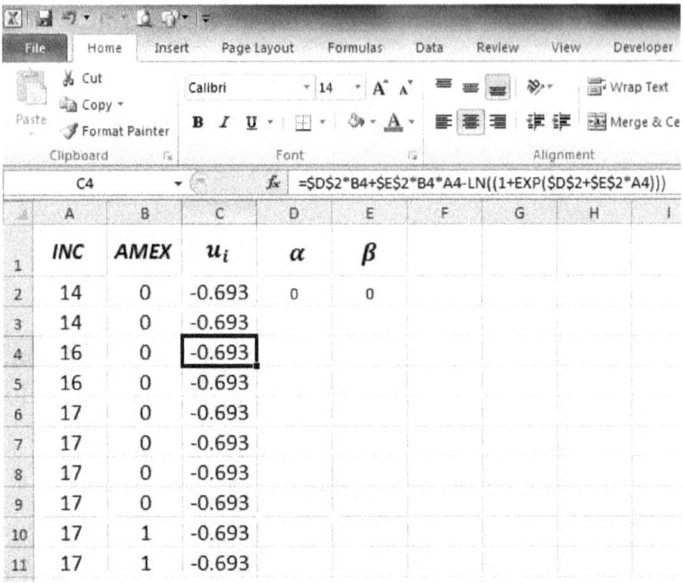

Figure 11.7. Preparing Excel Data for Solver

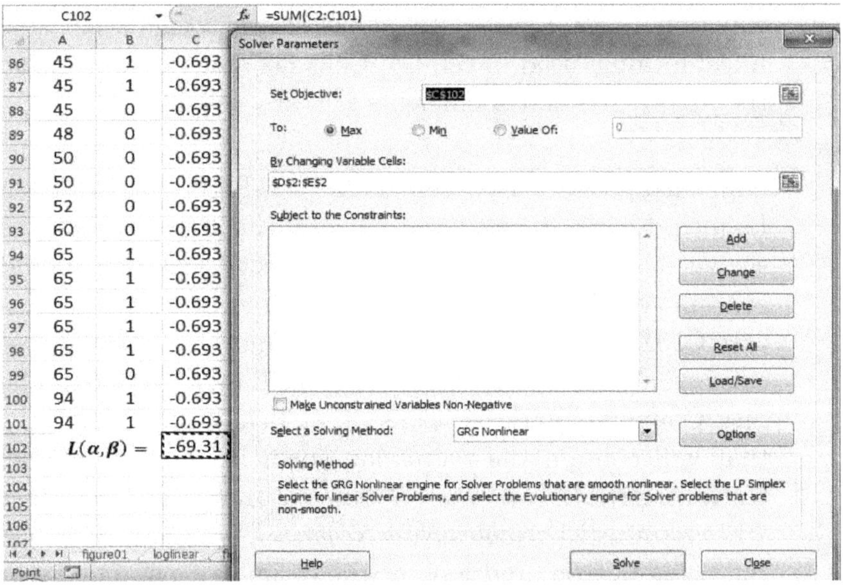

Figure 11.8. Launching Excel's Solver

11.3.2 Logistic Regression with 2 Independent Variables

Suppose you want to evaluate the effectiveness of a new SAT preparation program. One possible approach is to use historical data on a sample of former students. You will need to have in your sample, students who participated in the program, and those who did not, then compare the performance of the two groups. Can you not use the t-test for proportions to test whether the success rate of the participants exceed that of non-participants? The answer to this question depends on how the study was designed and what data you have at your disposal.

One reason why the t-test may not work here is that one group may have better students than the other group, which will make it more difficult to attribute its performance to the SAT program or to the quality of students. This will typically be the case in a retrospective study. In this section, I propose a solution to this problem using logistic regression with the data in Table 11.7.

Let y_i be a binary variable defined as $y_i = 1$ if student i obtained a passing score, and $y_i = 0$ otherwise. Let π_i be the probability for student i to obtain a passing grade, x_{1i} the GPA of student i, and x_{2i} a 0-1 dichotomous variable that flags students who participated in the SAT preparation program (i.e. $x_{2i} = 1$ if student i is a participant, and $x_{2i} = 0$ otherwise). The logistic regression model linking π_i to the 2 independent variables x_{1i} and x_{2i} is formulated as follows:

$$\log\left(\frac{\pi_i}{1 - \pi_i}\right) = \alpha + \beta_1 x_{1i} + \beta_2 x_{2i}. \qquad (11.30)$$

To fit this logistic regression model to Table 11.7 data using the R package, I first created a 3-variable csv file containing that data. The 3 variables in the data file are gpa, participation, and sat. Then I ran the R script file named `chapter11.logit2.r` the content of which is shown in Program 11.2. This file is executed from within R as shown in Figure 11.9.

Table 11.7: SAT Scores, GPA, and SAT Preparation Program
Participation of 32 Students

GPA	Participation	SAT	GPA	Participation	SAT
2.66	0	0	2.75	0	0
2.89	0	0	2.83	0	0
3.28	0	0	3.12	1	0
2.92	0	0	3.16	1	1
4	0	1	2.06	1	0
2.86	0	0	3.62	1	1
2.76	0	0	2.89	1	0
2.87	0	0	3.51	1	0
3.03	0	0	3.54	1	1
3.92	0	1	2.83	1	1
2.63	0	0	3.39	1	1
3.32	0	0	2.67	1	0
3.57	0	0	3.65	1	1
3.26	0	1	4	1	1
3.53	0	0	3.1	1	0
2.74	0	0	2.39	1	1

It follows from Figure 11.9 that the 3 logistic regression coefficients of interest are estimated by, $\widehat{\alpha} = -11.602$, $\widehat{\beta}_1 = 3.063$, and $\widehat{\beta}_2 = 2.338$. Therefore, if student i has a GPA of x_i, then the probability $\widehat{\pi}_i$ of her receiving a passing SAT score depends on whether she participated in the SAT preparation program or not, and is estimated as,

$$\widehat{\pi}_i = \begin{cases} \dfrac{\exp(-11.602 + 3.063x_i)}{1 + \exp(-11.602 + 3.063x_i)} & \textit{if non-participant,} \\[3mm] \dfrac{\exp(-9.264 + 3.063x_i)}{1 + \exp(-9.264 + 3.063x_i)} & \textit{if participant.} \end{cases}$$

$$(11.31)$$

You may have noted that $-9.264 = -11.602 + 2.338$. These probabilities can be interpreted in terms of Odds Ratios as done in the previous section.

```
R RGui
File  Edit  View  Misc  Packages  Windows  Help

R R Console

> source("chapter11.logit2.r")
[1] "factor(sat) ~ gpa + factor(participation)"

Call:
glm(formula = model, family = binomial(link = "logit"), data

Deviance Residuals:
    Min       1Q    Median       3Q      Max
-1.8396  -0.6282   -0.3045   0.5629   2.0378

Coefficients:
                         Estimate Std. Error z value Pr(>|z|)
(Intercept)               -11.602      4.213  -2.754  0.00589
gpa                         3.063      1.223   2.505  0.01224
factor(participation)1      2.338      1.041   2.246  0.02470
---
Signif. codes:  0 '***' 0.001 '**' 0.01 '*' 0.05 '.' 0.1 ' '

(Dispersion parameter for binomial family taken to be 1)

    Null deviance: 41.183  on 31  degrees of freedom
Residual deviance: 26.253  on 29  degrees of freedom
AIC: 32.253
```

Figure 11.9. Logistic Regression with R on Table 11.7 Data

```
01  input.data <- read.csv("chap11.tab11.7.csv",
header=T)
02  model <- "factor(sat)    gpa +
factor(participation)"
03  print(model)
04  logitfit <- glm(model, family = binomial(link
= "logit"), data=input.data)
05  print(summary(logitfit))
```

Program 11.2. R Script for fitting logistic regression to Table 11.7 data

Using Excel

Note that the logistic regression model 11.30 can be fitted to Table 11.7 data using Excel. But with Excel, you need the log likelihood function associated with the model you are currently using. That function is given by:

$$
L(\alpha, \beta) = \\
\alpha\left(\sum_{i=1}^{n} y_i\right) + \beta_1\left(\sum_{i=1}^{n} y_i x_{1i}\right) + \beta_2\left(\sum_{i=1}^{n} y_i x_{2i}\right) \\
- \sum_{i=1}^{n} \log\left(1 + e^{\alpha + \beta_1 x_{1i} + \beta_2 x_{2i}}\right)
\tag{11.32}
$$

Subject i's contribution to this function is given by:

$$
u_i = \alpha y_i + \beta_1 y_i x_{1i} + \beta_2 y_i x_{2i} - \log(1 + \exp(\alpha + \beta_1 x_{1i} + \beta_2 x_{2i}))
\tag{11.33}
$$

If you are an Excel user, you may want to attempt obtaining the 3 coefficients shown in Figure 11.9 using Excel as an exercise.

In section 11.4, I like to briefly discuss Poisson (or log linear) regression mentioned in the introductory section.

11.4. Poisson Regression

Table 11.8 shows data on the number of c-sections, number of births, and hospital type (public or private) for 20 hospitals. The problem as discussed in section 11.1 is to find out whether the frequency of c-sections differ between private and public hospitals. I already mentioned a few reasons why a straight comparison of c-section rates between the two hospital types was not recommended. Another reason why comparing proportions of c-sections in private and public hospitals may not work is if the data came from a retrospective study where the hospital type is determined only after the sample of hospitals has been selected. In this case, neither the public nor the private hospital samples will be representative of their respective populations.

You may also think about using the number of c-sections as the dependent variable in an ordinary linear regression model as those discussed previously. But linear regression models work well only when the data is not skewed, and works best when the data is normally distributed. In many situations where the response variable represents counts, neither condition is met. Poisson regression, will generally be the most viable option at your disposal.

The data to be analyzed is in the form of a series of k counts $n_1, n_2, \cdots, n_k$, where n_i is the count associated with object i (in our example of Table 11.8, $k = 20$, and the counts are the 20 numbers of c-sections $n_1, n_2, \cdots, n_{20}$) . Count n_i is a random variable whose expected value is m_i. Since I want to explain the number of c-sections with the number of births, and the hospital type, the resulting log linear model will be formulated as follows:

$$\log(m_i) = \beta_0 + \beta_b x_{bi} + \beta_t x_{ti}, \qquad (11.34)$$

where x_{bi} =*the number of births in hospital i*, x_{ti} =*hospital i's type,* β_0, β_b, and β_t are the parameters to be estimated from data.

Fitting the Log Linear Model with R

(*a*) I created a csv file named chap11.c.sections.csv, which contains the data in Table 11.8.

(*b*) I created an R script file named chap11.poisson.csections.r containing the following lines of code:

```
input.data<-read.csv("chap11.c.sections.csv")

results<-glm(caesareans~births+factor(h.type),
    family=poisson,data = input.data)
print(summary(results))
```

(*c*) I then submitted these statements as shown in Figure 11.10.

It follows from Figure 11.10 and equation 11.34, that for a given hospital i, number of births x_{bi} and hospital type x_{ti}, the predicted number of c-sections denoted by $\widehat{n}_i$ is implicitly defined as, $\log\left(\widehat{n}_i\right) = 1.351 + 0.0003261 x_{bi} + 1.045 x_{ti}$, which is equivalent to $\widehat{n}_i = e^{1.351} e^{0.0003261 x_{bi}} e^{1.045 x_{ti}}$. Therefore,

$$\widehat{n}_i = \begin{cases} e^{1.351} e^{0.0003261 x_{bi}}, & \textit{if private hospital,} \\ e^{1.351} e^{1.045} e^{0.0003261 x_{bi}}, & \textit{if public hospital} \end{cases} \quad (11.35)$$

It follows from equation 11.35 that for the same number of births x_{bi}, the ratio of the predicted number of c-sections in public hospitals to the predicted number of c-sections in private hospitals in $e^{1.045} = 2.8$. That is, c-sections are almost 3 times more common in public hospitals than they are in private hospitals.

Equation 11.35 also indicates that for each 100 additional births, you can expect the number of c-sections to be multiplied by a factor of $\exp(0.0003261 \times 10) = 1.033$, which is an increase of 3.3%.

```
R Console
> source("chap11.poisson.csections.r")

Call:
glm(formula = caesareans ~ births + factor(h.type), family
    data = input.data)

Deviance Residuals:
     Min       1Q   Median       3Q      Max
  -2.3270  -0.6121  -0.0899   0.5398   1.6626

Coefficients:
                  Estimate Std. Error z value Pr(>|z|)
(Intercept)      1.351e+00  2.501e-01   5.402 6.58e-08 ***
births           3.261e-04  6.032e-05   5.406 6.45e-08 ***
factor(h.type)1  1.045e+00  2.729e-01   3.830 0.000128 ***
---
Signif. codes:  0 '***' 0.001 '**' 0.01 '*' 0.05 '.' 0.1 '

(Dispersion parameter for poisson family taken to be 1)

    Null deviance: 99.990  on 19  degrees of freedom
Residual deviance: 18.039  on 17  degrees of freedom
AIC: 110.80

Number of Fisher Scoring iterations: 4
```

Figure 11.10. Log Linear Regression with R on Table 11.8 Data

Fitting the Log Linear Model using Excel

Using Table 11.8, it is possible to compute the log linear regression parameters of equation 11.34 using Excel's Solver. The procedure for doing so is very similar to the one used to compute the logistic regression coefficients in the previous section. Before you even consider opening Excel, you need to have the log likelihood function associated with the regression model 11.34. This function[10] is defined as,

$$L(\beta_0, \beta_b, \beta_t) = \sum_{i=1}^{k} u_i, \ where$$

$$u_i = n_i(\beta_0 + \beta_b x_{bi} + \beta_t x_{ti}) - \exp(\beta_0 + \beta_b x_{bi} + \beta_t x_{ti}).$$

$$(11.36)$$

The coefficients you want are those specific values of β_0, β_b, and β_t that will yield the maximum value for the function $L(\beta_0, \beta_b, \beta_t)$ as defined above. It is this maximization process that Excel's Solver will carry out.

Compute the coefficients as follows:

(a) Set up your Excel spreadsheet as shown in Figure 11.11. The first 3 columns A, B, and C contain your input data. Column D contains the u_i values (see 11.36), and cells E3, F3, and G3 will hold the coefficients you are interested in. For now these 3 cells are empty (i.e. hold each a value of 0).

Expression 11.37 shows how the u_i values must be computed, by displaying the formula used to compute cell D4. *Sum all u_i values into cell D23.* Cell D23 hold the value to be maximized.

D4 =C4*($E$3+$F$3*A4+$G$3*B4) - EXP($E$3+$F$3*A4+$G$3*B4)

$$(11.37)$$

(b) Launch solver and fill out the form as shown in Figure 11.12.

(c) I also advise to improve the precision with which the parameters are calculated. This is accomplished by clicking the "Options" button on Solver's main form, so that the

---

[10]I referred to this function as the log likelihood. But to be accurate, it is actually the portion of the log likelihood function that is relevant for the purpose of computing the coefficients

options form of Figure 11.13 can be displayed. Change the "Constraint Precision:" text box to a smaller number such as 0.000001. *Reducing this number may be necessary whenever you think that some of the coefficients will be very small. For most applications I do not expect that you will need to use a number that is smaller than 0.000001.*

(d) Click OK to go back to the main form of Figure 11.12, then press "Solve." After a few seconds, the coefficients you want will appear in cells E3, F3, and G3.

Table 11.8 : Distribution of c-sections by hospital type and number of births

births	h.type[a]	caesareans
236	0	8
739	1	16
970	1	15
2,371	1	23
309	1	5
679	1	13
26	0	4
1,272	1	19
3,246	1	33
1,904	1	19
357	1	10
1,080	1	16
1,027	1	22
28	0	2
2,507	1	22
138	0	2
502	1	18
1,501	1	21
2,750	1	24
192	1	9

[a] *Hospital Type*: 0=*Private hospital*, 1=*Public hospital*

	D4		▾ (	f_x	=C4*($E$3 + $F$3*A4 + $G$3*B4) - EXP($E$3 + $F$3*A4 + $G$3*B4)						
▲	A	B	C	D	E	F	G	H	I	J	K
1											
2	x_{bi}	x_{ti}	n_i	u_i	β_0	β_b	β_t				
3	236	0	8	-1							
4	739	1	16	-1							
5	970	1	15	-1							
6	2,371	1	23	-1							
7	309	1	5	-1							
8	679	1	13	-1							
9	26	0	4	-1							
10	1,272	1	19	-1							
11	3,246	1	33	-1							
12	1,904	1	19	-1							
13	357	1	10	-1							
14	1,080	1	16	-1							
15	1,027	1	22	-1							
16	28	0	2	-1							
17	2,507	1	22	-1							
18	138	0	2	-1							
19	502	1	18	-1							
20	1,501	1	21	-1							
21	2,750	1	24	-1							
22	192	1	9	-1							
23	$L(\beta_0,\beta_b,\beta_t) =$			-20							
24											

Figure 11.11. Setting Up Excel

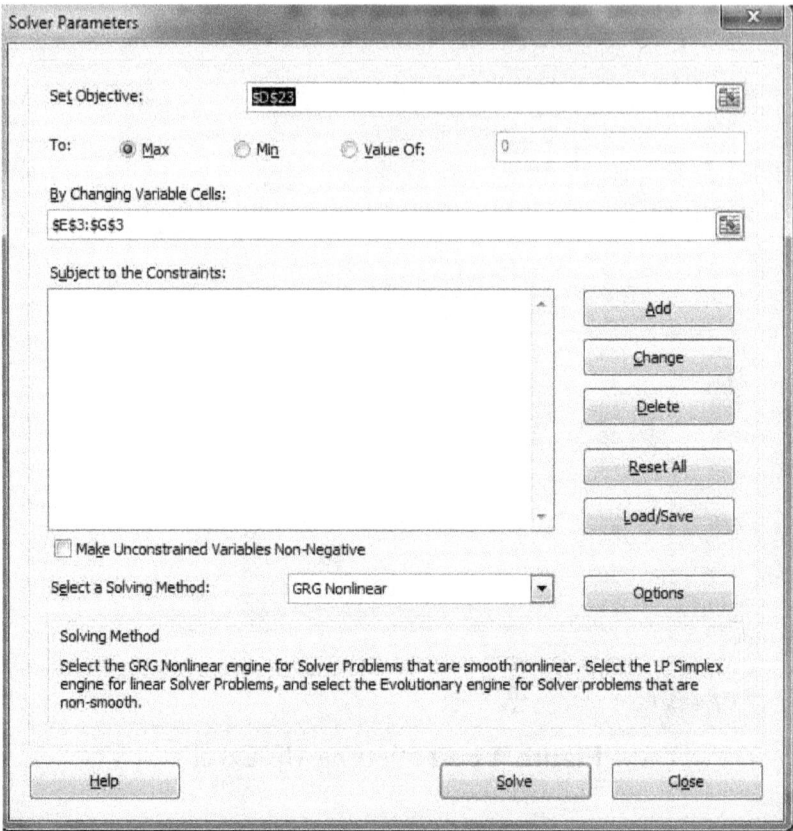

Figure 11.12. Specifying Solver's Parameters

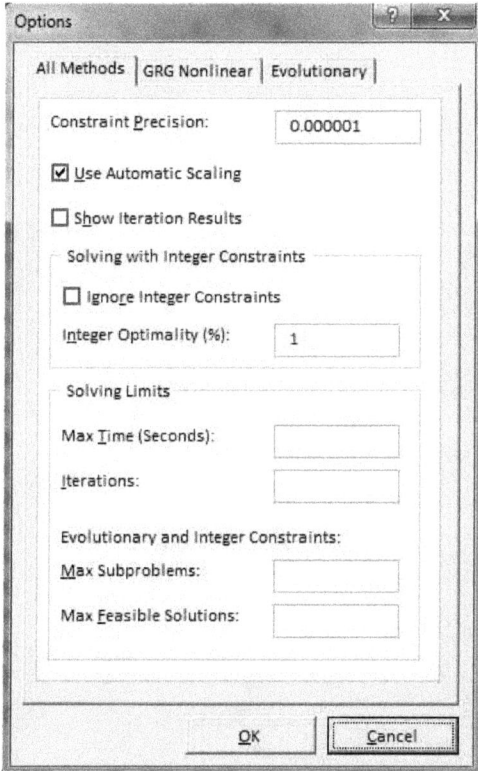

Figure 11.13. Increasing the precision level

CHAPTER $\boxed{12}$

Survey Sampling

OBJECTIVE

This chapter provides you with a broad overview of some important survey sampling concepts. You will learn why survey data should not be analyzed with the methods of classical statistics discussed in the past few chapters. You will also learned about the principles of a good sample design.

CONTENTS

12.1. Genesys

Statistical inference as discussed throughout the past few chapters is essentially about calculating (or approximating) the parameters of a theoretical probability distribution (such as the mean of a Normal distribution), calculating associated precision measures (such as the standard error of the mean), and testing hypotheses about the values of these parameters. However, this framework for doing statistical inference is inappropriate for analyzing data collected with a statistical survey that has targeted a specific and well-defined population of individuals or objects. The framework that you need to analyze survey data is known as *Finite Population Sampling and Inference*, also called *Survey Sampling*.

For an unknown reason, this very important area of statistics is often overlooked in many introductory statistics textbooks, and is generally unknown to many researchers. If you are going to use data from a statistical survey, or if you are going to conduct such a survey yourself, then you need to have some exposure to the principles of finite population sampling. A statistical survey, also called scientific survey, is essentially a survey based on a random sample of subjects that is representative of the finite population it was selected from. All surveys sponsored by the US Federal government for example are statistical surveys. Therefore to analyze data produced by government-sponsored surveys, you will need to use the right framework, which is that of survey sampling.

What is the issue?

Consider a fictitious city that is divided into low-income and high-income areas. Each of the 1,000 low-income households has the same annual income of $35,000, while each of the 50 high-income households has annual income of $250,000 as summarized in Table 12.1. Assume that all you know about this city is its population size (i.e. the population size is $N = 1,050$ households), and the distribution of households by income level (low versus high). However, you know nothing about the actual households income. Your goal is to survey a sample of households from the city so that you can estimate the city overall household mean income.

Table 12.1 : Population Income statistics

Variable	Low-Income Area	High-Income Area
Population Size	1,000	50
Household Income	$35,000	$250,000

To conduct this survey, you decide to make the following two decisions:

(i) Based on your limited budget, you decide that you can only afford to sample a total of 100 households (i.e. the sample size in this case is $n = 100$)

(ii) Since high-income households live in a secured area, and there are only 50 of them, you decide to select them all. Although there are many more low-income households, they live in a unsafe area, and you decide to select a random sample of 50 such households and collect their annual income.

Based on the data collected with your survey, you decide to compute the city mean household income as,

$$\text{Estimated Mean} = \frac{50 \times \$35,000 + 50 \times \$250,000}{100} = \$142,000.$$

Does $142,000 really represent the income level in this city? If you compute the actual overall mean income from Table 12.1 (I am cheating here because I am using income I am not supposed to know), then you will get the following "true" population mean income:

$$\text{"True" Mean} = \frac{1,000 \times \$35,000 + 50 \times \$250,000}{1050} = \$45,238.$$

The difference between your estimated mean and the true mean you are interested in is striking. The estimated mean has overestimated the true mean by a wide margin. What went wrong?

The main problem that caused this poor estimation of the household population's mean income stems from the fact that the oversampling[1] of the high-income area and the undersampling[2] of the low-income were not taken into consideration, when calculating the estimated mean income. I am going to mention two ways this problem could have been resolved:

(a) *Changing the Estimation Method.* Selecting 50 out of 1,000 low-income households means that each sampled low-income household represents 20 others (note that $20 = 1,000/50$). That is, each sampled low-income household carries a *Weight* of 20 (i.e. $W_L = 20$). On the other hand, selecting all 50 high-income households means that each selected high-income household represents only 1 high-income household, which is itself. Each high-income household carries a *Weight* of 1 (i.e. $W_H = 1$). The estimated population mean, should

---

[1]I am talking about oversampling here because 50 high-income households represent 50% of the sample, while representing only 4.76% of the total population of 1,050 households

[2]I am talking about undersampling here because 50 low-income households represent only 50% of the sample, when in fact low-income households represent over 95% of the population of 1,050 households

now be computed as a weighted mean:

$$\text{Estimated Mean} = \frac{W_{\text{L}}(50 \times \$35000) + W_{\text{H}}(50 \times \$250000)}{50W_{\text{L}} + 50W_{\text{H}}},$$

$$= \frac{20(50 \times \$35000) + 1(50 \times \$250000)}{50 \times 20 + 50 \times 1},$$

$$= \$45,238.$$

As you can see, this weighted method yields the exact same value of the population mean. Using this method will not always yields the exact number you are interested in. In this case, the low-income and high-income populations are homogeneous with respect to their annual income. Even a single individual will be representative of each of these 2 populations, providing all the information about the entire population. Still, the weighted approach produces the exact solution under ideal circumstances. *Because of its focus on a specific "finite" population of interest, survey sampling puts considerable emphasis on weighting. Always remember that in sample surveys, weighting is a big deal. It is done every time, always.*

(b) *Changing the Sample Selection Method (i.e. the Sampling Design).* In (a), I suggested to modify the standard un-weighted method for computing the mean in order to in-corporate into the procedure, the specific way the data was obtained, or the way the sampling was designed. You may however modify your sampling so that you do not have to use the weighted method. The solution is to select low-income and high-income households proportionally to their representation in the whole population. That is, since high-income households represent 4.76% of all households, about the same percentage of the 100 households in the sample must be made up of high-income households. Therefore, I would select in my sample of 100 households, 5 high-income households and 95 low-income households. The unweighted mean will the be,

$$\text{Estimated Mean} = \frac{95 \times \$35000 + 5 \times \$250000}{100} = \$45,750.$$

Although I did not get the exact same value of the population mean, my estimation of $45750 remains very close to it. This way of selecting the sample is known as *Proportional Allocation* of the sample, or the *Self-weighted* design.

The discussion of Table 12.1 data shows the nature and the strength of the link between the sampling design and the estimation procedure. Also tied to these 2 components of survey sampling, is the the finite population inference involving the calculation of expectations and variances. In the next 3 sections, I will discuss successively about different possible sampling designs, some estimation methods in survey sampling, and will provide a glimpse into the theory of finite population inference.

12.2. Sampling Design

One of the primary objectives in the development of a sample survey is to obtain a sample that is as representative as possible of the target population of inference. The term representative as used here is rather vague. In practice, the sample is a representation of the target population of inference only with respect to a limited number of characteristics such as gender, income level, educational attainment and other such variables known for having a big influence on individuals' interests and behaviors. The term sampling design refers to a broad range of activities including,

▶ Defining the physical medium that will be used to carry out the actual selection of the sample. This medium is called the *Sampling Frame*

▶ Deciding whether to divide the survey population into subgroups (subgroups could be strata or clusters) in order to obtain a better representation of the population (e.g. separating low-income from high-income households).

▶ Allocating the sample to each subgroup. Sample allocation amounts to deciding how many units will be selected from each population subgroup.

▶ Determining the selection probabilities. I will further elaborate on this later on.

Note that the terms *Sample design* and sampling design are often used interchangeably in many texts.

How you will design the sampling depends on the difficulty for reaching members of the survey population, the availability of the sampling frame, the heterogeneity of the population units with respect to the variable of interest, the precision you want for your estimations. The area of sampling design is vast, and if it gets very complex, you may want to seek assistance from a sampling statistician. My goal in this chapter, is to raise some of the core issues I believe you should care about as a data user, or as a developer of small-scale surveys.

For the rest of this chapter, I will assume that your target or survey population U has N units, of which a smaller number n will be randomly selected to be included in the sample labeled as s. A specific population or sample unit will be denoted by i, and the value of the analytic variable y associated with i is denoted by y_i.

12.2.1 Simple Random Sampling

The simple random sampling (SRS) is the simplest sampling design that exists. It will work well for regular survey populations that do not exhibit a lot of disparities among its members, and therefore do not require any particular precautions to ensure representativity.

Selecting a Simple Random Sample

Consider the list of 20 hospitals in Table 12.2, which contains the number of births, the hospital type (private or public), and the number of c-sections performed. Suppose that you want to select a simple random sample of 5 hospitals out of the 20 in the population. This task can be performed as follows:

(a) Assign a random number between 0 and 1 to each of the 20 hospitals. Such random numbers can be generated from Excel using the function =RAND() (with no argument). *Because these numbers change each time the spreadsheet is modified, you may want to copy and paste the values only into other cells.*

(b) Sort the entire dataset by the random numbers, and take the first 5 hospitals in the sorted file as your simple random sample

Characteristics of Simple Random Samples

Simple random samples are easy to select, and possess the following basic characteristics that you need to know, if you are to use them.

(a) *Total Number of Possible Simple Random Samples*

If I consider Table 12.2 as representing the population of interest, then the population size is $N = 20$. The total number of samples of size $n = 5$ you can select is calculated as $\binom{N}{n} = \binom{20}{5} = 15,504$ *This is the number of combinations of 20 taken 5 at a times. You may compute this number using the Excel function =COMBIN(20,5).* As you may see, the

number of different samples of size 5 from that population
can be quite high.

(b) *Probability of Selecting a Specific Sample s*

All 15,504 samples of size 5 have the exact same chance
of being selected during the selection process. That is, any
particular sample s you can think of, will have the same
probability $P(s) = 1/15,504 = 0.0000645$ of being selected.
That means, if you select 2 samples successively, there is
near to 0 chance that you will end up selecting the same
sample.

(c) *Inclusion Probabilities*

If you are about to select a sample, then any particular
subject i among the 20 in the target population will have
the exact same chance n/N of being included in the sample.
That is, the *Inclusion Probability* of subject i is $\pi_i = n/N = 5/20 = 0.25$.

The primary reason why I am showing interest in the inclusion probabilities stems from the fact that the weighted sample
mean will be used later to estimate the value of the population
mean, and the weight associated with an individual i is calculated as the inverse of the selection probability π_i. Furthermore,
the probability $P(s)$ for selecting a specific sample s is a key
building block of the finite population inference as I will show in
section 12.4.

Limitations of Simple Random Sampling

As you may have noticed, selecting a simple random sample
requires two things, which are the knowledge of the target population size N, and an easy access to the population units.
Some population units are not always easily identifiable. As an
example, suppose that you want to review 10 invoices to be se-

lected randomly from a batch of 2,000 invoices organized in chronological order and perhaps located in different places. If your simple random sample includes the 2 invoices #55, and #1,433, how would you know which physical invoice is associated with the numbers 55, and 1,433? In many instances, this will be an impossible task.

Table 12.2: Distribution of c-sections by hospital type and number of births

Hospital#	births	h.type[a]	caesareans
1	236	0	8
2	739	1	16
3	970	1	15
4	2,371	1	23
5	309	1	5
6	679	1	13
7	26	0	4
8	1,272	1	19
9	3,246	1	33
10	1,904	1	19
11	357	1	10
12	1,080	1	16
13	1,027	1	22
14	28	0	2
15	2,507	1	22
16	138	0	2
17	502	1	18
18	1,501	1	21
19	2,750	1	24
20	192	1	9

[a] *Hospital Type: 0=Private hospital, 1=Public hospital*

If the population size N is known, but there is no easy access to the population units, it is generally recommend to use *Systematic Sampling* in place of the simple random sampling. When the population size is unknown, the recommendation is to use *Bernoulli Sampling*.

In the next two sections 12.2.2 and 12.2.3, I will review the systematic and the Bernoulli sampling methods.

12.2.2 Systematic Sampling

Suppose that you want to select a random sample of size n from a population of N units. Let $a = N/n$, a ratio that may not be an integer. a is called the sampling interval. The selection of the systematic sample is carried out as follows:

(a) Randomly chose a random number ε between 0 and a then compute n integer values by rounding up the following n values to the nearest integer:

$$\varepsilon + (j-1)a, \ for \ j \ varying \ from \ 1 \ to \ n \qquad (12.1)$$

Suppose that $N = 20$ and you want to select a systematic sample of size $n = 7$. Therefore, $a = 20/7 = 2.857143$. Suppose that you have generated $\varepsilon = 0.8423$ as your random number between 0 and a. The 7 numbers of equation 12.1 are 0.842, 3.699, 6.557, 9.414, 12.271, which leads to the 7 integer values, 1, 4, 7, 10, 13.

(b) Select all population units corresponding to the integer values obtained in step (a).

Systematic samples are practical when sampling populations with units that are not clearly labeled nor easily accessible. The sample units are selected in a sequential manner, avoiding the undesirable situation where the units must be located by their numbers.

Inclusion Probabilities

If you are about to select a systematic sample, then any particular subject i among the N in the target population will have the exact same chance $1/a$ of being included in the sample. That is, the inclusion probability of subject i is $\pi_i = n/N$.

12.2.3   Bernoulli Sampling

Bernoulli sampling is recommended when you want to select a random sample from a finite population whose size is unknown. The Internal Revenue Service (IRS), which is the U.S. government agency responsible for tax collection and tax law enforcement, often selects samples of tax returns for audit as they are filed. This selection of tax returns on a flow basis is carried out without the knowledge of the population size, since the population of tax returns is still incomplete at the time of sampling. What the Bernoulli and the simple random sampling (SRS) approaches have in common is that all units are selected with the same probability. However, with the SRS approach you will know upfront what the sample size will be, while the size of the Bernoulli sample will only be known at the end of the sample selection process. You would implement Bernoulli sampling as follows:

(a) Decide about the probability π with which each unit will be selected. How do you determine π? One way to look at it is to realize that in the end of the selection process, you can expect a sample of size n determined as $n = N\pi$, where N is the population size. Since N is unknown, you may want to have an educated guess of its value based on past data or based on the experience of subject-matter experts. A value for π could then be the ratio n_0/N_0 of the desired sample size n_0 to the population size N_0.

(b) Each time a population unit i presents itself, you must

generate a uniform random number ε between 0 and 1, then
decide to include that unit into the sample if π exceeds ε.

Characteristics of Bernoulli Samples

Bernoulli sampling is one of the simplest of all sampling plans,
and is practical when the sample selection must begin while the
survey population is still incomplete.

(a) *Total Number of Possible Bernoulli Samples*

Because the size of a Bernoulli sample is random, knowing
the number of samples that can be selected carries no interest.

(b) *Probability of Selecting a Specific Sample s*

Any Bernoulli sample s of a given size n has the exact same
probability $P(s) = \pi^n (1 - \pi)^{N-n}$ of being selected.

(c) *Inclusion Probabilities*

Any particular subject i has by definition the same chance
π of being included into the sample. That is, the inclusion
probability of subject i is $\pi_i = \pi$.

12.2.4 Poisson Sampling

Bernoulli sampling is too simple, and you should not
expect to meet the task requirements in a serious professional
environment which such a soft tool. In reality, the probability
with which you will select a particular population unit i will not
be the same for all units. This differential selection probability is
often imposed by business considerations. The IRS for example
will certainly be more interested in tax returns with larger annual
revenues than those with smaller annual revenues. This interest
will translate into higher selection probabilities for large-revenue
tax returns, during the sample selection phase.

Poisson sampling generalizes Bernoulli sampling by defining blocks of population units within which Bernoulli sampling is carried out. Here is how Poisson sampling could be implemented:

(a) For each population unit k, decide about the probability π_k with which it will be included in the sample. In practice, these probabilities are calculated by first categorizing the population units into a number of groups according the magnitude of a characteristic of interest.

Regarding the audit of tax returns for example, you may decide that all returns with annual revenues in excess of $250,000 will be selected with certainty; that is with proba-bility 1 (i.e. $\pi_k = 1$), and that returns with annual below $100,000 will be selected with probabi-lity $\pi_k = 0.005$. Note that all returns included in your target survey population must have a non-zero selection probabi-lity, otherwise you will induce a bias into your estimation procedure as I will show in section 12.3.

(b) To decide whether a given population unit k must be in-cluded in the sample or not, generate a uniform random number ε_k between 0 and 1. Then include unit k in the sample only if the selection probability π_k exceeds ε_k.

Characteristics of Poisson Samples

Poisson sampling is a very practical sampling plans often used in practice. It generalizes Bernoulli sampling and possesses the following characteristics:

(a) *Total Number of Possible Poisson Samples*

Just like the Bernoulli sample, the Poisson sample has a random size. Therefore, knowing the number of samples that can be selected carries no interest.

(b) *Probability of Selecting a Specific Sample s*

The probability of selecting a specific Poisson sample s of size n is obtained as follows:

$$P(s) = \textit{(Product of the } \pi_k\text{'s for sample units)} \times$$
$$\textit{(Product of the } (1 - \pi_k)\text{'s for non-sample units)}$$

(c) *Inclusion Probabilities*

Any particular unit i has by definition a π_i chance of being selected in the sample.

12.2.5 Stratified Sampling

Stratified sampling is to simple random sampling, what Poisson sampling is to Bernoulli sampling. That is, a generalization of the basic SRS procedure to the more realistic situation where the population is first segmented, and the SRS carried out independently within each segment. Stratified random sampling consists of dividing the survey population into a number of sub-populations called strata, and selecting a simple random sample of a certain size independently within each stratum. Why do this?

There are a few practical reasons that justify the use of this popular sampling method. Suppose that the 20 hospitals in Table 12.2 represent the population of hospitals that you want to survey for the purpose of estimating the average number of c-sections performed per hospital. Also assume that the information on the number of births as well as the hospital type (public or private) is also available and accessible for each of the 20 hospitals targeted. However, the number of c-sections (our study variable) is not available, and will have to be collected for the hospitals in the selected sample. Here are two concerns that you may have regarding the proper selection of hospitals:

(a) If it is suspected that private and public hospitals may have different c-section rates, then you may end up overestimating the c-section average if the hospitals in the group with the highest c-section rate are over-represented in the sample. Note that the over-representation of public or private hospitals may well occur if you select a simple random sample from the entire population of hospitals. After all, there are far more public hospitals than private hospitals.

(b) The second problem stems from the fact large hospitals in terms of number of births are likely to have more c-sections than small hospitals. Therefore you may also want to have a better representation of hospitals in your (small) sample with respect to their size.

You can resolve problem (a) by creating strata based on the hospital type, before selecting a simple random sample independently from each of the 2 strata to ensure an adequate overall representation of hospitals in your sample. You can resolve problem (b) by categorizing the "births" to create one stratum for large hospitals (e.g. these could be hospitals with the number of births of 1,000 or more). If you are concerned about both the hospital type and size, then you could even cross the "Type" and "Size" strata. You would try to avoid having strata that are too small.

12.2.6 Cluster Sampling

If the purpose of stratification is to create within-group homogeneity, clustering is recommended in the following:

(i) You want to ensure geographical proximity among the sampled units to reduce data collection costs, when the population units are scattered over a wide geographical area.

(ii) You need to create a sampling frame when none is available.

Suppose you want to conduct a household survey in a given city and that a comprehensive list of houses and their street addresses is not available to you. Therefore, you will not have a sampling frame to select a sample from. An often recommended option is to divide the city into several smaller geographical areas called *clusters*, before selecting a simple random sample of clusters as described earlier. All households within the selected clusters will then be surveyed. To be more specific, this sampling method is known as the single-stage cluster sampling.

More complex sampling approaches are often used in practice where after the selection of an initial sample of clusters, the units in each the selected clusters are listed to build the sampling frame. Simple random sampling will then be conducted within each sampled cluster using the newly-built sampling frame. This method is called *two-stage cluster sampling*.

12.3. Estimation in Survey Sampling

In this section, I want to show how the population mean must be approximated with sample data, and stress out the important role that the sampling design plays in this estimation process. In the previous chapters on classical statistical inference, you had a single way of selecting the sample, and a single way of computing the mean from the selected sample. You dealt exclusively with abstract populations considered to be infinitely large with no specific structure of interest. In survey sampling, you deal solely with concrete populations that are structured in a specific way that you want to see replicated in the sample that serves as basis for estimation. Consequently, each sampling procedure you use will dictate the specific way the population mean must be estimated. This is the main reason why data coming from a statistical survey should never by analyzed unless you have in your possession all the information regarding the sampling design. In the survey sampling jargon, sampling statisticians use the term *Design Information.*

Let N be the population size, n the sample size, and y the variable of interest the mean of which you want to estimate. The mean calculated using all N population data is called population mean and denoted by $\overline{Y}$ (read Y bar, Y being capitalized). This population will be unknown and must be approximated using sample data. If on the other hand, the mean is calculated using the n sample observations only, it is referred to as the **sample mean** and denoted by $\overline{y}$ (read small y bar).

The general principle guiding the calculation of $\overline{y}$ is to always see it as a weighted mean where the weight w_i associated with sample unit i is the inverse of the selection probability π_i. That is,

$$\overline{y} = \frac{\displaystyle\sum_{i=1}^{n} w_i y_i}{\displaystyle\sum_{i=1}^{n} w_i}, \quad where \ w_i = 1/\pi_i. \tag{12.2}$$

The value of $\overline{y}$ is random since any random sample will lead to a different estimation of the population mean $\overline{Y}$. $\overline{y}$ is often called an *estimator* of $\overline{Y}$. To emphasize its link to the population mean, $\overline{y}$ is sometimes denoted by $\widehat{\overline{Y}}$ in some textbooks on survey sampling. For the sake of simplicity, I will stick with one notation, which $\overline{y}$ to designate any estimator of the population mean regardless of the sampling design used.

Using the weighted mean as opposed to the simple arithmetic mean as done in the past few chapters is motivated by the desire to have an estimator whose expected value[3] always matches the population parameter $\overline{Y}$ being estimated. This property is referred to as *Unbiasedness*, and will be further discussed in section 12.4.

Estimator of Mean under SRS, Systematic and Bernoulli Sampling

The inclusion probability π_i is constant under the SRS, systematic, and Bernoulli sampling approaches. That is all population units will be selected with the exact same probability $\pi_i = \pi$, which leads to a unique sampling weight $w_i = 1/\pi$. If follows from equation 12.2 that the mean estimator will reduce to the simple and familiar arithmetic mean $\overline{y}$ given by,

$$\overline{y} = \frac{1}{n} \sum_{i=1}^{n} y_i. \tag{12.3}$$

Although the arithmetic mean is a simple and intuitive estimator of the population mean, it often has unsatisfactory properties in the context of survey sampling. To see this, consider Table 12.2 data, and suppose that you must estimate the mean number of c-sections based on a simple random sample of 7 hospitals. Your mean number of c-sections based on your sample could vary considerably in its magnitude depending on the relative number of large hospitals that happen to be selected in your sample. A statistical procedure that sends mixed messages is considered unreliable. If the number of births is available for all hospitals in a computerized database, then a different and more efficient estimation procedure can be used. It is known as the *ratio estimator*.

---

[3]Note that the expected value in the context of survey sampling is not calculated with respect to a theoretical law of probability such as the Normal distribution. Instead, it is calculated with respect to the law of probability induced by the sampling as will be seen in section 12.4

Ratio Estimator

Referring the Table 12.2 data, in addition to the variable of interest y_i = *number of c-sections in hospital i*, consider what I call the *Auxiliary variable* x_i = *number of births in hospital i*. The ratio estimator of the population mean $\overline{Y}$ is defined as follows:

$$\overline{y}_{\mathrm{R}} = \frac{\overline{X}}{\overline{x}}\overline{y}, \qquad (12.4)$$

where $\overline{X}$ is the average number of births per hospital based on all 20 hospital numbers of births in the population, and $\overline{x}$ the average number of births based on the hospitals selected in the sample. Equation 12.4 is very appealing in that if you select small hospitals in your sample, $\overline{x}$ will underestimate $\overline{X}$, and the ratio $\overline{X}/\overline{x}$ will exceed 1, and be used as a correction factor for increasing the value of the arithmetic mean $\overline{y}$. This method is based on the premise that if $\overline{x}$ underestimates $\overline{X}$, then $\overline{y}$ also will underestimate the parameter of interest $\overline{Y}$ by a proportional amount.

Estimator of Mean under Poisson Sampling

Under Poisson sampling, there will be no particular problem applying equation 12.2 after the selection probabilities and associated weights w_i have be defined for all subjects in the sample.

Estimator of Mean under Stratified Sampling

Suppose that you divide your survey population into H strata. Each stratum h contains N_h population units, n_h of which will be selected using the simple random sampling approach. That is each unit i belonging to stratum h will be selected with the same probability $\pi_h = n_n/N_h$, which leads to the stratum weight $w_h = N_h/n_h$ to be assigned to all units within the stratum. It then follows for equation 12.2 that the mean estimator under stratified sampling will be:

$$\overline{y}_{str} = \sum_{h=1}^{H} W_h \overline{y}_h, \qquad (12.5)$$

where $W_h = N_h/N$ is the relative number of units is stratum h, and $\overline{y}_h$ is the simple arithmetic mean calculated using stratum h values only.

In stratified sampling, you would select the units into your sample independently within each stratum. In practice, after you determine the total sample size n, you will then have to decide how to distribute these n sample units across the H strata previously defined. This important activity is referred to as *Sample Allocation*. There is an extensive literature on this topic alone that explores various sample allocation techniques. A simple and popular allocation technique is called *Proportional allocation*, and consists of allocating to each stratum, n_h units proportionally to the stratum size. That is $n_h = nN_h/N$.

Remember that in stratified sampling, different units are generally selected with different selection probabilities. Consequently, not using the correct mean estimator may result in a severe bias in your estimations.

Ratio Estimator and Stratified Sampling

Equation 12.5 shows that the mean estimator under stratified sampling is a weighted average of all H stratum-level sample means $\overline{y}_h$. Moreover, I showed previously that under the simple random sampling design, the use of the ratio estimator could lead to a vastly improved estimation of the population mean, provided an auxiliary variable x is available for all sample units, and that the population mean of the auxiliary variable is known. Likewise, the estimation of the population mean under stratified sampling can be improved by using a ratio estimator for each stratum in place of the simple arithmetic mean $\bar{y}_h$.

The ratio estimator of the population mean under stratified sampling is given by:

$$\bar{y}_{\text{SR}} = \sum_{h=1}^{H} W_h \left(\frac{\overline{X}_h}{\overline{x}_h} \bar{y}_h \right), \qquad (12.6)$$

Equation 12.6 describes what is known as the *Separate Ratio Estimator* because the ratio estimation technique is applied separately within each stratum. However, this technique can be applied at the global level as it was in equation 12.4 with the only exception that $\bar{y}$ and $\bar{x}$ will not be simple arithmetic means as in equation 12.3. Instead these will be stratified means calculated as in equation 12.5. This approach will lead to the *Combined Ratio Estimator* of the population mean defined as,

$$\bar{y}_{\text{CR}} = \frac{\overline{X}}{\overline{x}_{str}} \bar{y}_{str} \qquad (12.7)$$

where $\bar{y}_{str}$ and $\bar{x}_{str}$ are calculated as shown in equation 12.5.

Estimator of Mean under Cluster Sampling

Under the single-stage cluster sampling, all population units have the exact same population of being selected in the sample. Therefore, each population unit i will be selected in the sample only if the cluster it belongs to is selected as well. Consequently the probability π_i of selecting unit i equals the probability of selecting its cluster of membership. That is, $\pi_i = m/M$, which leads to the weight $w_i = 1/\pi_i = M/m$. As a result of this and following the general equation 12.2, the population is estimated with the usual arithmetic mean $\bar{y}$.

Note that, although the simple arithmetic mean $\bar{y}$ is acceptable under the single-stage cluster sampling design because it is unbiased, it will have a higher variance under cluster sampling than under the simple random sampling design. This problem will get worse if the clusters vary substantially in size. Remember that the number n of analytic units under cluster sampling is a random variable. To see this, suppose that you have defined a total of M clusters, and that each cluster c contains N_c units (or individuals). The total number of units in the target population is $N = N_1 + N_2 + \cdots + N_M$. Following a simple random sampling design, you will select m out of M clusters, and all units of the selected clusters are surveyed (remember this is single-stage cluster sampling). The number n of units in the selected sample is calculated as,

$$n = N_1 + N_2 + \cdots + N_m. \tag{12.8}$$

Consequently, the magnitude of n may vary considerably if cluster sizes N_c are very different.

12.4.   A Glimpse into Finite Population Inference

Because finite population sampling provides a framework for statistical inference different from that of classical statistics, it is essential to compare the sources of randomness and the meaning of the notion of probability in both contexts. Consider a sampling plan that calls for the selection of a sample of size n from a finite population of size N. Suppose that the total number of different samples of size n that could possibly be selected is B. The goal is to estimate the population mean $\overline{Y}$. I now want to explain what the expected value of $\bar{y}$ and its variance mean within the current context.

Expectation

Each individual sample s of size n will lead to an estimation $\bar{y}_s$. The expected value of $\bar{y}$ represents the mean of the estimations $\bar{y}_s$ calculated over all B samples that can possibly be selected. Mathematically, this can be expressed as,

$$E(\bar{y}) = \sum_{s=1}^{B} \bar{y}_s P(s), \qquad (12.9)$$

where $P(s)$ is the probability that the specific sample s is selected. Because the probability $P(s)$ of selecting sample s depends on the sampling design, the expected value $E(\bar{y})$ is expected to be very dependent on the sampling design. For the SRS design, one can prove that $E(\bar{y}) = \overline{Y}$, meaning that $\bar{y}$ is an unbiased estimator of $\overline{Y}$ under this design.

Variance

The variance of $\bar{y}$ is naturally defined as the expected value of the squared difference between $\bar{y}_s$ values and their overall mean $E(\bar{y})$. This is formulated mathematically as,

$$V(\bar{y}) = \sum_{s=1}^{B} (\bar{y}_s - E(\bar{y}))^2 P(s). \qquad (12.10)$$

While equations 12.9 and 12.10 are useful for defining the concepts of expectation and variance in the context of finite population sampling, they are not very useful for performing the actual computations. This is due to the fact that all possible samples will never be available in the real word. Other techniques are generally used to derive more useful expressions for population variances, which can be approximated more conveniently with sample data. Below are some variance expressions that can give you a sense of what you may find in a typically survey sampling textbook.

▶ **Simple Random Sampling** The variance of the sample mean $\overline{y}$ is given by:

$$V(\overline{y}) = \frac{1-f}{n(N-1)} \sum_{i=1}^{N} (y_i - \overline{Y})^2, \quad (12.11)$$

where $f = n/N$ is the sampling fraction. The variance of the ratio estimator is given by:

$$V(\overline{y}_{\mathrm{R}}) = \frac{1-f}{n(N-1)} \sum_{i=1}^{n} (y_i - Rx_i)^2, \quad (12.12)$$

where $R = \overline{Y}/\overline{X}$ and $f = n/N$

▶ **Cluster Sampling** The variance of $\overline{y}$ under the single-stage cluster sampling design is given by,

$$V(\overline{y}) = \frac{1-f}{m(M-1)} \sum_{k=1}^{M} N_k^2 (\overline{y}_k - \overline{Y})^2, \quad (12.13)$$

where $f = m/M$.

In practice, only the sample data will generally be available. In this case, the summations in equations 12.11, 12.12, and 12.13 will be calculated based on the sample data only. For more on survey sampling, I advise that you read Cochran, W.G. (1977).

Bibliography

[1] Agresti, A., Coull, B.A. (1998). "Approximate is better than "Exact" for interval estimation of binomial proportions." *American Statistician*, **52**, 119-126.

[2] Aleksandrov, A.D., Kolmogorov, A.N., and Lavrent'ev, M.A. (1999). *Mathematics: Its Content, Methods and Meaning*, Dover Publications.

[3] Birmingham, C.L., Muller, J.L., Palepu, A., Spinelli, J.J., and Anis, A.H. (1999). "The Cost of Obesity in Canada," *Canadian Medical Association Journal*, **160**(4), 483-488.

[4] Brown, L.D., Cai, T.T., and DasGupta, A. (2001). "Interval Estimation for a Binomial Proportion." *Statistical Science*, **16**(2), 101-133.

[5] Cochran, W.G. (1952). "The Chi-Square Goodness-of-fit Test." *Annals of Mathematical Statistics*, **23**, 315-345.

[6] Cochran, W.G. (1977). *Sampling Techniques*, John Wiley & Sons, Inc. : New York.

[7] Conover, W.J. (1980). *Practical Nonparametric Statistics (2nd ed.)*. New York: John Wiley & Sons, Inc.

[8] Conover, W.J. (1999). *Practical Nonparametric Statistics (3nd ed.)*. New York: John Wiley & Sons, Inc.

[9] Connett, J.E., Sith, J.A., and McHugh, R.B. (1987). "Sample size and power for pair-matched case-control studied." *Statistics in Medicine*, **6**, 53-59

[10] Coombs, W.T., Algina, J., Oltman, D. (1996). "Univariate and multivariate omnibus hypothesis tests selected to control type I error rates when population variances are not necessarily equal." *Review of Educational Research*, **66**, 137-179.

[11] Cornfield, J. (1962). "Joint dependence of risk of coronary heart disease on serum cholesterol and systolic blood pressure: a discriminant function analysis." *Federal Proceedings*, **21**, Supplement No. 11: 58-61.

[12] Daniel, W.W. (1990). *Applied Nonparametric Statistics* (2nd ed.). Boston: PWS-Kent Publishing Company.

[13] Dunn, O.J. (1961). "Multiple comparisons among means." *Journal of the American Statistical Association*, **56**, 52-64.

[14] Efron, B. (1982). *The jackknife, the bootstrap, and other resampling plans.* Society of Industrial and Applied Mathematics CBMS-NSF Monographs.

[15] Efron, B. (1987). "Better Bootstrap Confidence Intervals." *Journal of the American Statistical Association*, **82**, 171-185.

[16] Efron, B., and Tibshirani, R.J. (1994), *An Introduction to the Bootstrap* (Chapman & Hall/CRC Monographs on Statistics & Applied Probability), Chapman and Hall/CRC.

[17] Fleiss, J.L., Levin, B., and Paik, M. C. (2003). *Statistical Methods for Rates and Proportions* (3rd Ed.), Wiley Series in Probability and Statistics, A John Wiley & Sons, Inc. : New York.

[18] Freedman, D. and Diaconis, P. (1981). "On this histogram as a density estimator : L2 theory." *Zeit. Wahr. ver. Geb.*, **57**, 453-476.

[19] Hartley, H.O. (1940). "Testing the homogeneity of a set of variances." *Biometrika*, **31**, 249-255.

[20] Hartley, H.O. (1950). "The maximum F-ratio as a shortcut test for heterogeneity of variance." *Biometrika, 37*, 308-312.

[21] Hogg, R.V., Craig, A., McKean, J.W. (2004). *Introduction to Mathematical Statistics* (6th edition), Publisher: Prentice Hall.

[22] Hosmer, D.W., and Lemeshow, S. (2000). *Applied logistic regression (Wiley Series in probability and statistics)* (2nd edition), Wiley-Interscience Publication.

[23] Howell, D.C. (2002). *Statistical Methods for Psychology* (5th ed.). Pacific Grove, CA: Duxbury Press.

[24] Kish, L. (1995). *Survey Sampling (Wiley Classics Library).* New York: John Wiley & Sons, Inc.

[25] Kruskal, W.H. (1952). "A nonparametric test for the several sample problem." *Annals of Mathematical Statistics, 23*, 525-540.

[26] Kruskal, W.H. and Wallis, W.A. (1952). "Use of ranks in one-criterion variance analysis." *Journal of the American Statistical Association, 47*, 583-621.

[27] Mann, H., and Whitney, D. (1947). "On a test of whether one of two random variables is stochastically larger than the other." *Annals of Mathematical Statistics, 18*, 50-60.

[28] Marascuilo, L.A. and McSweeney, M. (1977). *Nonparametric and Distribution-Free Methods for the Social Sciences.* Monterey, CA : Brooks/Cole Publishing Company.

[29] McNemar, Q. (1947), "Note on the sampling error of the difference between correlated proportions or percentages." *Psychometrika, 12*, 153-157.

[30] Moser, B.K., Stevens, G.R. (1992). "Homogeneity of variance in the two-sample means test." *The American Statistician, 46*, 19-21.

[31] Moser, B.K., Stevens, G.R., Watts, C.L. (1989). "The two-sample t-test versus Satterwaite's approximate F test." *Communications in Statistics: Theory and Methods,* **18**, 3963-3975.

[32] Neyman J. (1934). "On the two different aspects of the representative method: the method of stratified sampling and the method of purposive selection." *Journal of the Royal Statistical Society* **97** 558-606.

[33] Särndal, C.E., Swensson, B., and Wretman, J. (2003). *Model Assisted Survey Sampling (Springer Series in Statistics).* Springer, New York.

[34] Satterwaite, F.E. (1946). "An approximate distribution of estimates of variance components." *Biometrics Bulletin,* **2**, 110-114.

[35] Scheffé, H.A. (1953). "A method for judging all possible contrasts in the analysis of variance." *Biometrika,* **40**, 87-104.

[36] Schenker, N. (1985), "Qualms About Bootstrap Confidence Intervals," *Journal of the American Statistical Association,* **80**, 360-361.

[37] Scott, D.W. (1979). "On optimal and data-based histograms." *Biometrika,* **66**, 605-610.

[38] Smith, H.F. (1936). "The problem of comparing the results of two experiments with unequal errors." *Journal of the Council for Scientific and Industrial Research,* **9**, 211-212.

[39] Student (1908a), "The Probable Error of a Mean." *Biometrika,* **6**, 1-25.

[40] Student (1908b), "Probable Error of a Correlation Coefficient." *Biometrika,* **6**, 302-310.

[41] Sturges, H. (1926). "The choice of a class-interval." *Journal of the American Statistical Association,* **21**, 65-66.

[42] Tukey, J.W. (1953). "The problem of multiple comparisons." *Unpublished Paper, Princeton University, Princeton, NJ.*

[43] Weinfurt, K.P. (2000). *Repeated measures analysis: ANOVA, MANOVA and HLM.* Grimm, L. G. and Yarnold, P.R. (Eds.), Reading and Understanding more multivariate statistics (pp. 317-361). Washington, D.C. : American Psychological Association.

[44] Welch, B.L. (1938). "The significance of the difference between two means when the population variances are unequal." *Biometrika*, **29**, 350-362.

[45] Welch, B.L. (1947). "The generalisation of students problem when several different population variances are involved." *Biometrika*, **34**, 23-35.

[46] Wilcoxon, F. (1949). *Some Rapid Approximate Statistical Procedures.* Stamford, CT: Stamford Research Laboratories, American Cyanamid Corporation.

[47] Yates, F. (1934). "On the association of the attributes in statistics: With illustrations from the material of the childhood society, &c." *Philosophical Transactions of the Royal Society*, Series A, **194**, 257-319.

[48] Zar, J.H. (1999). *Biostatistical Analysis*(4th ed.). Upper Saddle River, NJ: Prentice Hall.

Appendix A

Statistical Tables

Table A.1 : Upper-Tail Critical Values and Probabilities for the Wilcoxon Signed-Rank Statistic S^+.
- *P in the table represents $P(S^+ \geq c)$* -

n = 3		n = 5		n = 6		n = 7		n = 8	
c	P	c	P	c	P	c	P	c	P
6	0.13	15	0.03	21	0.02	28	0.01	36	0
5	0.25	14	0.06	20	0.03	26	0.02	34	0.01
4	0.38	13	0.09	19	0.05	25	0.04	33	0.02
3	0.63	12	0.16	18	0.08	24	0.05	32	0.03
2	0.75	11	0.22	17	0.11	23	0.08	31	0.04
1	0.88	10	0.31	16	0.16	22	0.11	30	0.05
0	1	9	0.41	15	0.22	21	0.15	29	0.07
n = 4		8	0.5	14	0.28	20	0.19	28	0.1
c	P	7	0.59	13	0.34	19	0.23	27	0.13
10	0.06	6	0.69	12	0.42	18	0.29	26	0.16
9	0.13	5	0.78	11	0.5	17	0.34	25	0.19
8	0.19	4	0.84	10	0.58	16	0.41	24	0.23
7	0.31	3	0.91	9	0.66	15	0.47	23	0.27
6	0.44	2	0.94	8	0.72	14	0.53	22	0.32
5	0.56	1	0.97	7	0.78	13	0.59	21	0.37
4	0.69	0	1	6	0.84	12	0.66	20	0.42
3	0.81			5	0.89	11	0.71	19	0.47
2	0.88			4	0.92	10	0.77	18	0.53
1	0.94			3	0.95	9	0.81	17	0.58
0	1			2	0.97	8	0.85	16	0.63
				1	0.98	7	0.89	15	0.68
				0	1	6	0.92	14	0.73

Table A.2: Upper-Tail Critical Values and Probabilities for the
Wilcoxon Signed-Rank Statistic S^+.
- *P in the table represents $P(S^+ \geq c)$ -*

\(n = 9\)		\(n = 10\)		\(n = 11\)		\(n = 12\)		\(n = 13\)	
c	P	c	P	c	P	c	P	c	P
44	0	52	0	61	0	71	0	82	0
41	0.01	49	0.01	58	0.01	67	0.01	77	0.01
40	0.02	47	0.02	56	0.02	65	0.02	74	0.02
39	0.03	46	0.03	54	0.03	63	0.03	72	0.03
38	0.04	45	0.04	53	0.04	62	0.04	71	0.04
37	0.05	44	0.05	52	0.05	60	0.05	69	0.05
36	0.06	43	0.07	51	0.06	59	0.06	68	0.06
35	0.08	42	0.08	50	0.07	58	0.08	67	0.07
34	0.10	41	0.10	49	0.09	57	0.09	66	0.08
33	0.13	40	0.12	48	0.10	56	0.10	65	0.10
32	0.15	39	0.14	47	0.12	55	0.12	64	0.11
31	0.18	38	0.16	46	0.14	54	0.13	63	0.12
30	0.21	37	0.19	45	0.16	53	0.15	62	0.14
29	0.25	36	0.22	44	0.18	52	0.17	61	0.15
28	0.29	35	0.25	43	0.21	51	0.19	60	0.17
27	0.33	34	0.28	42	0.23	50	0.21	59	0.19
26	0.37	33	0.31	41	0.26	49	0.23	58	0.21
25	0.41	32	0.35	40	0.29	48	0.26	57	0.23
24	0.46	31	0.38	39	0.32	47	0.28	56	0.25
23	0.50	30	0.42	38	0.35	46	0.31	55	0.27
22	0.54	29	0.46	37	0.38	45	0.34	54	0.29
21	0.59	28	0.50	36	0.42	44	0.37	53	0.32

Table A.3 : Upper-Tail Critical Values and Probabilities for the Wilcoxon Signed-Rank Statistic S^+.

- *P in the table represents $P(S^+ \geq c)$* -

$n = 14$		$n = 15$		$n = 16$		$n = 17$		$n = 18$	
c	P	c	P	c	P	c	P	c	P
93	0	105	0	117	0	130	0	144	0
87	0.01	99	0.01	110	0.01	123	0.01	136	0.01
84	0.02	95	0.02	107	0.02	119	0.02	131	0.02
82	0.03	93	0.03	104	0.03	116	0.03	128	0.03
81	0.04	91	0.04	102	0.04	113	0.04	125	0.04
79	0.05	89	0.05	100	0.05	111	0.05	123	0.05
78	0.06	88	0.06	98	0.06	110	0.06	121	0.06
77	0.07	87	0.07	97	0.07	108	0.07	120	0.07
76	0.08	85	0.08	96	0.08	107	0.08	118	0.08
75	0.09	84	0.09	95	0.09	106	0.09	117	0.09
74	0.1	83	0.1	94	0.1	104	0.1	116	0.1
73	0.11	82	0.11	93	0.11	103	0.11	114	0.11
72	0.12	81	0.13	92	0.12	102	0.12	113	0.12
71	0.13	80	0.14	91	0.13	101	0.13	112	0.13
70	0.15	79	0.15	90	0.14	100	0.14	111	0.14
69	0.16	78	0.17	89	0.15	99	0.15	110	0.15
68	0.18	77	0.18	88	0.16	98	0.16	109	0.16
67	0.2	76	0.19	87	0.17	97	0.18	108	0.17
66	0.21	75	0.21	86	0.19	96	0.19	107	0.18
65	0.23	74	0.23	85	0.2	95	0.2	106	0.2
64	0.25	73	0.24	84	0.22	94	0.22	105	0.21
63	0.27	72	0.26	83	0.23	93	0.23	104	0.22

Table A.4 : Upper-Tail Critical Values and Probabilities for the Wilcoxon Signed-Rank Statistic S^+.

- *P in the table represents $P(S^+ \geq c)$* -

$n = 19$		$n = 20$		$n = 21$		$n = 22$		$n = 23$	
c	P	c	P	c	P	c	P	c	P
158	0	173	0	189	0	205	0	222	0
149	0.01	163	0.01	178	0.01	194	0.01	210	0.01
144	0.02	158	0.02	173	0.02	188	0.02	203	0.02
141	0.03	154	0.03	169	0.03	183	0.03	198	0.03
138	0.04	151	0.04	165	0.04	180	0.04	195	0.04
136	0.05	149	0.05	163	0.05	177	0.05	192	0.05
134	0.06	147	0.06	160	0.06	174	0.06	189	0.06
132	0.07	145	0.07	158	0.07	172	0.07	187	0.07
130	0.08	143	0.08	156	0.08	170	0.08	184	0.08
129	0.09	141	0.09	155	0.09	168	0.09	182	0.09
128	0.10	140	0.10	153	0.10	166	0.10	181	0.10
126	0.11	139	0.11	151	0.11	165	0.11	179	0.11
125	0.12	137	0.12	150	0.12	163	0.12	177	0.12
124	0.13	136	0.13	149	0.13	162	0.13	176	0.13
123	0.14	135	0.14	147	0.14	161	0.14	174	0.14
122	0.15	134	0.15	146	0.15	159	0.15	173	0.15
121	0.16	132	0.16	145	0.16	158	0.16	171	0.16
120	0.17	131	0.17	144	0.17	157	0.17	170	0.17
119	0.18	130	0.18	143	0.18	156	0.18	169	0.18
118	0.19	129	0.19	142	0.19	154	0.19	168	0.19
117	0.20	128	0.20	141	0.20	153	0.20	167	0.20
116	0.21	127	0.22	140	0.21	152	0.21	165	0.21

Table A.5 : Upper-Tail Critical Values and Probabilities for the Wilcoxon Signed-Rank Statistic S^+.

- *P in the table represents* $P(S^+ \geq c)$ -

$n = 24$		$n = 25$		$n = 26$	
c	P	c	P	c	P
239	0	257	0	276	0
226	0.01	243	0.01	261	0.01
219	0.02	236	0.02	253	0.02
214	0.03	231	0.03	248	0.03
210	0.04	227	0.04	243	0.04
207	0.05	223	0.05	240	0.05
204	0.06	220	0.06	236	0.06
202	0.07	217	0.07	233	0.07
199	0.08	215	0.08	231	0.08
197	0.09	213	0.09	228	0.09
195	0.10	210	0.10	226	0.10
193	0.11	209	0.11	224	0.11
192	0.12	207	0.12	222	0.12
190	0.13	205	0.13	220	0.13
188	0.14	203	0.14	219	0.14
187	0.15	202	0.15	217	0.15
186	0.16	200	0.16	215	0.16
184	0.17	199	0.17	214	0.17
183	0.18	197	0.18	212	0.18
182	0.19	196	0.19	211	0.19
180	0.20	195	0.20	209	0.20
179	0.21	193	0.21	208	0.21

Table A.6: Upper-Tail Probabilities for Wilcoxon Rank-Sum Statistic W
- *Table entries represent* $P(W \geq c)$ -

c	$(3,3)$	$(3,4)$	$(3,5)$	$(3,6)$	$(3,7)$	$(3,8)$	$(3,9)$	$(3,10)$	$(4,4)$	$(4,5)$
36	0.000	0.000	0.000	0.000	0.000	0.000	0.000	0.003	0.000	0.000
35	0.000	0.000	0.000	0.000	0.000	0.000	0.000	0.007	0.000	0.000
34	0.000	0.000	0.000	0.000	0.000	0.000	0.000	0.014	0.000	0.000
33	0.000	0.000	0.000	0.000	0.000	0.000	0.005	0.024	0.000	0.000
32	0.000	0.000	0.000	0.000	0.000	0.000	0.009	0.038	0.000	0.000
31	0.000	0.000	0.000	0.000	0.000	0.000	0.018	0.056	0.000	0.000
30	0.000	0.000	0.000	0.000	0.000	0.006	0.032	0.080	0.000	0.008
29	0.000	0.000	0.000	0.000	0.000	0.012	0.050	0.108	0.000	0.016
28	0.000	0.000	0.000	0.000	0.000	0.024	0.073	0.143	0.000	0.032
27	0.000	0.000	0.000	0.000	0.008	0.042	0.105	0.185	0.000	0.056
26	0.000	0.000	0.000	0.000	0.017	0.067	0.141	0.234	0.014	0.095
25	0.000	0.000	0.000	0.000	0.033	0.097	0.186	0.287	0.029	0.143
24	0.000	0.000	0.000	0.012	0.058	0.139	0.241	0.346	0.057	0.206
23	0.000	0.000	0.000	0.024	0.092	0.188	0.300	0.406	0.100	0.278
22	0.000	0.000	0.000	0.048	0.133	0.248	0.364	0.469	0.171	0.365
21	0.000	0.000	0.018	0.083	0.192	0.315	0.432	0.531	0.243	0.452
20	0.000	0.000	0.036	0.131	0.258	0.388	0.500	0.594	0.343	0.548
19	0.000	0.000	0.071	0.190	0.333	0.461	0.568	0.654	0.443	0.635
18	0.000	0.029	0.125	0.274	0.417	0.539	0.636	0.713	0.557	0.722
17	0.000	0.057	0.196	0.357	0.500	0.612	0.700	0.766	0.657	0.794
16	0.000	0.114	0.286	0.452	0.583	0.685	0.759	0.815	0.757	0.857
15	0.050	0.200	0.393	0.548	0.667	0.752	0.814	0.857	0.829	0.905
14	0.100	0.314	0.500	0.643	0.742	0.812	0.859	0.892	0.900	0.944
13	0.200	0.429	0.607	0.726	0.808	0.861	0.895	0.920	0.943	0.968
12	0.350	0.571	0.714	0.810	0.867	0.903	0.927	0.944	0.971	0.984
11	0.500	0.686	0.804	0.869	0.908	0.933	0.950	0.962	0.986	0.992
10	0.650	0.800	0.875	0.917	0.942	0.958	0.968	0.976	1.000	1.000
9	0.800	0.886	0.929	0.952	0.967	0.976	0.982	0.986	1.000	1.000
8	0.900	0.943	0.964	0.976	0.983	0.988	0.991	0.993	1.000	1.000
7	0.950	0.971	0.982	0.988	0.992	0.994	0.995	0.997	1.000	1.000
6	1.000	1.000	1.000	1.000	1.000	1.000	1.000	1.000	1.000	1.000

Table A.7 : Upper-Tail Probabilities for Wilcoxon Rank-Sum Statistic W
Note: $P = P(W \geq c)$ and $(\cdot, \cdot) = (n_1, n_2)$

c	(4,6)	(4,7)	(4,8)	c	(4,9)	(4,10)	c	(5,5)	c	(5,6)	c	(5,7)
44	0.000	0.000	0.000	51	0.000	0.000	40	0.004	45	0.002	50	0.001
43	0.000	0.000	0.000	50	0.000	0.001	39	0.008	44	0.004	49	0.003
42	0.000	0.000	0.002	49	0.000	0.002	38	0.016	43	0.009	48	0.005
41	0.000	0.000	0.004	48	0.000	0.004	37	0.028	42	0.015	47	0.009
40	0.000	0.000	0.008	47	0.000	0.007	36	0.048	41	0.026	46	0.015
39	0.000	0.000	0.014	46	0.001	0.012	35	0.075	40	0.041	45	0.024
38	0.000	0.003	0.024	45	0.003	0.018	34	0.111	39	0.063	44	0.037
37	0.000	0.006	0.036	44	0.006	0.027	33	0.155	38	0.089	43	0.053
36	0.000	0.012	0.055	43	0.010	0.038	32	0.210	37	0.123	42	0.074
35	0.000	0.021	0.077	42	0.017	0.053	32	0.210	36	0.165	41	0.101
34	0.005	0.036	0.107	41	0.025	0.071	31	0.274	35	0.214	40	0.134
33	0.010	0.055	0.141	40	0.038	0.094	31	0.274	34	0.268	39	0.172
32	0.019	0.082	0.184	39	0.053	0.120	30	0.345	33	0.331	38	0.216
31	0.033	0.115	0.230	38	0.074	0.152	30	0.345	32	0.396	37	0.265
30	0.057	0.158	0.285	37	0.099	0.187	29	0.421	32	0.396	36	0.319
29	0.086	0.206	0.341	36	0.130	0.227	29	0.421	31	0.465	35	0.378
28	0.129	0.264	0.404	35	0.165	0.270	28	0.500	31	0.465	34	0.438
27	0.176	0.324	0.467	27	0.587	0.682	28	0.500	30	0.535	32	0.562
26	0.238	0.394	0.533	26	0.645	0.730	27	0.579	30	0.535	31	0.622
25	0.305	0.464	0.596	25	0.698	0.773	27	0.579	29	0.604	30	0.681
24	0.381	0.536	0.659	24	0.748	0.813	26	0.655	29	0.604	29	0.735
23	0.457	0.606	0.715	23	0.793	0.848	26	0.655	28	0.669	28	0.784
22	0.543	0.676	0.770	22	0.835	0.880	25	0.726	27	0.732	27	0.828
21	0.619	0.736	0.816	21	0.870	0.906	25	0.726	26	0.786	26	0.866
20	0.695	0.794	0.859	20	0.901	0.929	24	0.790	25	0.835	25	0.899
19	0.762	0.842	0.893	19	0.926	0.947	24	0.790	24	0.877	24	0.926
18	0.824	0.885	0.923	18	0.947	0.962	23	0.845	23	0.911	23	0.947
17	0.871	0.918	0.945	17	0.962	0.973	22	0.889	22	0.937	22	0.963
16	0.914	0.945	0.964	16	0.975	0.982	21	0.925	21	0.959	21	0.976
15	0.943	0.964	0.976	15	0.983	0.988	20	0.952	20	0.974	20	0.985
14	0.967	0.979	0.986	14	0.990	0.993	19	0.972	19	0.985	19	0.991
13	0.981	0.988	0.992	13	0.994	0.996	18	0.984	18	0.991	18	0.995
12	0.990	0.994	0.996	12	0.997	0.998	17	0.992	17	0.996	17	0.997
11	0.995	0.997	0.998	11	0.999	0.999	16	0.996	16	0.998	16	0.999
10	1.000	1.000	1.000	10	1.000	1.000	15	1.000	15	1.000	15	1.000

Table A.8: Upper-Tail Probabilities for Wilcoxon Rank-Sum Statistic W

Note: $P = P(W \geq c)$ and $(\cdot, \cdot) = (n_1, n_2)$

c	(5,8)	c	(5,9)	c	(5,10)	c	(6,6)	c	(6,7)	c	(6,8)	c	(6,9)
55	0.001	60	0.000	65	0.000	57	0.001	63	0.001	69	0.000	74	0.000
54	0.002	59	0.001	64	0.001	56	0.002	62	0.001	68	0.001	73	0.001
53	0.003	58	0.002	63	0.001	55	0.004	61	0.002	67	0.001	72	0.001
52	0.005	57	0.003	62	0.002	54	0.008	60	0.004	66	0.002	71	0.002
51	0.009	56	0.006	61	0.004	53	0.013	59	0.007	65	0.004	70	0.004
50	0.015	55	0.009	60	0.006	52	0.021	58	0.011	64	0.006	69	0.006
49	0.023	54	0.014	59	0.010	51	0.032	57	0.017	63	0.010	68	0.009
48	0.033	53	0.021	58	0.014	50	0.047	56	0.026	62	0.015	67	0.013
47	0.047	52	0.030	57	0.020	49	0.066	55	0.037	61	0.021	66	0.018
46	0.064	51	0.041	56	0.028	48	0.090	54	0.051	60	0.030	65	0.025
45	0.085	50	0.056	55	0.038	47	0.120	53	0.069	59	0.041	64	0.033
44	0.111	49	0.073	54	0.050	46	0.155	52	0.090	58	0.054	63	0.044
43	0.142	48	0.095	53	0.065	45	0.197	51	0.117	57	0.071	62	0.057
42	0.177	47	0.120	52	0.082	44	0.242	50	0.147	56	0.091	61	0.072
41	0.218	46	0.149	51	0.103	43	0.294	49	0.183	55	0.114	60	0.091
40	0.262	45	0.182	50	0.127	42	0.350	48	0.223	54	0.141	59	0.112
39	0.311	44	0.219	49	0.155	41	0.409	47	0.267	53	0.172	58	0.136
32	0.689	32	0.781	33	0.815	38	0.591	38	0.733	39	0.793	40	0.836
31	0.738	31	0.818	32	0.845	37	0.650	37	0.777	38	0.828	39	0.864
30	0.782	30	0.851	31	0.873	36	0.706	36	0.817	37	0.859	38	0.888
29	0.823	29	0.880	30	0.897	35	0.758	35	0.853	36	0.886	37	0.909
28	0.858	28	0.905	29	0.918	34	0.803	34	0.883	35	0.909	36	0.928
27	0.889	27	0.927	28	0.935	33	0.845	33	0.910	34	0.929	35	0.943
26	0.915	26	0.944	27	0.950	32	0.880	32	0.931	33	0.946	34	0.956
25	0.936	25	0.959	26	0.962	31	0.910	31	0.949	32	0.959	33	0.967
24	0.953	24	0.970	25	0.972	30	0.934	30	0.963	31	0.970	32	0.975
23	0.967	23	0.979	24	0.980	29	0.953	29	0.974	30	0.979	31	0.982
22	0.977	22	0.986	23	0.986	28	0.968	28	0.983	29	0.985	30	0.987
21	0.985	21	0.991	22	0.990	27	0.979	27	0.989	28	0.990	29	0.991
20	0.991	20	0.994	21	0.994	26	0.987	26	0.993	27	0.994	28	0.994
19	0.995	19	0.997	20	0.996	25	0.992	25	0.996	26	0.996	27	0.996
18	0.997	18	0.998	19	0.998	24	0.996	24	0.998	25	0.998	26	0.998
17	0.998	17	0.999	18	0.999	23	0.998	23	0.999	24	0.999	25	0.999
16	0.999	16	1.000	17	0.999	22	0.999	22	0.999	23	0.999	24	0.999
15	1.000	15	1.000	16	1.000	21	1.000	21	1.000	22	1.000	23	1.000

Table A.9 : Upper-Tail Probabilities for Wilcoxon Rank-Sum Statistic W
Note: $P = P(W \geq c)$ *and* $(\cdot, \cdot) = (n_1, n_2)$

(6, 10)		(7, 7)		(7, 8)		(7, 9)		(7, 10)	
c	P	c	P	c	P	c	P	c	P
79	0.000	77	0.000	83	0.000	89	0.000	92	0.002
78	0.001	76	0.001	82	0.001	88	0.001	91	0.002
77	0.001	75	0.001	81	0.001	87	0.001	90	0.003
76	0.002	74	0.002	80	0.002	86	0.002	89	0.005
75	0.004	73	0.003	79	0.003	85	0.003	88	0.007
74	0.005	72	0.006	78	0.005	84	0.004	87	0.009
73	0.008	71	0.009	77	0.007	83	0.006	86	0.012
72	0.011	70	0.013	76	0.010	82	0.008	85	0.017
71	0.016	69	0.019	75	0.014	81	0.011	84	0.022
70	0.021	68	0.027	74	0.020	80	0.016	83	0.028
69	0.028	67	0.036	73	0.027	79	0.021	82	0.035
68	0.036	66	0.049	72	0.036	78	0.027	81	0.044
67	0.047	65	0.064	71	0.047	77	0.036	80	0.054
66	0.059	64	0.082	70	0.060	76	0.045	79	0.067
65	0.074	63	0.104	69	0.076	75	0.057	78	0.081
64	0.090	62	0.130	68	0.095	74	0.071	77	0.097
63	0.110	61	0.159	67	0.116	73	0.087	50	0.903
41	0.868	46	0.809	47	0.860	48	0.895	49	0.919
40	0.890	45	0.841	46	0.884	47	0.913	48	0.933
39	0.910	44	0.870	45	0.905	46	0.929	47	0.946
38	0.926	43	0.896	44	0.924	45	0.943	46	0.956
37	0.941	42	0.918	43	0.940	44	0.955	45	0.965
36	0.953	41	0.936	42	0.953	43	0.964	44	0.972
35	0.964	40	0.951	41	0.964	42	0.973	43	0.978
34	0.972	39	0.964	40	0.973	41	0.979	42	0.983
33	0.979	38	0.973	39	0.980	40	0.984	41	0.988
32	0.984	37	0.981	38	0.986	39	0.989	40	0.991
31	0.989	36	0.987	37	0.990	38	0.992	39	0.993
30	0.992	35	0.991	36	0.993	37	0.994	38	0.995
29	0.995	34	0.994	35	0.995	36	0.996	37	0.997
28	0.996	33	0.997	34	0.997	35	0.997	36	0.998
27	0.998	32	0.998	33	0.998	34	0.998	35	0.998
26	0.999	31	0.999	32	0.999	33	0.999	34	0.999
25	0.999	30	0.999	31	0.999	32	0.999	33	0.999
24	1.000	29	1.000	30	1.000	31	1.000	32	1.000

Table A.10: Upper-Tail Probabilities for Wilcoxon Rank-Sum Statistic W

Note: $P = P(W \geq c)$ and $(\cdot, \cdot) = (n_1, n_2)$

	(8, 8)				(8, 9)				(8, 10)		
c	P	c	P	c	P	c	P	c	P	c	P
108	0.000	73	0.323	110	0.000	75	0.407	111	0.000	76	0.517
107	0.000	72	0.360	109	0.000	74	0.444	110	0.001	75	0.552
106	0.000	71	0.399	108	0.000	73	0.481	109	0.001	74	0.586
105	0.000	70	0.439	107	0.000	72	0.519	108	0.002	73	0.619
104	0.000	69	0.480	106	0.000	71	0.556	107	0.002	72	0.652
103	0.000	68	0.520	105	0.000	70	0.593	106	0.003	71	0.683
102	0.000	67	0.561	104	0.000	69	0.629	105	0.004	70	0.714
101	0.000	66	0.601	103	0.001	68	0.664	104	0.006	69	0.743
100	0.000	65	0.640	102	0.001	67	0.697	103	0.008	68	0.770
99	0.000	64	0.677	101	0.002	66	0.729	102	0.010	67	0.796
98	0.000	63	0.713	100	0.003	65	0.760	101	0.013	66	0.820
97	0.001	62	0.747	99	0.004	64	0.788	100	0.017	65	0.842
96	0.001	61	0.779	98	0.006	63	0.815	99	0.022	64	0.863
95	0.001	60	0.809	97	0.008	62	0.839	98	0.027	63	0.882
94	0.002	59	0.836	96	0.010	61	0.862	97	0.034	62	0.898
93	0.003	58	0.861	95	0.014	60	0.882	96	0.042	61	0.914
92	0.005	57	0.883	94	0.018	59	0.900	95	0.051	60	0.927
91	0.007	56	0.903	93	0.023	58	0.916	94	0.061	59	0.939
90	0.010	55	0.920	92	0.030	57	0.931	93	0.073	58	0.949
89	0.014	54	0.935	91	0.037	56	0.943	92	0.086	57	0.958
88	0.019	53	0.948	90	0.046	55	0.954	91	0.102	56	0.966
87	0.025	52	0.959	89	0.057	54	0.963	90	0.118	55	0.973
86	0.032	51	0.968	88	0.069	53	0.970	89	0.137	54	0.978
85	0.041	50	0.975	87	0.084	52	0.977	88	0.158	53	0.983
84	0.052	49	0.981	86	0.100	51	0.982	87	0.180	52	0.987
83	0.065	48	0.986	85	0.118	50	0.986	86	0.204	51	0.990
82	0.080	47	0.990	84	0.138	49	0.990	85	0.230	50	0.992
81	0.097	46	0.993	83	0.161	48	0.992	84	0.257	49	0.994
80	0.117	45	0.995	82	0.185	47	0.994	83	0.286	48	0.996
79	0.139	44	0.997	81	0.212	46	0.996	82	0.317	47	0.997
78	0.164	43	0.998	80	0.240	45	0.997	81	0.348	46	0.998
77	0.191	42	0.999	79	0.271	44	0.998	80	0.381	45	0.998
76	0.221	41	0.999	78	0.303	43	0.999	79	0.414	44	0.999
75	0.253	40	0.999	77	0.336	42	0.999	78	0.448	43	0.999
74	0.287	39	1.000	76	0.371	41	1.000	77	0.483	42	1.000

Table A.11 : Upper-Tail Probabilities for Wilcoxon Rank-Sum Statistic W
Note: $P = P(W \geq c)$ and $(\cdot, \cdot) = (n_1, n_2)$

(9, 9)				(9, 10)				(10, 10)			
c	P	c	P	c	P	c	P	c	P	c	P
121	0.000	85	0.534	128	0.000	87	0.610	147	0.000	98	0.711
120	0.001	84	0.568	127	0.001	86	0.640	146	0.001	97	0.736
119	0.001	83	0.602	126	0.001	85	0.670	145	0.001	96	0.759
118	0.001	82	0.635	125	0.001	84	0.698	144	0.001	95	0.782
117	0.002	81	0.667	124	0.002	83	0.726	143	0.001	94	0.803
116	0.003	80	0.698	123	0.003	82	0.752	142	0.002	93	0.824
115	0.004	79	0.727	122	0.004	81	0.777	141	0.003	92	0.843
114	0.005	78	0.755	121	0.005	80	0.800	140	0.003	91	0.860
113	0.007	77	0.782	120	0.007	79	0.822	139	0.004	90	0.876
112	0.009	76	0.807	119	0.009	78	0.842	138	0.006	89	0.891
111	0.012	75	0.830	118	0.011	77	0.861	137	0.007	88	0.905
110	0.016	74	0.851	117	0.014	76	0.879	136	0.009	87	0.917
109	0.020	73	0.871	116	0.017	75	0.894	135	0.012	86	0.928
108	0.025	72	0.889	115	0.022	74	0.909	134	0.014	85	0.938
107	0.031	71	0.905	114	0.027	73	0.922	133	0.018	84	0.947
106	0.039	70	0.919	113	0.033	72	0.933	132	0.022	83	0.955
105	0.047	69	0.932	112	0.039	71	0.944	131	0.026	82	0.962
104	0.057	68	0.943	111	0.047	70	0.953	130	0.032	81	0.968
103	0.068	67	0.953	110	0.056	69	0.961	129	0.038	80	0.974
102	0.081	66	0.961	109	0.067	68	0.967	128	0.045	79	0.978
101	0.095	65	0.969	108	0.078	67	0.973	127	0.053	78	0.982
100	0.111	64	0.975	107	0.091	66	0.978	126	0.062	77	0.986
99	0.129	63	0.980	106	0.106	65	0.983	125	0.072	76	0.988
98	0.149	62	0.984	105	0.121	64	0.986	124	0.083	75	0.991
97	0.170	61	0.988	104	0.139	63	0.989	123	0.095	74	0.993
96	0.193	60	0.991	103	0.158	62	0.991	122	0.109	73	0.994
95	0.218	59	0.993	102	0.178	61	0.993	121	0.124	72	0.996
94	0.245	58	0.995	101	0.200	60	0.995	120	0.140	71	0.997
93	0.273	57	0.996	100	0.223	59	0.996	119	0.157	70	0.997
92	0.302	56	0.997	99	0.248	58	0.997	118	0.176	69	0.998
91	0.333	55	0.998	98	0.274	57	0.998	117	0.197	68	0.999
90	0.365	54	0.999	97	0.302	56	0.999	116	0.218	67	0.999
89	0.398	53	0.999	96	0.330	55	0.999	115	0.241	66	0.999
88	0.432	52	0.999	95	0.360	54	0.999	114	0.264	65	0.999
87	0.466	51	1.000	94	0.390	53	1.000	113	0.289	64	1.000

Appendix B

Power Curves for t-Tests

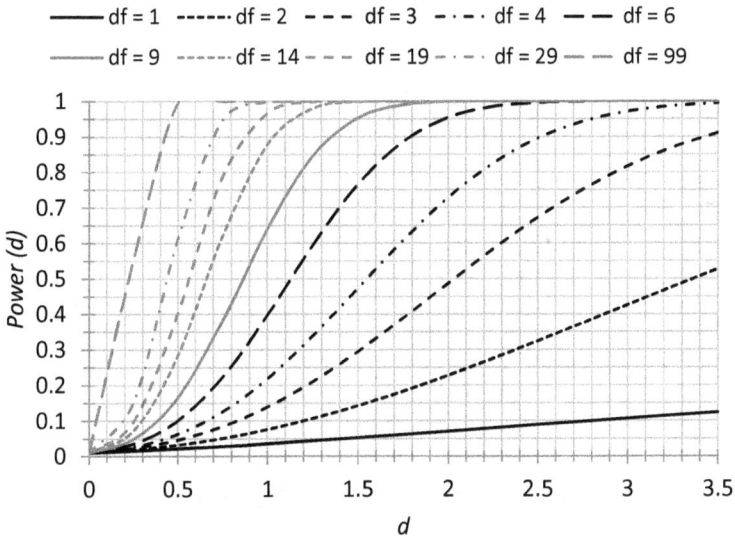

Figure B.1. Power curve of the one-tailed t-test at the 1% significance level ($\alpha = 0.01$)

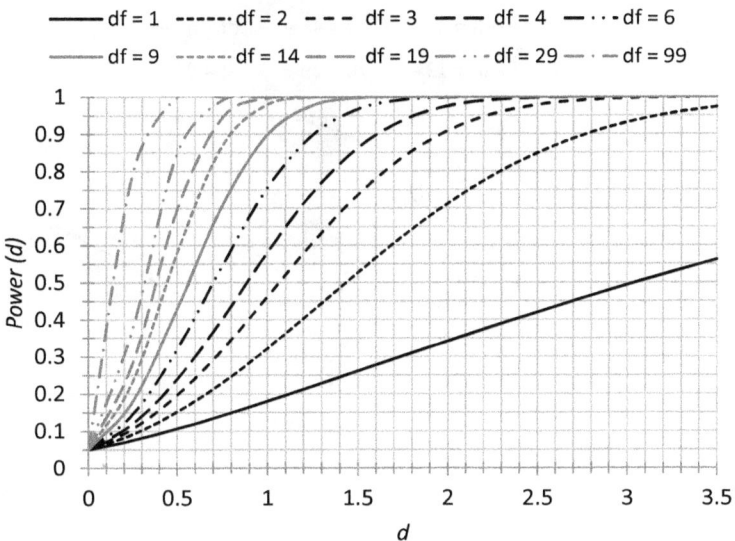

Figure B.2. Power curve of the one-tailed t-test at the 5% significance level ($\alpha = 0.05$)

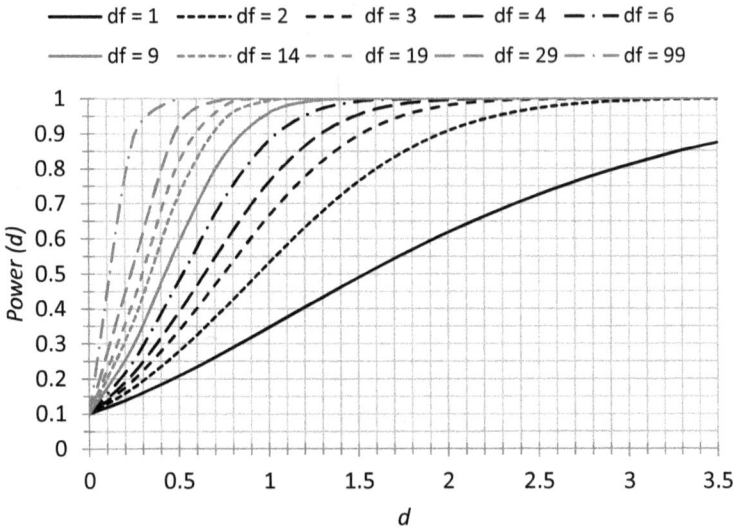

Figure B.3. Power curve of the one-tailed t-test at the 10% significance level ($\alpha = 0.10$)

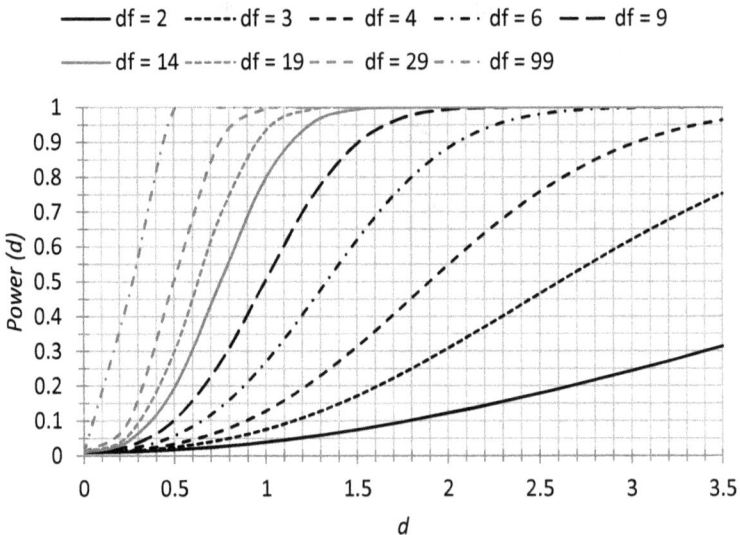

Figure B.4. Power curve of the two-tailed t-test at the 1% significance level ($\alpha = 0.01$)

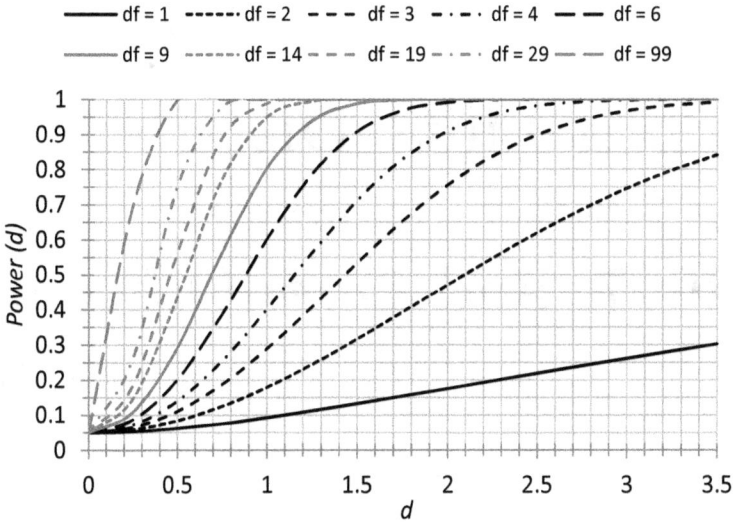

Figure B.5. Power curve of the two-tailed t-test at the 5% significance level ($\alpha = 0.05$)

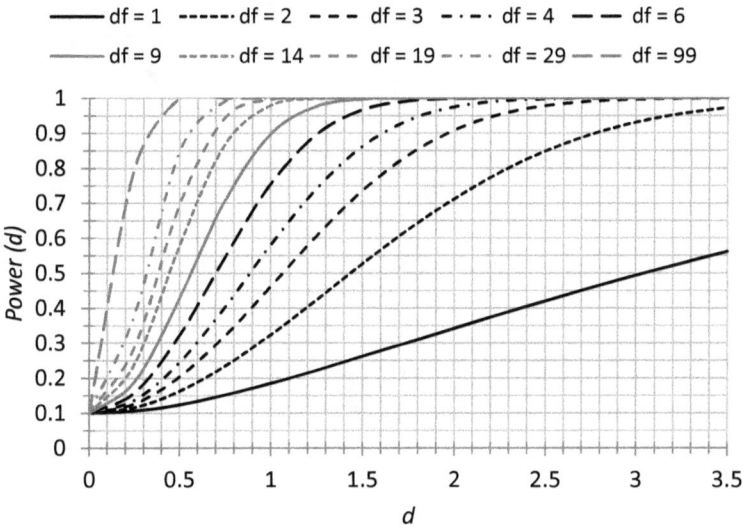

Figure B.6. Power curve of the two-tailed t-test at the 10% significance level ($\alpha = 0.10$)

Appendix C

Probability Distributions with R, Excel 2010 & 2007, and OpenOffice.org Calc

Table C.1 : Using Excel 2007 & 2010 to obtain Cumulative Probabilities
for various Probability Distributions[a]

Distribution	Excel 2007	Excel 2010
Binomial $\mathcal{B}(n,p)$	BINOMDIST(x_0,n,p,TRUE)	BINOM.DIST(x_0,n,p,TRUE)
Chi-Square χ_n^2	1-CHIDIST(x_0,n)	CHISQ.DIST(x_0,n,TRUE)
Exponential $\mathcal{E}(\lambda)$	EXPONDIST(x_0,λ,TRUE)	EXPON.DIST(x_0,λ,TRUE)
F $F(df_1,df_2)$	1-FDIST(x_0,df_1,df_2)	=F.DIST(x_0,df_1,df_2,TRUE)
Gamma $\mathcal{G}(\alpha,\lambda)$	GAMMADIST[b]	GAMMA.DIST[c]
Hypergeometric[d]	N/A[e]	HYPGEOM.DIST[f]
Normal $\mathcal{N}(\mu,\sigma^2)$	NORMDIST(x_0,μ,σ,TRUE)	NORM.DIST(x_0,μ,σ,TRUE)
Poisson $\mathcal{P}(\lambda)$	POISSON(x_0,λ,TRUE)	POISSON.DIST(x_0,λ,TRUE)
Stand. Normal	NORMSDIST(x_0)	NORM.S.DIST(x_0,TRUE)
Student's t	TDIST(x_0,df,1)	T.DIST(x_0,df,TRUE)
NC Student's[g] t	N/A	N/A

  [a]The probabilities are in the in the form $P(X \leq x_0)$ for a given random
variable X and a predetermined number x_0

  [b]GAMMADIST(x_0,α,1/λ,TRUE)

  [c]GAMMA.DIST(x_0,α,1/λ,TRUE)

  [d]$\mathcal{H}(n, N_s, N)$ where n=sample size, N_s=number of population successes,
and N=population size

  [e]In Excel 2007, you may compute the probability mass function at x_0 as
follows: HYPGEOMDIST(x_0,n,N_s,N). Repeating this for all values preceding
x_0, and summing all the probabilities will produce $P(X \leq x_0)$

  [f]HYPGEOM.DIST(x_0,n,N_s,N,TRUE)

  [g]The Noncentral Student's t distribution is not implemented in Excel

Table C.2 : Using R & OpenOffice.org Calc to obtain Cumulative Probabilities for various Probability Distributions[a]

Distribution	OpenOffice Calc 3.2.0	R
Binomial $\mathcal{B}(n, p)$	BINOMDIST$(x_0\,;n\,;p\,;1)$	<-pbinom(x_0,n,p)
Chi-Square χ_n^2	1-CHIDIST$(x_0\,;n)$	<-pchisq(x_0,n)
Exponential $\mathcal{E}(\lambda)$	EXPONDIST$(x_0\,;\lambda\,;1)$	<-pexp$(x_0,\text{rate}=\lambda)$
F $F(df_1, df_2)$	1-FDIST$(x_0\,;df_1\,;df_2)$	<-pf(x_0,df_1,df_2)
Gamma $\mathcal{G}(\alpha, \lambda)$	GAMMADIST[b]	<-pgamma[c]
Hypergeometric[d]	N/A[e]	<-phyper[f]
Normal $\mathcal{N}(\mu, \sigma^2)$	NORMDIST$(x_0\,;\mu\,;\sigma\,;1)$	<-pnorm(x_0,μ,σ)
Stand. Normal	NORMSDIST(x_0)	<-pnorm(x_0)
Poisson $\mathcal{P}(\lambda)$	POISSON$(x_0\,;\lambda\,;1)$	<-ppois$(x_0,\text{ lambda})$
Student's t	TDIST$(x_0;df;1)$	<-pt(x_0,df)
NC Student's t	N/A[g]	<-pt(x_0,df,δ)[h]

[a]The probabilities are in the in the form $P(X \leq x_0)$ for a given random variable X and a predetermined number x_0

[b]GAMMADIST$(x_0\,;\alpha\,;1/\lambda\,;1)$

[c]<-pgamma$(x_0,\ \alpha,\ [\text{rate} = \lambda],\ [\text{scale} = 1/\lambda])$

[d]$\mathcal{H}(n, N_s, N)$ where n=sample size, N_s=number of population successes, and N=population size

[e]In Calc 2007, you may compute the probability mass function at x_0 as follows: HYPGEOMDIST$(\ x_0\,;n\,;N_s\,;N)$. Repeating this for all values preceding x_0, and summing all the probabilities will produce $P(X \leq x_0)$

[f]<-phyper$(x_0,N_s,N - N_s,n)$

[g]The noncentral t distribution is not implemented in Calc 3.2.0

[h]δ is the noncentrality parameter

Table C.3: Using Excel 2007 & 2010 to obtain $(1 - \alpha)^{th}$ Quantile x_α of a random variable X for various probability distributions[a]

Distribution	Excel 2007	Excel 2010
Binomial $\mathcal{B}(n, p)$	N/A	BINOM.INV($n,p,1 - \alpha$)
Chi-Square χ^2_n	CHIINV(α,n)	CHISQ.INV.RT(α,n)
Exponential $\mathcal{E}(\lambda)$	N/A	N/A
F $F(df_1, df_2)$	FINV(α,df_1,df_2)	F.INV.RT(α,df_1,df_2)
Gamma $\mathcal{G}(\theta, \lambda)$	GAMMAINV()[b]	GAMMA.INV()[c]
Hypergeometric[d]	N/A	N/A
Normal $\mathcal{N}(\mu, \sigma^2)$	NORMINV($1 - \alpha,\mu,\sigma$)	NORM.INV($1 - \alpha,\mu,\sigma$)
Poisson $\mathcal{P}(\lambda)$	N/A	N/A
Standard Normal	NORMSINV($1 - \alpha$)	NORM.S.INV($1 - \alpha$)
Student's t	TINV($2 * \alpha$,df)	T.INV($1 - \alpha$,df)
NC Student's[e] t	N/A	N/A

[a]The $(1 - \alpha)^{th}$ quantile x_α is implicitly defined by $P(X > x_\alpha) = \alpha$
[b]GAMMAINV($1 - \alpha,\theta,1/\lambda$)
[c]GAMMA.INV($1 - \alpha,\theta,1/\lambda$)
[d]$\mathcal{H}(n, N_s, N)$ where n=sample size, N_s=number of population successes, and N=population size
[e]The Noncentral Student's t distribution is not implemented in Excel

Table C.4 : Using R & OpenOffice.org Calc for obtaining $(1 - \alpha)^{th}$ Quantile x_α of a random variable X for various probability distributions[a]

Distribution	OpenOffice Calc 3.2.0	R
Binomial $\mathcal{B}(n, p)$	N/A	<-qbinom$(1 - \alpha, n, p)$
Chi-Square χ_n^2	CHISQINV$(1 - \alpha ; n)$	<-qchisq$(1 - \alpha, n)$
Exponential $\mathcal{E}(\lambda)$	N/A	<-qexp()[b]
F $F(df_1, df_2)$	FINV$(\alpha ; df_1 ; df_2)$	<-qf()[c]
Gamma $\mathcal{G}(\theta, \lambda)$	GAMMAINV()[d]	<-qgamma()[e]
Hypergeometric[f]	N/A	<-qhyper()[g]
Normal $\mathcal{N}(\mu, \sigma^2)$	NORMINV$(1 - \alpha ; 0 ; 1)$	<-qnorm$(1 - \alpha, \mu, \sigma)$
Normal Standard	NORMSINV$(1 - \alpha)$	<-qnorm$(1 - \alpha)$
Poisson $\mathcal{P}(\lambda)$	N/A	<-qpois$(1 - \alpha, \lambda)$
Student's t	TINV$(2 * \alpha ; df)$	<-qt$(1 - \alpha, df)$
NC Student's t	N/A[h]	<-qt$(1 - \alpha, df, \delta)$[i]

[a]The $(1 - \alpha)^{th}$ quantile x_α is implicitly defined by $P(X > x_\alpha) = \alpha$
[b]<-qexp$(1 - \alpha$, rate $= \lambda$,lower.tail $=$ TRUE)
[c]<-qf$(\alpha, df_1, df_2$,lower.tail=FALSE)
[d]GAMMAINV$(1 - \alpha ; \theta ; 1/\lambda)$
[e]¡- qgamma$(1 - \alpha, \theta$, [rate=λ], [scale=$1/\lambda$])
[f]$\mathcal{H}(n, N_s, N)$ where n=sample size, N_s=number of population successes, and N=population size
[g]<-qhyper$(1 - \alpha, N_s, N - N_s, n)$
[h]The noncentral t distribution is not implemented in Calc 3.2.0
[i]δ is the noncentrality parameter

Appendix D

Sample Size Determination in the Two-Sample Test of Hypothesis for Proportions

Table D.1: Sample size as a function of d and p_1 for $\alpha = 0.05$ and Power $= 0.90$

d	p_1								
	0.05	0.1	0.2	0.3	0.5	0.7	0.8	0.9	0.95
0.01	7,359	14,724	26,884	35,619	42,813	36,305	27,912	16,094	8,900
0.02	1,642	1,114	6,588	8,815	10,699	9,157	7,102	4,190	2,413
0.03	641	470	2,867	3,876	4,752	4,104	3,210	1,935	1,155
0.04	310	250	1,578	2,156	2,670	2,327	1,835	1,129	695
0.05	165	151	987	1,364	1,707	1,501	1,193	747	473
0.06	N/A	98	669	936	1,183	1,050	841	536	348
0.07	N/A	67	480	679	868	777	627	406	270
0.08	N/A	47	358	513	663	599	486	320	217
0.09	N/A	34	275	400	522	476	389	260	180
0.1	N/A	25	216	319	422	388	319	216	152
0.15	N/A	N/A	81	131	184	177	150	108	81
0.2	N/A	N/A	37	67	101	101	88	67	53
0.25	N/A	N/A	N/A	38	62	65	58	46	38
0.3	N/A	N/A	N/A	22	41	45	41	34	29
0.35	N/A	N/A	N/A	N/A	29	33	31	26	22
0.4	N/A	N/A	N/A	N/A	21	25	24	21	18
0.45	N/A	N/A	N/A	N/A	15	19	19	17	15
0.5	N/A	N/A	N/A	N/A	11	15	15	14	12

Table D.2 : Sample size as a function of d and p_1 for $\alpha = 0.05$
and **Power** $= 0.80$

d	p_1								
	0.05	0.1	0.2	0.3	0.5	0.7	0.8	0.9	0.95
0.01	5,313	10,630	19,409	25,715	30,909	26,210	20,151	11,619	6,426
0.02	1,186	1,114	4,756	6,364	7,724	6,611	5,127	3,025	1,742
0.03	463	470	2,070	2,799	3,431	2,963	2,318	1,397	834
0.04	224	250	1,140	1,557	1,928	1,681	1,325	815	502
0.05	120	151	713	985	1,232	1,084	861	540	342
0.06	N/A	98	484	676	855	758	607	387	252
0.07	N/A	67	347	490	627	561	453	294	195
0.08	N/A	47	259	371	479	433	351	232	157
0.09	N/A	34	199	289	378	344	281	188	130
0.1	N/A	25	157	231	305	280	231	157	110
0.15	N/A	N/A	59	95	133	128	109	78	59
0.2	N/A	N/A	27	48	73	73	64	48	38
0.25	N/A	N/A	N/A	27	45	47	42	33	27
0.3	N/A	N/A	N/A	16	30	33	30	25	21
0.35	N/A	N/A	N/A	N/A	21	24	23	19	17
0.4	N/A	N/A	N/A	N/A	15	18	17	15	13
0.45	N/A	N/A	N/A	N/A	11	14	14	12	11
0.5	N/A	N/A	N/A	N/A	8	11	11	10	9

Table D.3: Sample size as a function of d and p_1 for $\alpha = 0.10$ and **Power** $= 0.90$

d	p_1								
	0.05	0.1	0.2	0.3	0.5	0.7	0.8	0.9	0.95
0.01	5,645	11,295	20,623	27,324	32,843	27,850	21,412	12,346	6,827
0.02	1,260	1,114	5,054	6,762	8,207	7,024	5,448	3,214	1,851
0.03	491	470	2,200	2,973	3,645	3,148	2,462	1,484	886
0.04	237	250	1,210	1,654	2,048	1,785	1,408	866	533
0.05	126	151	757	1,046	1,309	1,151	915	573	363
0.06	N/A	98	513	718	908	805	645	411	267
0.07	N/A	67	368	521	665	596	480	311	207
0.08	N/A	47	274	393	508	459	373	246	166
0.09	N/A	34	211	307	401	365	298	199	138
0.1	N/A	25	166	245	324	297	245	166	117
0.15	N/A	N/A	62	100	141	135	115	83	62
0.2	N/A	N/A	28	51	77	77	67	51	40
0.25	N/A	N/A	N/A	29	48	50	44	35	29
0.3	N/A	N/A	N/A	17	32	34	32	26	22
0.35	N/A	N/A	N/A	N/A	22	25	23	20	17
0.4	N/A	N/A	N/A	N/A	16	19	18	16	14
0.45	N/A	N/A	N/A	N/A	11	14	14	13	11
0.5	N/A	N/A	N/A	N/A	8	11	11	10	9

Table D.4 : Sample size as a function of d and p_1 for $\alpha = 0.10$
and Power $= 0.80$

d					p_1				
	0.05	0.1	0.2	0.3	0.5	0.7	0.8	0.9	0.95
0.01	3,874	7,750	14,152	18,750	22,536	19,110	14,692	8,472	4,685
0.02	865	1,114	3,468	4,640	5,632	4,820	3,738	2,206	1,270
0.03	337	470	1,509	2,040	2,501	2,161	1,690	1,019	608
0.04	163	250	831	1,135	1,406	1,225	966	594	366
0.05	87	151	520	718	898	790	628	394	249
0.06	N/A	98	352	493	623	553	443	282	183
0.07	N/A	67	253	357	457	409	330	214	142
0.08	N/A	47	188	270	349	315	256	169	114
0.09	N/A	34	145	211	275	251	205	137	95
0.1	N/A	25	114	168	222	204	168	114	80
0.15	N/A	N/A	43	69	97	93	79	57	43
0.2	N/A	N/A	19	35	53	53	46	35	28
0.25	N/A	N/A	N/A	20	33	34	31	24	20
0.3	N/A	N/A	N/A	12	22	24	22	18	15
0.35	N/A	N/A	N/A	N/A	15	17	16	14	12
0.4	N/A	N/A	N/A	N/A	11	13	13	11	10
0.45	N/A	N/A	N/A	N/A	8	10	10	9	8
0.5	N/A	N/A	N/A	N/A	6	8	8	7	7

Appendix E

Excel and Statistical Analysis

This appendix summarizes some Excel features that I have deemed important for performing statistical analysis. One of these features is the Data Analysis ToolPak, an built-in Excel Add-In that provides a useful collection of modules for performing various statistical analyzes. This appendix contains detailed instructions for installing this add-in. Additionally, I describe steps for using the ToolPak to implement specific techniques described in the body of the book. Also included in this appendix is a description of special Excel functions that I found useful for performing specific tasks. These Excel functions are organized by rubric rather than alphabetically for practical reasons.

E.1. Analysis ToolPak

The Data Analysis ToolPak is an Excel Add-In for statistical analysis. It includes a number of modules that implement various statistical techniques ranging from producing basic descriptive statistics to performing linear regression analysis. These modules are referred to in the body of this book as they are needed. Although this Add-In generally ships with new versions of Microsoft Office for Windows, it must be activated before you can use it. The purpose of this Appendix is to present a step-by-step procedure for activating it.

The steps for installing the ToolPak for Excel 2007 and Excel 2010 are very similar. For Excel 2003, the installation instructions are substantially different. I will show installation instructions for Excel 2007 only, since they will be equally valid for Excel 2010.

E.2. Activating the Analysis ToolPak

1. Once in Excel, **Click on the Office button** on the top left side (see Figure E.1)

2. Then **Click on the Excel Options Button** *(near the "Exit Excel" button on Figure E.1).* This action will open a new window

3. On the window opened in step 2, **Click the Add-Ins Button** to open the window shown in Figure E.2.

4. From the window of step 3, **Select Analysis ToolPak** as shown in Figure E.2.

5. Then **Click on the <u>Go</u>... Button**[1]

---

[1]Be careful not to click the OK button ; because then nothing will happen

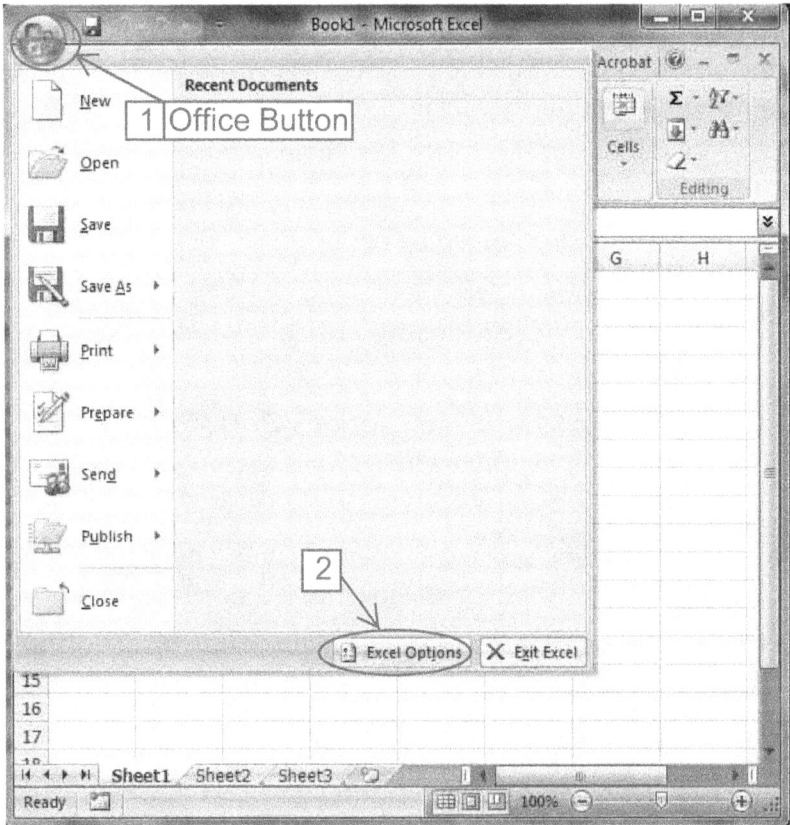

Figure E.1. Displaying the Excel Options

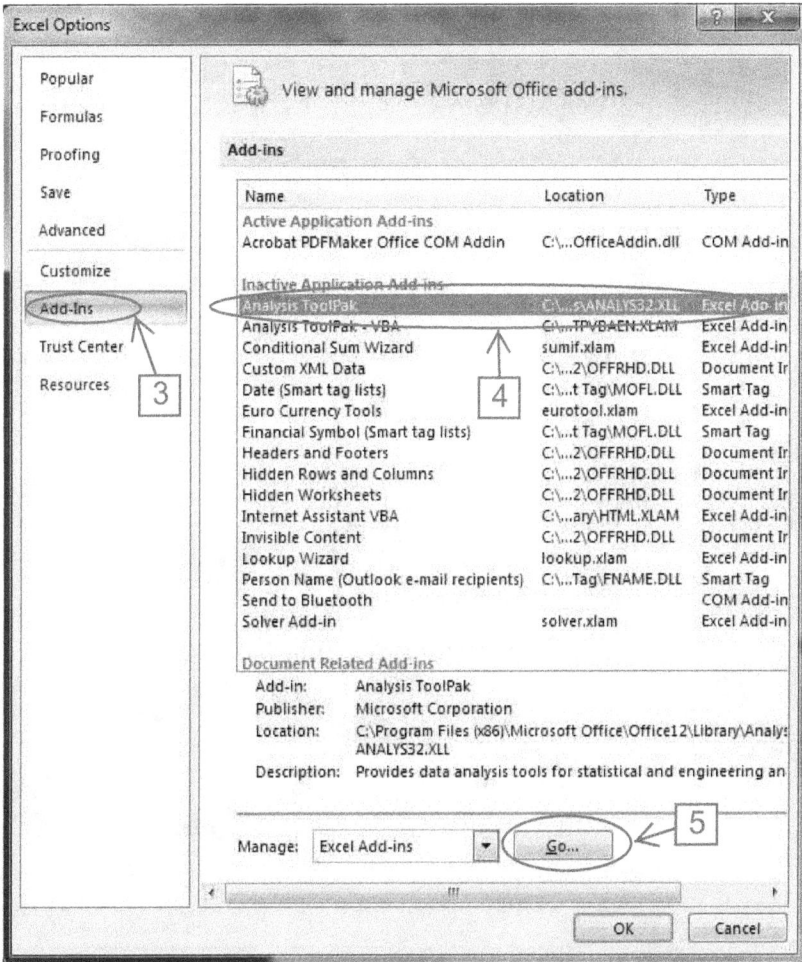

Figure E.2. Selecting the Analysis ToolPak for Installation

[6] The execution of step 5 will open the window of Figure E.3.
Check the Analysis ToolPak Box as shown in Figure E.3.

[7] **Click on the OK Button**[2]

---

[2]Occasionally, some Office installations require that you insert an MS Office DVD at this stage. It has never happened to me, but some of my students at the University of Phoenix brought this to my attention

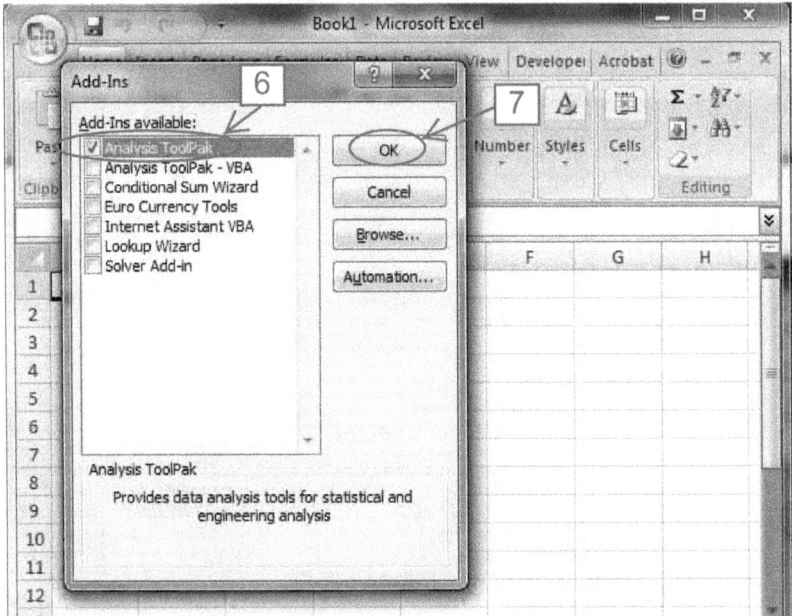

Figure E.3. Activating the Analysis ToolPak

8 The execution of step 7 will activate the Analysis ToolPak Add-In. To verify that this activation was successful, go back to Excel's main menu system and **Select the "Data" Tab from the Menu Bar on the Ribbon.** Data Analysis should appear as the rightmost icon on the ribbon (see Figure E.4). You may or may not have noticed that; but this icon was not there before.

9 **Click on the Data Analysis Icon.** You should see a long list of **Analysis Tools.** These are actually separate programs that implement specific data analysis techniques. I will use a few of them.

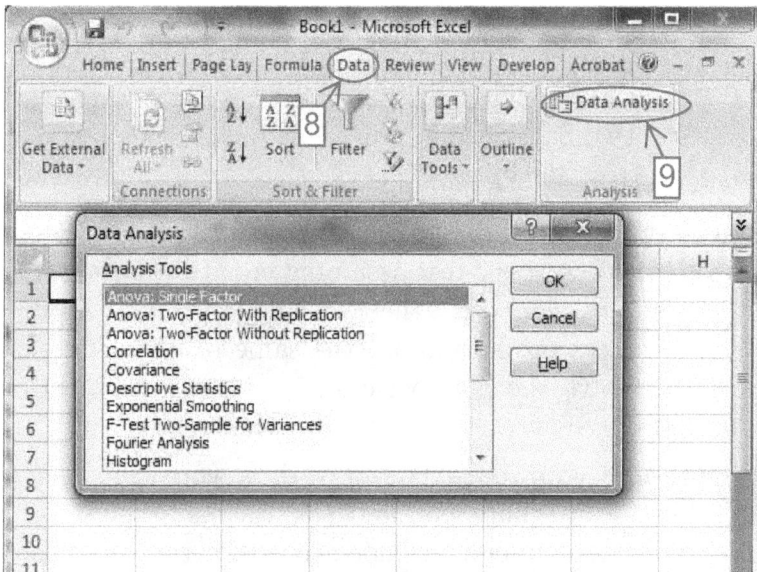

Figure E.4. Verifying the Installation

E.3. Example 8.2: *t*-Test using Excel's ToolPak

In this section, I describe how the *t*-test can be implemented with the hourly wage rate data of plumbers and electricians shown in Table 8.4, to test the hypothesis that plumbers are earning more than electricians. Note that the two sample sizes ($n_{\mathrm{P}} = 33$ and $n_{\mathrm{E}} = 31$) are sufficiently large to conduct the *z*-test. But Excel's implementation of the *z*-test requires that I supply the two "true" population variances, which are generally assumed known for this test, and which I do not have[3]. Nevertheless, Excel offers four options for comparing two populations means, which are the following:

▶ *z*-test: Two Sample for Means

▶ *t*-test: Two-Sample Assuming Equal Variances

▶ *t*-test: Two-Sample Assuming Unequal Variances

▶ *t*-test: Paired Two Sample for Means

Since I do not have any true population variances to supply, I will use the two-sample *t*-test assuming equal variances. In fact, for large samples, the *z*-test, the two-sample *t*-test for equal and unequal variances produce about the same results. Therefore, I decided to resolve the plumber-electrician wage problem using the two-sample *t*-test assuming equal variance.

[1] **Organize your data.** The only data you will need from Table 8.4 are plumbers' and electricians' wage data. The observation numbers are not needed for this analysis. The first thing to do is to have all plumber's data in a single worksheet's column, and all electricians' data in an adjacent column. It is essential that both columns be adjacent.

---

[3]The situations in practice where you would know the true population variances are very rare. You may encounter such a situation with simulated data, where the experimenter can set experimental parameters. In business or social statistics, it is unlikely that you ever have to deal with this situation

2. **Open the ToolPak.** Check Figure E.4 to see how you can display the list of statistical modules contained in the Excel Analysis ToolPak.

3. **Select the Statistical Module.** Scroll down the list of statistical until the two-sample *t*-test assuming equal variances is highlighted as shown in Figure E.5. Then click on OK to launch the *t*-test.

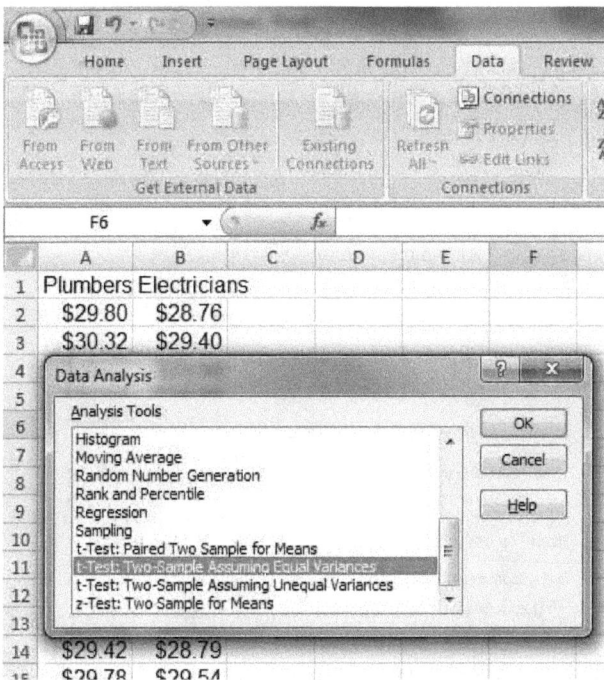

Figure E.5. Selecting the two-sample *t*-test assuming equal variances

4. **Input your data.** The execution of step 3 will open the window shown in Figure E.6, where you will input your data, set the test parameters, and specify the output location. "Variable 1 Range" and "Variable 2 Range" are the two places where you will describe plumber and electrician

wage data respectively. Note that when capturing the data, I captured column labels as well.

"Hypothesized Mean Difference" is what I referred to in chapter 8 as the hypothetical mean difference D_0. In this case $D_0 = 0$ as is often the case. This is a straight comparison between two means.

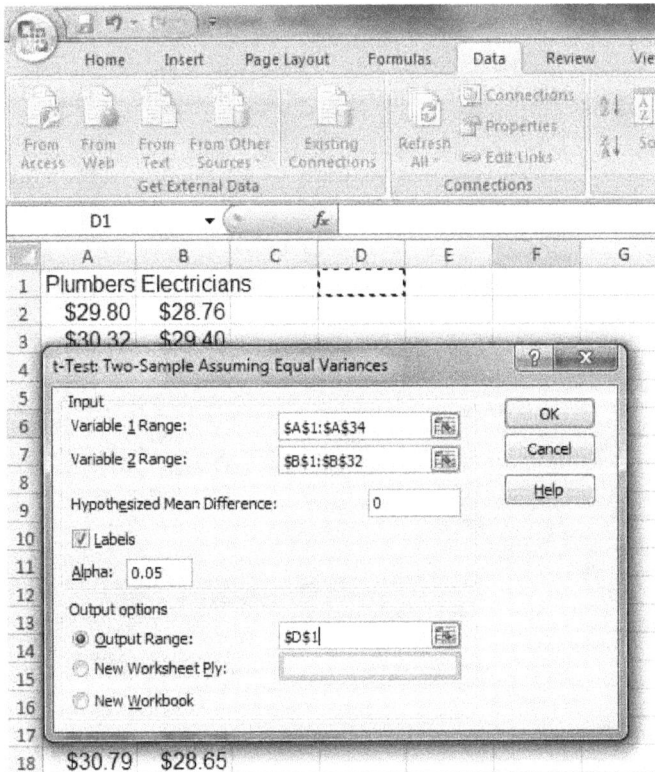

Figure E.6. Describing your Input Data

The "Labels" checkbox must be checked only if you decide to include column labels in your data range. Otherwise the box must be left unchecked. Note that if you check this box without including column labels among your data, you will lose the first data point of each range that Excel will treat as labels.

The "Alpha" text box is the place where you will supply the significance level. If your chosen significance level is 5%, do not enter 5 or 5%. Instead, you should enter 0.05.

Regarding the "Output options" here is the place where you will instruct Excel about the specific location to output the results. You can choose to display the results on the same worksheet containing your data (my preferred option). In this case you would select the "Output Range" radio button as shown in Figure E.6 before clicking inside the associated RefEdit control, and clicking on the specific cell where Excel would begin displaying the results. Alternatively, you may create a new worksheet or even an entirely new workbook that will only contain the results. I find these 2 later options unattractive because then you may end up toggling back and forth between your data and your results.

5 **Viewing the Results**When you are satisfied with your input, then click on OK, and view the results as shown in Figure E.7.

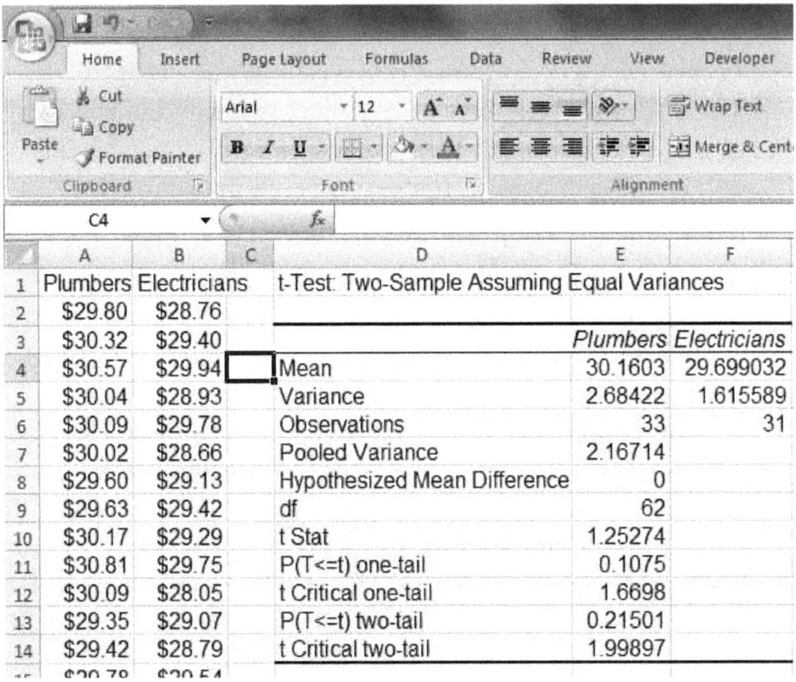

	A	B	C	D	E	F
1	Plumbers	Electricians		t-Test: Two-Sample Assuming Equal Variances		
2	$29.80	$28.76				
3	$30.32	$29.40			*Plumbers*	*Electricians*
4	$30.57	$29.94		Mean	30.1603	29.699032
5	$30.04	$28.93		Variance	2.68422	1.615589
6	$30.09	$29.78		Observations	33	31
7	$30.02	$28.66		Pooled Variance	2.16714	
8	$29.60	$29.13		Hypothesized Mean Difference	0	
9	$29.63	$29.42		df	62	
10	$30.17	$29.29		t Stat	1.25274	
11	$30.81	$29.75		P(T<=t) one-tail	0.1075	
12	$30.09	$28.05		t Critical one-tail	1.6698	
13	$29.35	$29.07		P(T<=t) two-tail	0.21501	
14	$29.42	$28.79		t Critical two-tail	1.99897	

Figure E.7. Viewing the Results

E.4. Example 10.1: Single-Factor ANOVA With Excel

This section shows the steps for performing the ANOVA analysis of Example 10.1 using Excel. These instructions are valid for both the 2010 and 2007 versions of Excel.

1. **Launching the ToolPak.** In Excel, capture the work-performance data of Peter, Jennifer, and Kosta as shown in Figure E.8. Then select "Data" for Excel's menu, and "Data Analysis" as shown in Figure E.8, in order to launch the ToolPak.

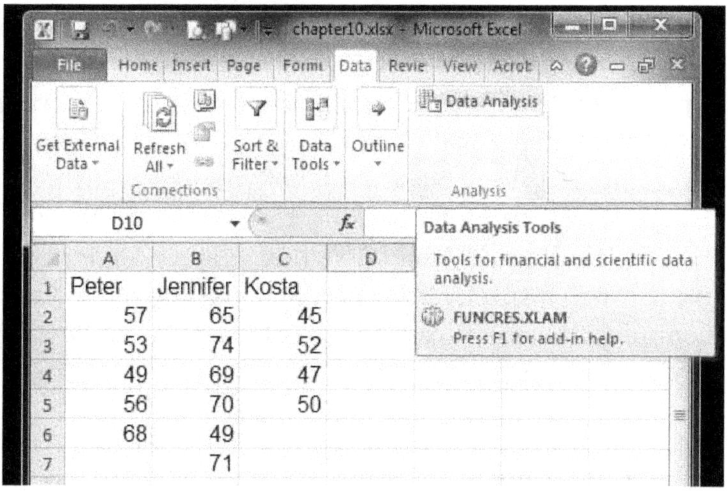

Figure E.8. Launching the ToolPak

[2] **Selecting the ANOVA Procedure.** Select "Anova: Single Factor" from the Analysis ToolPak menu (the first menu item) as shown in Figure E.9. The Analysis ToolPak offers 3 modules for ANOVA. These modules are the "Single Factor," "Two-Factor With Replication," and "Two-Factor Without Replication." Only the single-factor ANOVA falls within the scope of this book.

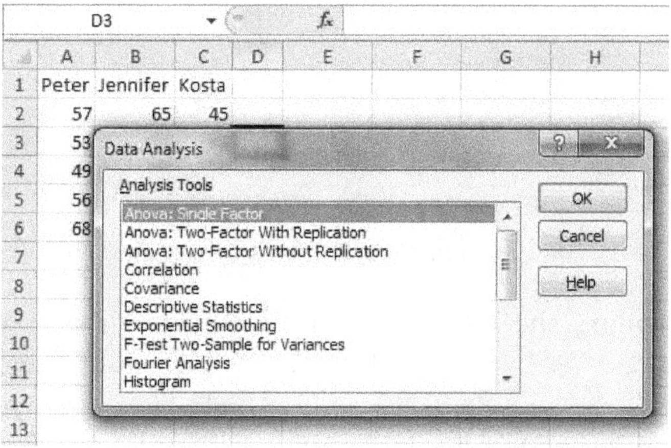

Figure E.9. Selecting the ANOVA Single-factor Module

[3] **Specifying the Parameters.** The selection of "Anova: Single Factor" in step 2 will open the form shown in Figure E.10. I recommend that you select all of your data (including the labels, which requires checking the label-in-first-row checkbox) in the Input Range text box. The Grouped By radio button helps specify whether your data is organized horizontally (Rows) or vertically (columns). Change "Alpha" is the significance level is not 0.05.

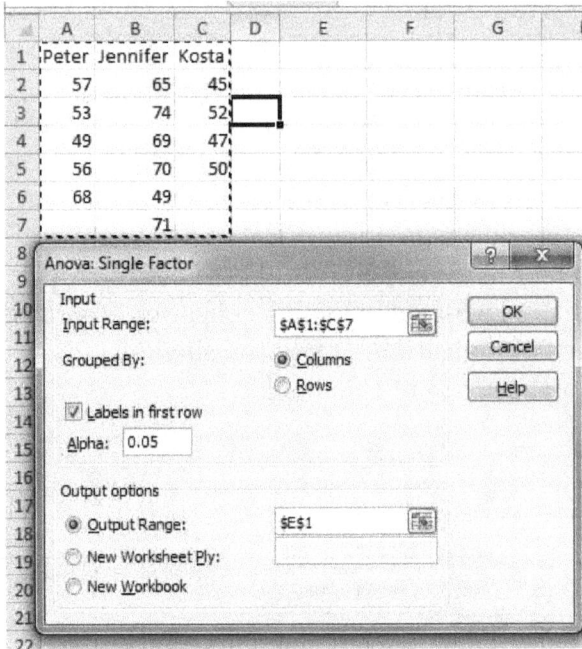

	A	B	C	D	E	F	G	H
1	Peter	Jennifer	Kosta					
2	57	65	45					
3	53	74	52					
4	49	69	47					
5	56	70	50					
6	68	49						
7		71						

Anova: Single Factor

Input
Input Range: $A$1:$C$7
Grouped By: ◉ Columns
⦿ Rows

☑ Labels in first row
Alpha: 0.05

Output options
◉ Output Range: $E$1
⦿ New Worksheet Ply:
⦿ New Workbook

OK
Cancel
Help

Figure E.10. Specifying Parameters of the ANOVA Module

[4] **Reading the Results.** Figure E.11 shows the outcome of the ANOVA analysis. The SUMMARY table presents basic statistics about the input dataset. The "Count" column shows the number of observations used by Excel, for each

employee, and may help verify that the data was read accurately. The ANOVA table contains the main results.

A	B	C	D	E	F	G	H	I	J	K
ter	Jennifer	Kosta		Anova: Single Factor						
57	65	45								
53	74	52		SUMMARY						
49	69	47		Groups		Count	Sum	Average	Variance	
56	70	50		Peter		5	283	56.6	50.3	
68	49			Jennifer		6	398	66.33	80.67	
	71			Kosta		4	194	48.5	9.67	
				ANOVA						
				Source of Variation	SS	df	MS	F	P-value	F crit
				Between Groups	785.8	2	392.9	7.44	0.0079	3.89
				Within Groups	633.53	12	52.79			
				Total	1419.33	14				

Figure E.11. The ANOVA Results

E.5. Example 10.4: Repeated-Measures ANOVA

This section shows the steps for performing the repeated-measures ANOVA analysis of Example 10.4 using Excel. These instructions are valid for both the 2010 and 2007 versions of Excel.

1. **Launching the ToolPak and Selecting ANOVA Module.** In Excel, *type in the return data of Table 10.7* as shown in Figure E.12. Whether the data is organized with Days as rows and Salespersons as columns or vice-versa is irrelevant. After capturing your data, *select "Data" for Excel's menu, then "Data Analysis" on far right of the Data menu*. Excel will then display "Analysis Tools" menu as shown in Figure E.12. *Select "Anova: Two-Factor Without Replication"*.

1	A	B	C	D	E	F	G	H	I	J	K
1		P1	P2	P3	P4	P5	P6	P7	P8	P9	
2	MON	2.8	5.9	3.3	4.4	1.7	3.8	6.6	3.1	0	
3	TUE	3.6	1.7	5.1	2.2	2.1	4.1	4.7	2.7	1.3	
4	WED	1.4	0.9	1.1	3.2	0.8	1.5	2.8	1.4	0.5	
5	THU	2	2.2	0.9	1.1	0.5	1.5	1.4	3.5	1.2	

Data Analysis

Analysis Tools

Anova: Single Factor
Anova: Two-Factor With Replication
Anova: Two-Factor Without Replication
Correlation
Covariance
Descriptive Statistics
Exponential Smoothing
F-Test Two-Sample for Variances
Fourier Analysis
Histogram

OK
Cancel
Help

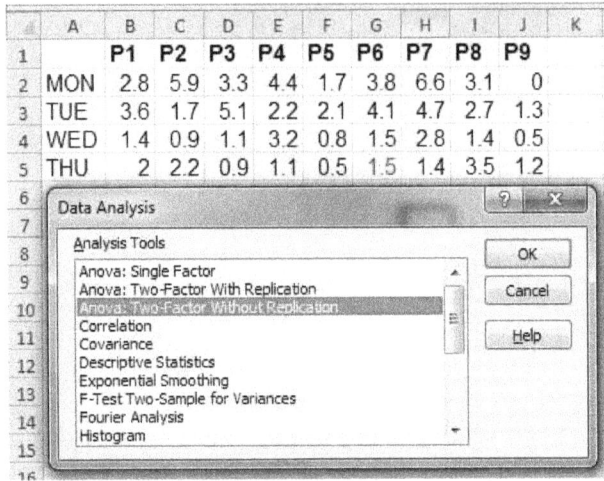

Figure E.12. Selected Module for Repeated-Measures ANOVA

2 **Filling out the ANOVA Form.** In the textbox Input Range, *(i) Use the cursor to select all of your data, including the labels, (ii) Check the Labels checkbox, (iii) Specify the significance level of the test (0.05 is the defaults value)*, and *(iv) in the Output Range textbox, specify the cell where you want Excel to output the result.* The completed form will look like Figure E.13. After you complete the entire form as indicated, *(v) Click OK*

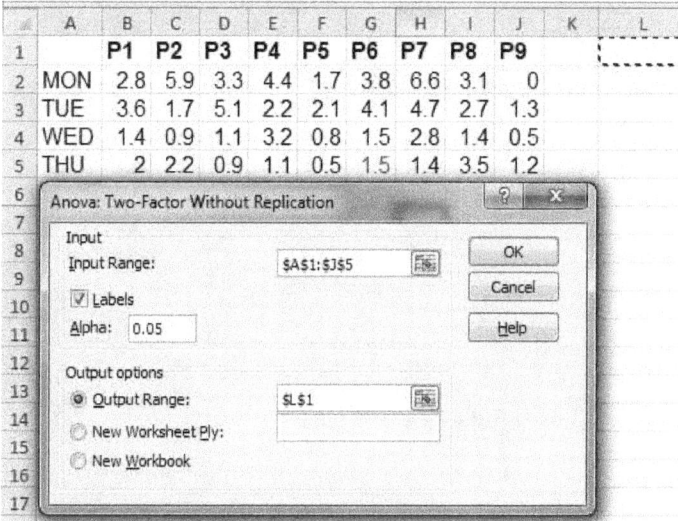

Figure E.13. Filling Out the ANOVA Two-Factor Without Replication Form

[3] **Interpret the Result.** Figure E.14 shows Excel's output. The summary table is followed with the ANOVA table, which contains all the ANOVA results you need. The factor you are studying is "Day of the week". Because day-of-the-week data is displayed row-wise, "Day of the week" is identified in the ANOVA table as Rows. Therefore, the observed F statistic you need is $F = 6.580$, which the critical value of 3.009. Therefore, the null hypothesis must be rejected.

L	M	N	O	P	Q	R
Anova: Two-Factor Without Replication						
SUMMARY	Count	Sum	Average	Variance		
MON	9	31.6	3.51	4.06		
TUE	9	27.5	3.06	1.87		
WED	9	13.6	1.51	0.83		
THU	9	14.3	1.59	0.79		
P1	4	9.8	2.45	0.92		
P2	4	10.7	2.68	4.91		
P3	4	10.4	2.60	3.96		
P4	4	10.9	2.73	1.98		
P5	4	5.1	1.28	0.56		
P6	4	10.9	2.73	2.02		
P7	4	15.5	3.88	5.13		
P8	4	10.7	2.68	0.83		
P9	4	3	0.75	0.38		
ANOVA						
Source of Variation	SS	df	MS	F	P-value	F crit
Rows	28.001	3	9.334	6.580	0.0021	3.009
Columns	26.265	8	3.283	2.315	0.0535	2.355
Error	34.044	24	1.418			
Total	88.310	35				

Figure E.14. Repeated-Measures ANOVA Results

E.6. Example 11.1: Obtaining the Regression Equation with Excel

This section shows the steps for obtaining the regression of Example 11.1 with Excel. These instructions are valid for both the 2010 and 2007 versions of Excel and will produce Figure 11.3.

1 **Creating the Scatterplot.** Select the 2 columns containing GPA.HS and GPA.CO (these 2 columns must be adjacent with GPA.HS (the independent variable) coming first. Select "Insert" from Excel menu, then "Scatter", then *Scatter*

with only Markers (the first picture). You may need to do a little formatting if you want the picture to look nice.

2 **Adding the Trendline.** After creating the scatterplot, right-click on one the markers to display the menu shown in Figure E.15, and select "Add Trendline ...".

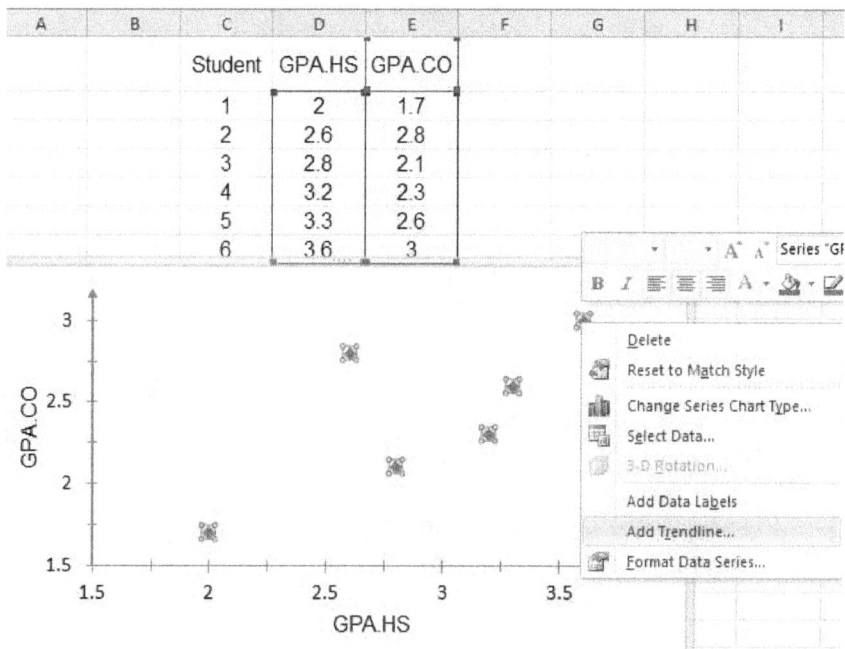

Figure E.15. Adding Trendline to a Scatterplot

3 **Displaying the Regression Equation.** After clicking on "Add Trendline ...", Excel will add the regression line to the scatterplot, before displaying the "Format Trendline" form shown in Figure E.16. You need to check the "Display Equation on chart" checkbox before the regression can be displayed as shown in Figure 11.3.

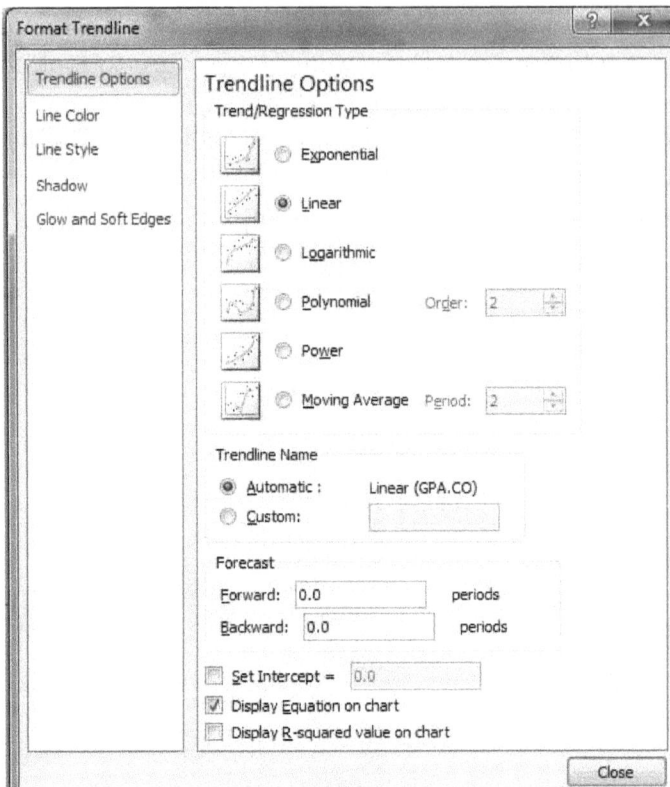

Figure E.16. Displaying the Regression Equation on the Scatterplot

E.7. Example 11.2: Multiple Regression Analysis with Excel

[1] **Organizing Input Data.** Create 3 columns of data in Excel representing the 3 variables of Table 11.4 and their values.

[2] **Launching the Regression Module.** Select "Menu" from Excel's main horizontal menu, then click on the "Data Analysis" icon to open the Analysis ToolPak. Excel will display

the list of Analysis Tools from which you will select "Regression" as shown in Figure E.17.

[3] **Defining Input Data on the Regression Form.** After selecting the "Regression" module and clicking OK, Excel will display the form shown in Figure E.18. The "Input Y Range" text box will always contain the independent variable (in this case Sales). *Select of the data in that column including the labels.* The "Input Y Range" text box contain all values of the independent variables. *Select both "POPULATION" and "PCI" columns, including the labels.* Then, *Fill out the rest of the form as shown in Figure E.18, then click OK.* Excel will then display the results shown in Example 11.2.

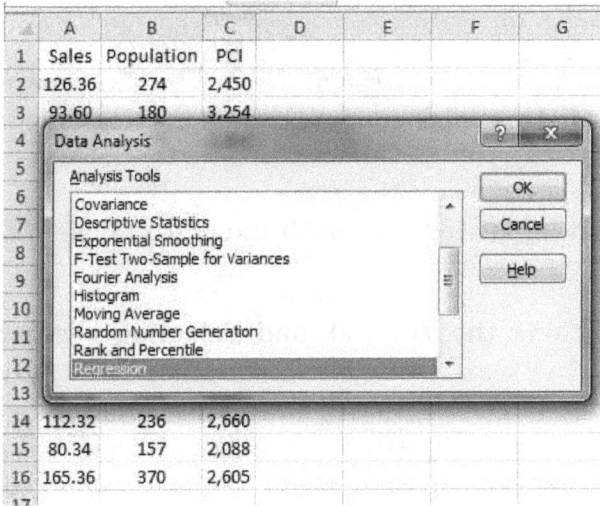

Figure E.17. Selecting the Regression Module of the ToolPak

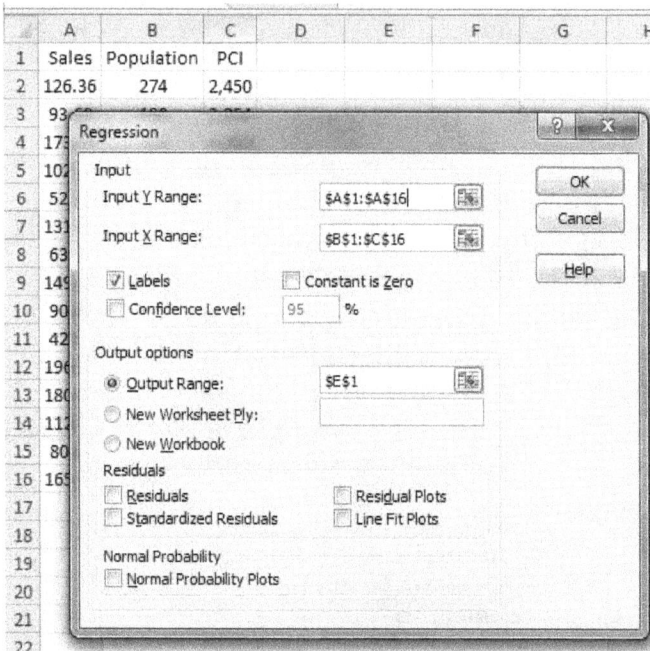

Figure E.18. Defining Input Data on the Regression Form

E.8.   Creating Histograms with Excel ToolPak

1 **Launching the ToolPak and Selecting Histogram Module.** In Excel, *type in the food expense data of Table 1.3* as shown in Figure E.19. All data is in column A and includes the label "Food". After capturing your data, *select "Data" for Excel's menu, then "Data Analysis" on far right of the Data menu.* Excel will then display the "Analysis Tools" menu as shown in Figure E.19. *Select "Histogram".*

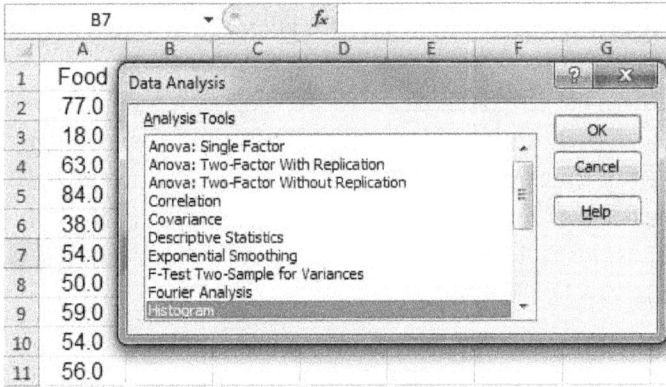

Figure E.19. Selecting the Histogram Module

[2] **Filling out the Histogram Dialog Form.** In the textbox Input Range, *(i) Use the cursor to select all of your data, including the label cell A1, (ii) Check the Labels checkbox, (iii) For your first run, leave the "Bin Range" text box blank, (iv) in the Output Range textbox, specify the cell where you want Excel to output the results. (v) Select the Chart Output checkbox.* The completed form will look like Figure E.20. After you complete the entire form as indicated, *(vi) Click OK*

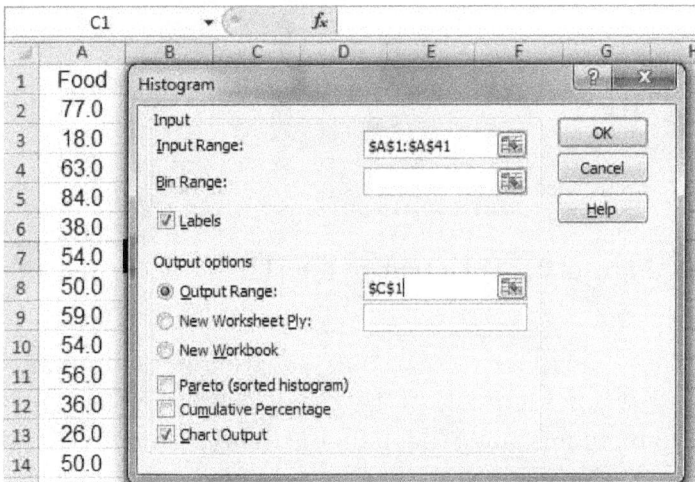

Figure E.20. Filling Out the Histogram Dialog Form

3 **Interpret the Result.** Figure E.21 shows Excel's output. Excel produces a frequency table and the histogram. Generally histograms do not have gap between its vertical bars. To remove the gap between bars, right-click on any bar, select "Format Data Series ...", select Series Options, then change the Gap Width to "No Gap."

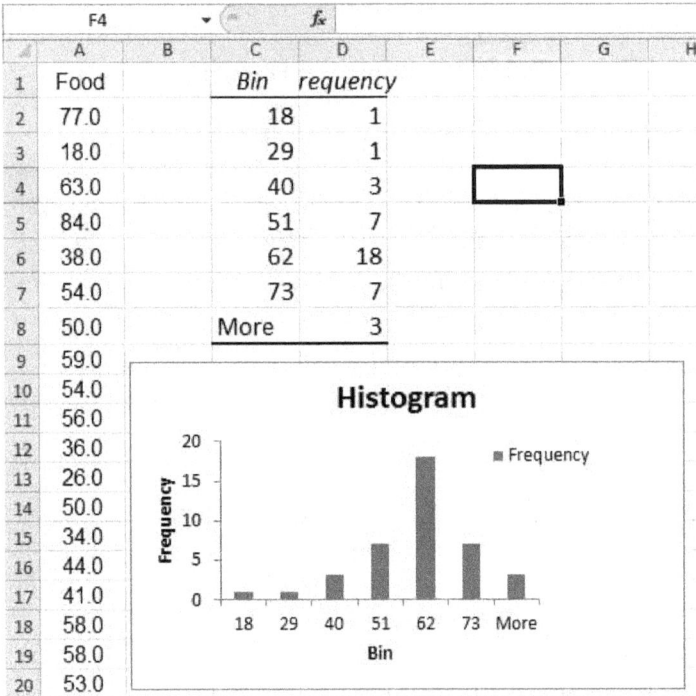

Figure E.21. Output of the Histogram Module

If you want to use your set of bins, create a data column similar to column C of Figure E.21 that will contain your bin values. All of your values must be numeric (there should not be any alphabetic value such as "More"). Note that a bin value such as

29 refers to the interval (18 to 29] that includes 29 but excludes 18.

E.9. Some Excel 2010 Statistical Functions

The statistical functions listed here are those of the 2010 version of Excel. Refer to Appendix C to find equivalent functions in Excel 2007 (in they exist), Calc 3.2.0, or R.

Binomial Distributions

▶ BINOM.DIST(n_s,n,p,cumulative). Returns the probability mass at point n_s when **cumulative=FALSE**, and will return the cumulative probability at that same point when **cumulative=TRUE**. n_s represents the number of successes out of n trials for a probability of success of p.

▶ BINOM.INV(n,p,α). Returns α^{th} quantile of the Binomial distribution, which is the smallest value n_s for which,
$$\text{BINOM.DIST}(n_s, n, p, \text{TRUE}) > \alpha$$

Chi-Square Distributions and Tests

▶ CHISQ.DIST(x,$deg\_freedom$,$cumulative$). Returns the cumulative chi-squared probability distribution function.

▶ CHISQ.DIST.RT(x,$deg\_freedom$). Returns the one-tailed probability of the chi-squared distribution.

▶ CHISQ.INV($probability$,$deg\_freedom$). Returns the cumulative chi-square probability distribution function.

▶ CHISQ.INV.RT($probability$,$deg\_freedom$). Returns the inverse of the one-tailed probability of the chi-squared distribution.

▶ CHISQ.TEST($actual\_range$,$expected\_range$). Returns the test for independence.

Hypergeometric Distribution

▶ HYPGEOM.DIST(k,n,N_s,N,FALSE). This calculates the probability mass of a Hypergeometric process as shown in equation 2.9.

▶ HYPGEOM.DIST(k,n,N_s,N,TRUE). This calculates the cumulative probability at mass point k of a Hypergeometric model, whose probability mass function is given by equation 2.9.

Normal Distributions

▶ NORM.DIST($x,\mu,\sigma,cumulative$). This function applies to a random variable X that follows the Normal probability distribution $\mathcal{N}(\mu, \sigma^2)$. If cumulative=TRUE then NORM.DIST is the Normal cumulative distribution function, which calculates the cumulative probability $P(X \leq x)$. For cumulative=FALSE, it represents the Normal probability density function, which calculates the density of probability at x.

▶ NORM.INV(p,μ,σ). This function is the inverse of the Normal cumulative distribution function. It produces the number x_p that satisfies the condition $P(X \leq x_p) = p$, when X follows the Normal distribution $\mathcal{N}(\mu, \sigma^2)$.

▶ NORM.S.DIST($z,cumulative$) is a special case of the function NORM.DIST($x,\mu,\sigma,cumulative$), where $\mu = 0$, and $\sigma = 1$.

▶ NORM.S.INV(probability) is a special case of the function NORM.INV(p,μ,σ) where $\mu = 0$ and $\sigma = 1$.

Poisson Distribution

▶ POISSON.DIST($x,\lambda,cumulative$). Returns the Poisson distribution. Use **cumulative=TRUE** for the cumulative distribution function, and **cumulative=FALSE** for the probability mass function.

Rank Tests

▶ RANK.AVG(number,ref,[order]). Excel 2010 function that returns the rank of a number in a list of numbers: its size relative to other values in the list; if more than one value has the same rank, the average rank is returned.

▶ RANK.EQ(number,ref,[order]). Excel 2010 function that returns the rank of a number in a list of numbers. Its size is relative to other values in the list; if more than one value has the same rank, the top rank of that set of values is returned. If you were to sort the list, the rank of the number would be its position.

Number is required and represents the number whose rank you want to find. **Ref** is required and represents an array of, or a reference to, a list of numbers. Nonnumeric values in ref are ignored. **Order** is optional and specifies how to rank **number**. If order is 0 (zero) or omitted, Microsoft Excel ranks number as if ref were a list sorted in descending order. If order is any nonzero value, Microsoft Excel ranks number as if ref were a list sorted in ascending order.

Student Distributions

The Excel 2010 functions under this rubric refer to the Student t-distribution with df degrees of freedom.

▶ T.DIST($x,df,$ cumulative). If *cumulative*=**FALSE** this function produces the probability density at a given point x. If *cumulative*=TRUE then it produces the cumulative probability at point x.

▶ T.DIST.2T(x,df). This function computes the probability $P(|T| > x)$ of the 2 tails defined by point x.

▶ T.DIST.RT(x,df). This function computes the probability $P(T > x)$ of the right tail at point x.

▶ T.INV(α,df). This function returns the αth percentile of the distribution.

▶ T.INV.2T(α,df). This function produces the number x that satisfies the condition $P(|T| > x) = \alpha$. It defines the tails of the distribution whose probability is α.

Survey Sampling

▶ RAND() Returns an evenly distributed random real number greater than or equal to 0 and less than 1. A new random real number is returned every time the worksheet is calculated, which at times can be very annoying. Excel help files suggest to enter this function in the formula bar before pressing F9 to obtain a random number that does not change. If you generate several numbers, copy them all and use "Paste special.." to paste the values only.

If you want to generate a random number between two arbitrary values a and b (with $b > a$), it could be obtained as follows:

$$=(b\text{-}a)\text{RAND}()+a$$

▶ RANDBETWEEN(bottom, top) Returns a random *integer number* between the "bottom", and the "top" numbers you specify.

▶ ROUNDUP(number, num_digits). This function can be very useful, and rounds the number *"number"* up to the smallest value above, which has *num_digits* digits after the decimal point. For example **ROUNDUP(3.43556,3)** yields 3.436, while **ROUNDUP(3.43556,0)** yields 4.

Appendix F

The R System for Statistical Computing

R is an integrated suite of software facilities for data manipulation, calculation and graphical display. Among other things it has the following features:

- ▶ an effective data handling and storage facility,
- ▶ a suite of operators for calculations on arrays, in particular matrices,
- ▶ a large, coherent, integrated collection of intermediate tools for data analysis,
- ▶ graphical facilities for data analysis and display either directly at the computer or on hardcopy, and
- ▶ a well developed, simple and effective programming language (called 'S') which includes conditionals, loops, user defined recursive functions and input and output facilities. (Indeed most of the system supplied functions are themselves written in the S language.)

To Input Your Data

Before you can analyze a dataset with an R function, you need to have it stored in an object such as a matrix. Although there are several options for achieving this, I will focus on one option I have personally found convenient and reliable. This option consist of first storing your data in a `csv` file before assigning it to a matrix x with the R command > `x=read.csv(file="data.csv")`.

One popular way to create a **csv** is to use spreadsheet to organize your data, then save it as a **csv** file using the spreadsheet **Save As** ... command. If you are already using Excel, you can use it to create **csv** files. If not, my advice is to download the free OpenOffice Suite, and use its spreadsheet called Calc. It will get the job done just fine. Alternatively, you will use any text editor you may find and type in your data separated with commas, or copy and paste your data into the editor from wherever it is stored.

F.1. Using R for *t*-Test

The **t.test** function of R allows you to easily perform the one- and two-sample *t*-tests on vectors of data. Although there are several arguments that may be supplied to this function, I will confine myself to the most commonly-used arguments, and will indicate their default values if you do not specify any. Do not hesitate to consult a R manual, if you want to perform more complex analyzes.

To produce the output shown in Figure F.4 using the plumber-electrician data of Table F.4, I first created a **csv** file containing this data, which is displayed in Notepad as in Figure F.1. Next, I assigned that file to a matrix **x** as follows:

```
> x<-read.csv(file="c:/advancedanalytics/chap8_plumbers_electricians.csv")
```

The R matrix **x** has 2 columns of data, which are the plumber data (can be accessed in R as **x[,1]**) and the electrician data (can be accessed in R as **x[,2]**). The last step was to call R's **t.test** function as follows:

```
> t.test(x[,1],y=x[,2],alternative="greater",var.equal=TRUE)
```

```
chap8_plumbers_electricians.csv - Notepad
File  Edit  Format  View  Help
Plumbers,Electricians
29.80,28.76
30.32,29.40
30.57,29.94
30.04,28.93
30.09,29.78
30.02,28.66
29.60,29.13
29.63,29.42
30.17,29.29
30.81,29.75
30.09,28.05
29.35,29.07
29.42,28.79
29.78,29.54
29.60,29.60
30.60,30.19
30.79,28.65
29.14,29.95
29.91,28.75
28.74,29.21
27.36,28.75
33.60,29.21
31.15,31.28
25.85,32.60
30.55,31.36
28.95,30.00
33.75,30.25
30.79,33.54
29.14,27.75
29.91,31.86
30.82,29.21
29.95,
35.00,
```

Figure F.1. The chap8_plumbers_electricians.csv Data File

More General Use of Function t.test

R's t.test function has the following more general form:

```
t.test(x, y = NULL,
       alternative = c("two.sided", "less", "greater"),
       mu = 0, paired = FALSE, var.equal = FALSE,
       conf.level = 0.95)
```

x	a (non-empty) numeric vector of data values.
y	an optional (non-empty) numeric vector of data values.
alternative	a character string specifying the alternative hypothesis, must be one of "two.sided" (default), "greater" or "less". You can specify just the initial letter.
μ	a number indicating the true value of the mean (or difference in means if you are performing a two sample test).
paired	a logical indicating whether you want a paired t-test.
var.equal	a logical variable indicating whether to treat the two variances as being equal. If TRUE then the pooled variance is used to estimate the variance otherwise the Welch (or Satterthwaite) approximation to the degrees of freedom is used.
conf.level	confidence level of the interval.

F.2. Probability Distributions in R

F.2.1 Binomial Distribution Family: $\mathcal{B}(n, p)$

The Binomial distribution is a discrete probability distribution, whose probability mass function is defined as follows:

$$f(x) = \begin{cases} \dbinom{n}{x} p^x (1-p)^{n-x}, & for\ x = 0, 1, 2, \cdots, n, \\ 0, otherwise \end{cases} \quad (F.1)$$

▶ **Probability mass**: $f(x)$

dbinom(x,n,p) .

▶ **Cumulative Probability**: $F(x)$

pbinom(x,n,p[,lower.tail = TRUE]) . Note that, pbinom(x,n,p,lower.tail = FALSE]) = 1-pbinom(x,n,p). When parameter lower.tail is not specified, it is assumed to be TRUE.

▶ α^{th} **Quantile**

qbinom(α,n,p) . This function returns the smallest value x whose cumulative probability exceeds α.

▶ **Simulated values**

rbinom(m, n, p) . This function generates m simulated Binomial values $\mathcal{B}(n,p)$.

F.2.2 Exponential Distribution Family: $\mathcal{E}(\lambda)$

Exponential is a continuous probability distribution, whose probability density function is defined as follows:

$$f_\lambda(t) = \begin{cases} \lambda e^{-\lambda t}, & \text{if } t \geq 0, \\ 0, & \text{otherwise.} \end{cases} \qquad (F.2)$$

▶ **Probability Density**: $f(x)$

dexp(x, rate $= \lambda$) . [rate $= \lambda$] is optional. If not specified then rate$=1$ will be assumed.

▶ **Cumulative Probability**: $F(x)$

pexp(x, rate $= \lambda$, lower.tail $=$ TRUE) . Note that $1 - F(x)$ may be calculated by replacing TRUE with FALSE.

▶ **Quantiles**: $F^{-1}(p)$

qexp(p, rate $= \lambda$, lower.tail $=$ TRUE)

▶ **Simulated Value**

rexp(*n*, rate = λ) . This function can be practical if you want to investigate specific aspects of the Exponential distribution. It allows you to randomly generate *n* data points from the Gamma distribution. Here is an example:

```
> rexp(10, rate = 0.5)
[1] 1.21813582 3.13028911 0.27266386 1.01837658
3.86885415 1.05177422
[7] 3.43713994 0.21975280 0.45795395 0.08082657
```

F.2.3   Gamma Distribution Family: $\mathcal{G}(\alpha, \lambda)$

Gamma is a continuous probability distribution, whose probability density function is defined as follows:

$$f(t|\alpha, \lambda) = \begin{cases} \dfrac{\lambda^\alpha}{\Gamma(\alpha)} t^{\alpha-1} e^{-\lambda t}, & \text{if } t \geq 0, \\ 0, & \text{otherwise.} \end{cases} \quad (F.3)$$

▶ **Probability Density**: $f(x)$

dgamma(x, α, [rate = λ], [scale = $1/\lambda$)] . If none of the parameters [rate = λ], [scale = $1/\lambda$)] is specified, then scale=1 will be assumed. If both are specified, then only [scale = $1/\lambda$)] will be considered. If only [rate = λ] is specified, it will be considered.

▶ **Cumulative Probability**: $F(x)$

pgamma(x, α, rate = λ, scale = $1/\lambda$)

▶ **Quantiles**: $F^{-1}(p)$

qgamma(p, α, [rate = λ], [scale = $1/\lambda$])

▶ **Simulated Value**

rgamma(*n*, α, [rate = λ], [scale = $1/\lambda$]) . This function can be practical if you want to investigate specific aspects of

the Gamma distribution. It allows you to randomly generate n data points from the Gamma distribution. Here is
an example:

```
> rgamma(10, 2, scale = 1/0.05)
[1] 22.024026 7.436632 7.505359 108.532917
61.207246 74.799016
[7] 52.794908 76.102660 65.976236 21.879415
```

F.2.4 Hypergeometric Distribution Family: $\mathcal{H}(n, N_s, N)$

Hypergeometric is a discrete probability distribution,
whose probability mass function is defined as follows:

$$h(k; n, N_s, N) = \begin{cases} \dfrac{\dbinom{N_s}{k}\dbinom{N-N_s}{n-k}}{\dbinom{N}{n}} & \text{if } k \in \mathcal{A}, \\ \\ 0 & \text{otherwise.} \end{cases} \quad (F.4)$$

▶ **Probability mass**: $h(k)$

dhyper(k, N_s, $N - N_s$, n). k is the mass point, N_s the
number of successes, N the population size (i.e. $N - N_s$
is the number of failures), and n the sample size.

▶ **Cumulative Probability**: $H(k)$

phyper(k, N_s, $N - N_s$, n). The parameters are defined as
above.

▶ **Quantiles**: $H^{-1}(p)$

qhyper(p, N_s, $N - N_s$, n) calculates the pth percentile of
the Hypergeometric distribution

▶ **Simulated Values**

rhyper(nn, N_s, $N - N_s$, n) . This function allows you to randomly generate nn data points from the Hypergeometric distribution.

F.2.5 Poisson Distribution Family: $\mathcal{P}(\lambda)$

The Poisson distribution is a discrete probability distribution, whose probability mass function is defined as follows:

$$p(k; \lambda) = \begin{cases} \lambda^k e^{-\lambda}/k! & \text{if } \lambda > 0 \text{ and } k = 0, 1, \cdots, \\ 0 & \text{otherwise.} \end{cases} \qquad (F.5)$$

▶ **Probability mass**: $p(k)$

dpois(k, λ) , k is the mass point.

▶ **Cumulative Probability**: $P(k, \lambda)$

ppois(k, λ) .

▶ **Quantiles**: $P^{-1}(p)$

qpois(p, λ) calculates the pth percentile of the Poisson distribution

▶ **Simulated Values**

rpois(n, λ) . This function allows you to randomly generate n data points from the Poisson distribution.

List of Notations

Author Index

Subject Index

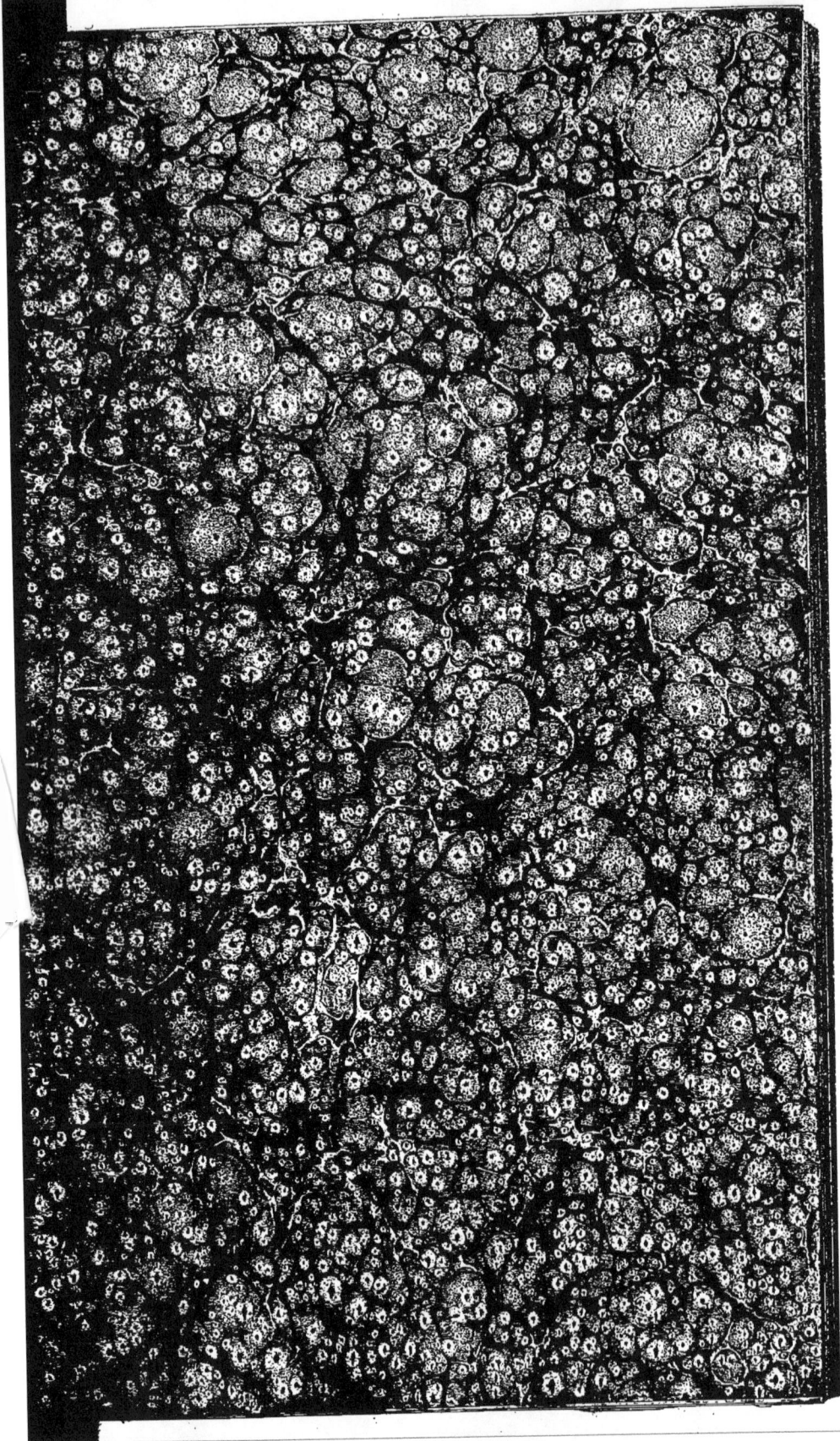

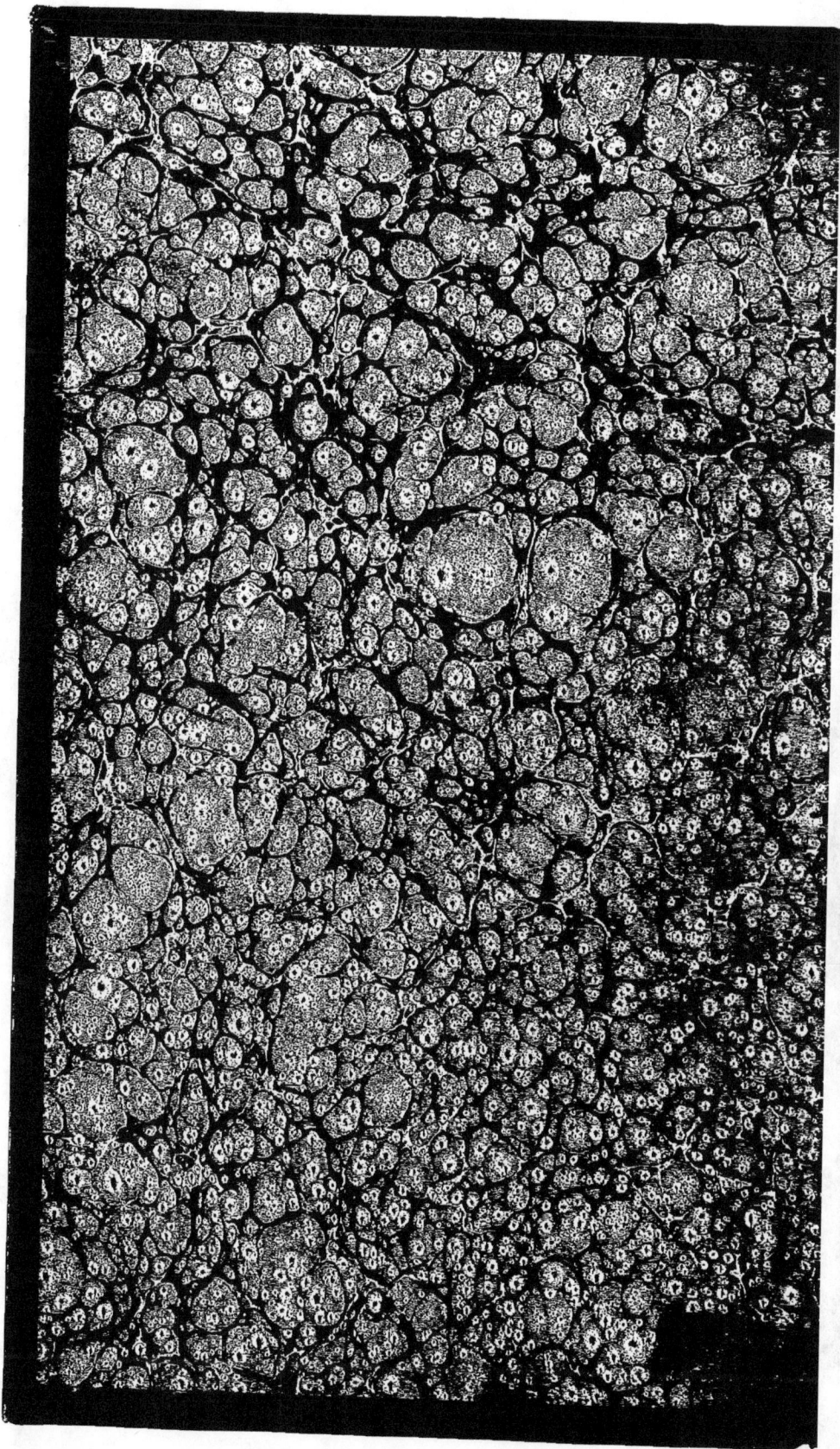

3 7531 0333463 3

BIBLIOTHEQUE NATIONALE DE FRANCE

www.ingramcontent.com/pod-product-compliance
Lightning Source LLC
Chambersburg PA
CBHW031354210326
41599CB00019B/2762